Advances in
Disease Vector Research

Advances in Disease Vector Research

Edited by

Kerry F. Harris
Virus-Vector Research Laboratory, Department of Entomology, Texas A&M
University, College Station, Texas 77843, USA

Advances in
Disease Vector Research

Volume 9

Edited by Kerry F. Harris

With Contributions by
John J. Cho Diana M. Custer Thomas L. German
Rainer Gothe U.B. Gunashinghe Peter J. Ham
Wayne B. Hunter Meir Klein Werner J. Kloft
Ronald F.L. Mau Robert G. Milne Gaylord I. Mink
Alexander S. Raikhel Jean Richardson
Edward S. Sylvester Diane E. Ullman
Daphne M. Westcot

With 84 Illustrations

Springer-Verlag
New York Berlin Heidelberg London Paris
Tokyo Hong Kong Barcelona Budapest

Kerry F. Harris
Virus-Vector Research Laboratory
Department of Entomology
Texas A&M University
College Station, TX 77843, USA

Volumes 1 and 2 of *Current Topics in Vector Research* were published by Praeger Publishers, New York, New York.

ISSN: 0934-6112

Printed on acid-free paper.

Production coordinated by Chernow Editorial Services, Inc. and managed by Christin R. Ciresi; manufacturing supervised by Jacqui Ashri.
Typeset by Asco Trade Typesetting Ltd., Hong Kong.

9 8 7 6 5 4 3 2 1

ISBN-13: 978-1-4612-7716-3   e-ISBN-13: 978-1-4612-2910-0
DOI: 10.1007/978-1-4612-2910-0

Preface

Volume 9 in this series consists of four chapters on vectors that affect human or animal health and six chapters on plant pathogens and their vectors.

In Chapter 1, Alex S. Raikhel discusses vitellogenesis in mosquitoes: the cornerstone of the reproductive cycle involving massive production of yolk precursors by the fat body and their accumulation in developing oocytes. In anautogeneous mosquitoes, vitellogenesis is dependent on the availability of a blood meal and, as a consequence, is linked to transmission of pathogens. Therefore, elucidation of mechanisms governing the mosquito vitellogenesis is critical for the successful development of novel strategies in vector and disease management. Previous reviews on mosquito vitellogenesis have dealt predominantly with hormonal control. The goal of this review, however, is to summarize significant progress which has been achieved in understanding mosquito vitellogenesis at the cellular, biochemical and molecular levels. It is with these disciplines that we expect to fully understand the mechanisms governing this key process in mosquito reproduction.

Studies at the cellular level have led to a realization of the complexity of events associated with vitellogenin synthesis in the mosquito fat body and its uptake by the oocyte. Regulation of vitellogenin synthesis is not limited to just turning on and off its genes, but is accompanied by a cycle of proliferation and degradation of biosynthetic machinery. The major genes involved in vitellogenic activities of the mosquito fat body have been identified and are currently under intensive investigation. Considerable progress also has been achieved in our understanding of the mechanism of specific accumulation of yolk protein in mosquito oocytes. The pathway of vitellogenin in oocytes has been mapped in detail by immunocytochemistry at the ultrastructural level. Properties of endocytosis of vitellogenin by developing oocytes and vitellogenin binding to ovarian membranes have been investigated. Finally, the vitellogenin receptor, solubilized from ovarian membranes, has been visualized and characterized.

In Chapter 2, Werner J. Kloft draws on more than thirty years of experience in applying radioisotope methodology to entomological research to explore how radioisotopes can be used as model substances for studying the transmission of disease agents by arthropod vectors. A knowledge of the different pathways of radioisotopes in vector organisms, as well as the time factors involved in traversing those pathways, allows one to discriminate between mechanical and so-called biological transmission. The author gives special emphasis to the phenomenon of arthropod vector ingestion–egestion behavior, especially in the context of his own recent research into the possible role of egestion (regurgitation) in the transmission of retroviruses by hematophagous insects. In spite of the controversy surrounding discussions about this mode of transmission, the author's latest results on survival and multiplication of egested retrovirus indicate a high probability for his hypothesis. His main experimental insect is the biting fly *Stomoxys calcitrans* (L.). The experiments involving egested retrovirus are briefly covered in the final section of Kloft's chapter.

Rainer Gothe, in Chapter 3, provides a comprehensive overview of the genus *Aegyptianella* with special emphasis on the epidemiology of *A. pullorum* relative to its argasid tick vectors and domestic poultry. The developmental biology of this obligatorily heteroxenous parasite in vertebrate hosts and vectors is included, as well as detailed outlines of pathogen-vertebrate host relationships and pathogen-vector interactions. From the investigations described and reviewed, the author concludes that *A. pullorum* possesses an exceptional biological plasticity and is epidemiologically extremely effective with respect to both vertebrate hosts and vector ticks. Both of the latter represent long-lasting reservoirs for the pathogen that complement and replenish one another continuously, thereby ensuring permanent natural circulation of the pathogen. The consequences of this very stable infection chain relative to the epidemiology of *A. pullorum* is discussed, as is the possibility of wild birds serving as yet another reservoir.

Immunity in hematophagous insect vectors of parasitic infection is discussed in Chapter 4. Peter J. Ham, the author, notes that since insects synthesize and secrete a large variety of molecules in response to infection by parasites, it is clear that the parasite does not have things all its own way and has to overcome an array of innate, as well as acquired, barriers. Some of these, such as the armatures in the foregut and the peritrophic membrane (PM), may be mechanical, at least in part. Also included in this is the midgut lumen and its epithelium. Many organisms attempt to occupy this area during development to the infective stage, both successfully and otherwise. However, several parasites develop away from the gut, in fat cells, hemocytes or flight musculature. These need to traverse the hemocoel with its attendant rapid response to non-self. The might of an array of cells and molecules are thrown at the parasite, as soon as it passes through the gut into the hemocoel. Some organisms survive, others fail. Some insects mount a strong response whereas others appear ineffective. Both parasite and host play a role in this process: the parasite by trying to evade the response and the host by overcoming the evasion. Carbohydrate specific lectins, antibacterial peptides, the phenoloxidase cascade and proteolytic enzymes are all involved in attempting to eliminate the parasite which, by definition, is detrimental to the existence of the insect intermediate host.

In addition to reviewing the literature and discussing the applications of this field, data are also provided for the system being studied in the author's laboratory: *Onchocerca* infections in *Simulium* spp. It is hoped that some of these data, which are new, will stimulate interest in this fascinating area of research.

In Chapter 5, Meir Klein reviews the role of the *Circulifer-Neoaliturus* group of leafhopper species in the transmission of plant pathogens. According to taxonomic experts, species of the genera *Circulifer* and *Neoaliturus* originated in the Old World. *Circulifer tenellus* is the only species belonging to these two genera to have been introduced in the New World, probably in plant crops carried by explorers. The species became a serious pest in the western United States by feeding on a large variety of plant hosts and being a vector of the beet curly top virus (BCTV), *Spiroplasma citri*, the agent of citrus stubborn disease, and the beet leafhopper-transmitted, virescence agent (BLTVA). The first two pathogens are present in the Mediterranean area and are probably transmitted by more than one species in the *Circulifer-Neoaliturus* complex. The safflower phyllody mycoplasma and perhaps other mycoplasma-like organisms (MLOs) are transmitted by leafhoppers of this group in the Mediterranean basin and the Near East. Disease spread in the United States is clearly related to the biology of *C. tenellus*, a connection not yet studied in the Old World. Molecular biology and genetic engineering techniques developed in the last few years enable us now to investigate relationships among similar pathogens in both continents. Also, such studies might reveal the locations of genes coding for both BCTV and mycoplasma vector specificity and multiplication in plants.

Epidemics of insect-transmitted plant viruses in agricultural ecosystems require the interaction of three basic components: the host plant of the virus, the insect vector, and the plant-pathogenic virus. In Chapter 6, Diane E. Ullman and associates provide us with a comprehensive overview of a multidisciplinary, interactive research approach aimed at unraveling the relationship between tomato spotted wilt virus (TSWV) and its thrips vector, and elucidating the epidemiology of this globally important plant disease. Information regarding thrips morphology, feeding behavior, biology on different plant hosts, and cellular thrips-TSWV interactions is integrated and presented as a pivotal foundation for understanding TSWV epidemiology and control.

The traditional classification of thrips feeding as rasping-sucking is challenged, and morphological and behavioral evidence is presented to suggest that thrips feed in a piercing-sucking manner. Furthermore, innervation of sensory structures on the thrips mouthcone support their classification as dual function receptors with probable sensitivity to both chemical and mechanical stimuli. Innervation of the stylets also suggests that thrips have directional control over stylet movements and may select feeding sites on the plant surface and within plant cells in a more complex and selective manner than previously thought. The significance of these findings to host choice and transmission of TSWV is discussed.

A theoretical framework is proposed for integrating TSWV epidemiology with the impact of thrips biology on various hosts, TSWV acquisition in particular developmental stages, and cell-TSWV interactions. Evidence that a midgut epithelial barrier prevents TSWV acquisition by adult thrips is presented, as well as data demonstrating how TSWV is retained in the midgut epithelium, an epidemiologically nonsignificant tissue, in adults feeding on TSWV-infected plants. In contrast, infection of the salivary glands, an epidemiologically significant thrips tissue, occurs when thrips acquire the virus as larvae. Finally, analogies and potential differences between TSWV and other members of the Bunyaviridae are discussed.

In Chapter 7, Thomas L. German, Diane E. Ullman, and U.B. Gunashinghe take a fresh look at the etiology of mealybug wilt disease of pineapple. Mealybug wilt is a serious problem in all pineapple growing regions of the world. The association of mealybug feeding with the disease is the source of the name "mealybug wilt" and for many years the disease was assumed to be caused by a toxin secreted during mealybug feeding on pineapple plants. A preponderance of biological data suggested that a virus may be involved in disease etiology, and recently a closterovirus was found in pineapple and mealybugs taken from fields where disease occurs. However, the virus is widespread and the complex epidemiology of mealybug wilt of pineapple makes it impossible to determine the exact cause of the disease.

In Chapter 8, Gaylord I. Mink critically examines all reports of ilarvirus transmission via pollen or various arthropods beginning with the work of George and Davidson in the late 1950's. While pollen itself can function as a vector for some ilarviruses by delivering them to embryos produced in fruit on healthy trees, there is no convincing evidence that any ilarvirus is transmitted plant-to-plant through pollen alone. There is also no evidence that any arthropod acts as an ilarvirus vector in the conventional sense. There is, on the other hand, an increasing body of evidence demonstrating that thrips plus pollen are at least two essential components required for plant-to-plant spread. However, as the author points out, specific details about the interaction of thrips, viruses, pollen, plant, and possibly pollinating insects remain to be determined.

In Chapter 9, Robert G. Milne discusses immunoelectron microscopy of plant viruses and plant-infecting mycoplasma-like organisms (MLOs). The emphasis is on techniques, but some of the major results are considered. Only transmission electron microscopy results are reviewed, and cryo-work is not considered. For the viruses, in vitro methods (those for handling preparations in suspension) are the first treated. The techniques reviewed include the classical antibody-virus mixture methods: leaf-dip serology, immunosorbent electron microscopy (ISEM), use of protein A, decoration, and gold labeling. Following this, thin-sectioning methods for viruses are reviewed: uses of gold- and ferretin-labeled antibody to trace antigens by pre- and post-embedding methods. The discussion on MLOs follows a similar format: in vitro methods and results, and then methods and results for pre- and post-embedding gold-labeling of MLOs in thin sections.

For fifty years now, electron microscopy and serology have continued a fruitful partnership, allying high resolution to high specificity. Now that gold labeling has become routine for virus coat protein antigens, and is rapidly developed for virus-specified noncapsid proteins and for MLOs, both *in vitro* and *in situ* work offer exciting prospects.

Our volume closes with Chapter 10, a treatise by Edward S. Sylvester and Jean Richardson on the aphid-borne rhabdoviruses. The viruses, as a group, are briefly characterized. Aphids, as alternate hosts for the viruses, provide a mechanism for both horizontal and vertical transmission. Information is given on ultrastructure studies of infected aphids and on the transmission parameters. Assay techniques used in vector studies are mentioned. The aphid and the plant host ranges are briefly discussed. The impact of virus infection on vectors is emphasized. Both longevity and reproduction can be reduced. Data are presented on the previously unreported phenomenon of sensitivity to carbon dioxide anaesthesia that is found in aphids infected with all examples of the viruses tested. The sensitivity is similar in some respects to that found in *Drosophila* infected with the vertically-transmitted sigma rhabdovirus. The epidemiology of three diseases caused by aphid-borne rhabdoviruses is discussed, and finally, it is suggested that long-term association of these viruses with invertebrates possibly has provided an evolutionary focus for the movement of this group of pathogens into plant as well as vertebrate hosts.

In conclusion, I thank the authors for their outstanding chapter contributions and patience in working with me to bring Volume 9 of *Advances in Disease Vector Research* to a most successful conclusion. I also greatly appreciate the continuing support and help of the excellent production staff at Springer-Verlag.

Kerry F. Harris

Contents

Contributors

John J. Cho
Department of Plant Pathology, Hawaiian Institute of Tropical Research and Human Resources-Maui Research Station, P.O. Box 269, Kula, Hawaii 96790, USA

Diana M. Custer
Department of Plant Pathology, Hawaiian Institute of Tropical Research and Human Resources-Maui Research Station, P.O. Box 269, Kula, Hawaii 96790, USA

Thomas L. German
Department of Plant Pathology, University of Wisconsin, 1630 Linden Drive, Madison, Wisconsin 53706, USA

Rainer Gothe
Institute for Comparative Tropical Medicine and Parasitology, University of Munich, D-8000 Munich 40, Germany

U.B. Gunashinghe
Department of Plant, Soil and Entomological Sciences, University of Idaho, Moscow, Idaho 83843, USA

Peter J. Ham
Department of Biological Sciences, Center for Vector Biology, University of Keele, Keele, Staffordshire, United Kingdom

Wayne B. Hunter
Department of Entomology, University of Hawaii, 3050 Maile Way, Honolulu, Hawaii 96822, USA

Meir Klein
Department of Entomology, Institute of Plant Protection, ARO, The Volcani Center, Bet Dagan 50250, Israel

Werner J. Kloft
Institut für Angewandte Zoologie, Universität Bonn, Germany

Ronald F.L. Mau
Department of Entomology, University of Hawaii, 3050 Maile Way, Honolulu, Hawaii 96822, USA

Robert G. Milne
Istituto di Fitovirologia Applicata, CNR, Strada delle Cacce 73, I-10135 Torino, Italy

Gaylord I. Mink
Department of Plant Pathology, Washington State University, Irrigated Agriculture Research and Extension Center, Prosser, Washington 99350, USA

Alexander S. Raikhel
Department of Entomology and Programs in Cell and Molecular Biology and Genetics, Michigan State University, East Lansing, Michigan 48824, USA

Jean Richardson
Department of Entomological Sciences, University of California, Berkeley, California 94720, USA

Edward S. Sylvester
Department of Entomological Sciences, University of California, Berkeley, California 94720, USA

Diane E. Ullman
Department of Entomology, University of Hawaii, 3050 Maile Way, Honolulu, Hawaii 96822, USA

Daphne M. Westcot
Department of Entomology, University of Hawaii, 3050 Maile Way, Honolulu, Hawaii 96822, USA

Contents for Previous Volumes

Volume 3

Volume 4

Volume 5

Volume 6

Volume 7

Volume 8

1
Vitellogenesis in Mosquitoes

Alexander S. Raikhel

Introduction

Malaria, dengue and other mosquito-borne infections are among the most devastating diseases (57, 144). The maintenance and dispersal of mosquito-borne disease depends upon the successful reproduction of the mosquito. The cornerstone of the reproductive cycle is vitellogenesis involving massive production of yolk protein precursors and their accumulation in developing oocytes. In anautogeneous mosquitoes, vitellogenesis is dependent on the availability of a blood meal and, as a consequence, is linked to transmission of pathogens. Pathogens acquired transovarially or from an infected host during an initial blood meal are transmitted during subsequent blood meals. Therefore, elucidation of vitellogenesis and other aspects of mosquito reproductive physiology is critical for the successful development of novel strategies in vector and disease management.

During the past two decades much attention has been paid to the hormonal control of mosquito vitellogenesis. This issue has been extensively reviewed (12, 20, 48, 60, 61, 62, 91). Recent progress in research on mosquito neuropeptide hormones have been also summarized (24, 93). Endocrine regulation of mosquito vitellogenesis remains an important and only partially resolved issue. The goal of this review, however, is to summarize significant progress which has been recently achieved in understanding of mosquito vitellogenesis at the cellular, biochemical and molecular levels. It is with these disciplines that we expect to fully understand the mechanisms governing this key process in mosquito reproduction.

Alexander S. Raikhel, Department of Entomology and Programs in Cell and Molecular Biology and Genetics, Michigan State University, East Lansing, Michigan 48824, USA.

Production of Yolk Protein Precursors in the Fat Body

The Cell Biology of the Mosquito Fat Body

The mosquito fat body, as in other insects, is a tissue of great functional diversity. It is responsible for the metabolism and storage of carbohydrates, lipids, and nitrogenous compounds and synthesis of hemolymph proteins. These functions are hormonally controlled and change successively according to the demands of the insect at different stages in the life cycle (149). In oogenic females, the primary important function of the fat body is the synthesis of the yolk protein precursor, vitellogenin (Vg), and other proteins crucial for oocyte maturation (see Figure 1.1). Hagedorn and Judson (65) were the first to establish that the fat body of the mosquito *Aedes aegypti* is the only site of synthesis of Vg.

In adult mosquito, the fat body is located in the head, thorax and abdomen. However, most of the fat body is in the abdomen, where it is organized as sheets or lobes of tissue, which are attached to the epidermis and surround the midgut (28). According to immunocytochemical analysis, all parts of the fat body are involved in Vg synthesis but, at least in the abdomen, the fat body exhibits a peripheral-visceral gradient for increasing intensity of Vg synthesis (126).

The majority of cells in the mosquito fat body are of a single type, the trophocyte, which is exclusively responsible for Vg synthesis (126). A second cell type, oenocytes, are far less numerous than trophocytes and are located at the surface of fat body lobes. Their function is still unclear, although they may be involved in metabolism of ecdysteroids (138).

A preliminary account of Vg synthesis in the fat body using radioimmunoassay (RIA) was presented by Hagedorn and Judson (65). Later, with the combined use of RIA and rocket immunoelectrophoresis, Hagedorn and co-workers (61, 64) have shown that Vg synthesis in the mosquito fat body starts at 3–4 hrs, reaches its peak at 24–28 hrs and then declines to back-

---

FIGURE 1.1. Summary of events during the first cycle of egg maturation in anautogeneous mosquito, *Aedes aegypti*. The first cycle proceeds through two developmental periods. The previtellogenic period begins at eclosion (E) of the adult female and is divided in a preparatory stage (preparation) and the state-of-arrest (arrest). During the preparatory stage, both the fat body and the ovary become competent for subsequent vitellogenesis. The female then enters a state-of-arrest; Vg is not synthesized during previtellogenic period if a female feeds on nectar. Arrest terminates when the female ingests vertebrate blood (BM) and vitellogenesis is initiated. Vitellogenic period proceeds for 48 hrs and is divided in a synthetic stage (synthesis) and a termination stage (termination). (A) Hormonal events. EDNH, egg development neurosecretory hormone; JH III, juvenile hormone III; NHRF, neurosecretory hormone-releasing

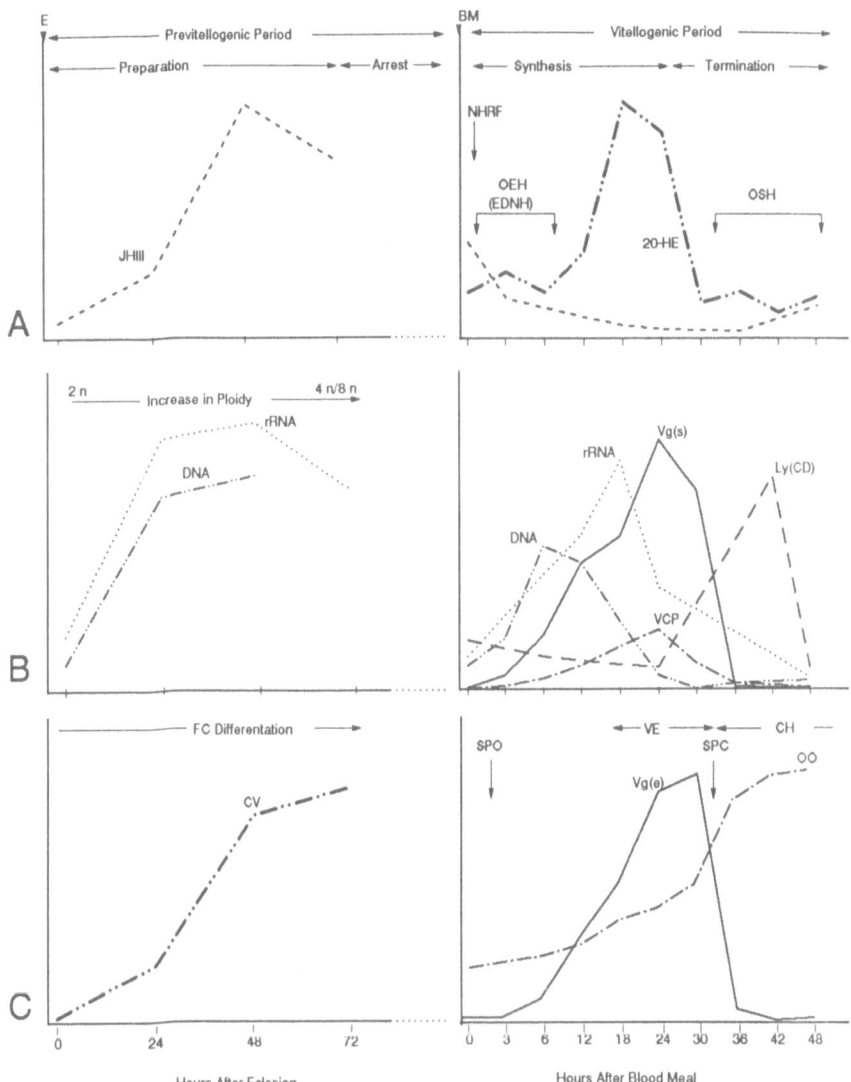

FIGURE 1.1 (*contd*)

factor; OEH, ecdysteroidogenic hormone; OSH, oostatic hormone; 20-HE, 20-hydroxyecdysone. (With modifications from 66, 89, 91, 94, 101, 105, 141.) (B) Events in the fat body. DNA, rate of DNA synthesis; Ly(CD), specific activity of a lysosomal enzyme, cathepsin D; rRNA, rate of ribosomal RNA synthesis; VCP, rate of synthesis of vitellogenic carboxypeptidase; Vg(s), rate of vitellogenin synthesis. (With modifications from 36, 64, 71, 73, 118, 130.) (C) Events in the ovarian primary follicle. CV, number of coated vesicles per unit of the oocyte cortex; CH, chorion deposition; FC, follicle cell; OO, oocyte yolk length; SPC, closing of interfollicular spaces; SPO, openning of interfollicular spaces; VE, vitelline envelope formation; Vg(e), rate of Vg endocytosis. (With modifications from 81, 127, 131.)

ground levels by 36–40 hrs post blood meal (PBM) (Figure 1.1B). Other measurements of Vg production in vitellogenic mosquitoes have confirmed this kinetics of Vg synthesis in the mosquito (95, 96, 97, 118, 148). An immunofluorescent study (126) has found, however, that Vg appears in trophocytes as early as one hour after a blood meal.

The ultrastructure of the mosquito fat body was described by several authors (6, 76, 145). These investigators, however, did not define the major changes in the fat body ultrastructure associated with onset, maintenance and termination of Vg synthesis. More recent ultrastructural analysis combined with cyto- and immunocytochemical studies of the *A. aegypti* fat body has demonstrated that during vitellogenesis trophocytes undergo three successive stages of activity: (1) proliferation of biosynthetic organelles; (2) the synthesis of vitellogenic proteins; and (3) termination of Vg production and breakdown of the biosynthetic organelles by lysosomes (118, 119, 120, 122 126, 130).

The signs of proliferation of biosynthetic organelles in trophocytes are first evident during the previtellogenic period. These changes include the enlargement of nucleoli, organelles producing ribosomes; a considerable increase in number of ribosomes and cisternae of rough endoplasmic reticulum (RER) in the cytoplasm; and the appearance of extensive foldings of the plasma membrane. A similar increase of ribosomes, RER cisternae and Golgi complexes in the trophocyte cytoplasm also occurs during the first part of the vitellogenic period. By 18–20 hrs post-blood meal, the trophocytes have transformed from a predominantly storage depot of lipid, proteins and glycogen to an efficient protein-producing factory; their cytoplasm is loaded with RER, Golgi complexes and electron-dense secretory granules, which are released from the cells (Figure 1.2A). At this time, while the rate of Vg synthesis still continues to rise, the trophocyte nucleoli show signs of termination in their activity by transforming from multilobed structures into compact bodies. The size of nucleoli progressively decreases until 42–48 hrs PBM, when it becomes similar to that of the trophocyte nucleoli at eclosion.

More obvious signs of the termination of Vg synthesis become evident in the trophocyte cytoplasm only by 30–32 hrs PBM. The most striking feature of this termination is the appearance of numerous autophagical vacuoles (autolysosomes) which clearly participate in degradation of biosynthetic organelles, i.e., RER, Golgi complexes and secretory granules (Figure 1.2B; 118, 119). By 42–48 hrs PBM, trophocytes contain a few biosynthetic organelles and numerous protein granules, lipid droplets, and glycogen in their cytoplasm (126).

High-resolution immunocytochemistry was utilized to localize the secretory pathway of Vg in mosquito trophocytes (119, 126). Vitellogenin is routed through compartments of rough endoplasmic reticulum to the Golgi complex, where it is packed into secretory granules and subsequently secreted (Figure 1.2C).

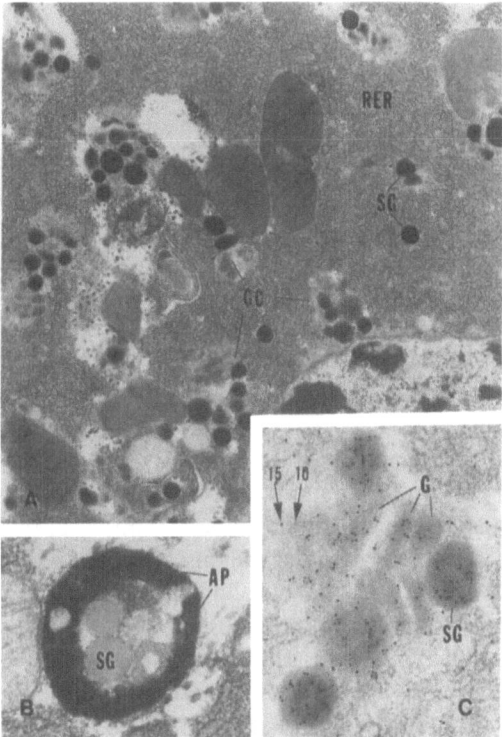

FIGURE 1.2. Ultrastructural analysis of Vg synthesis in the mosquito fat body. (A) Electron microscope image of a trophocyte at the peak of Vg synthesis, 27 hrs after a blood meal; note numerous cisternae of rough endoplasmic reticulum (RER), Golgi complexes (GC) and secretory granules (SG). X18,000. (From 117, with permission from publisher.) (B) An acid-phosphatase-contaning secondary lysosome with mature secretory granules (SG). Cytochemical reaction for acid phosphatase (AP). X80,000. (From 117, with permission from publisher.) (C) Double immunolabeling of the vitel-logenic trophocyte using monoclonal antibodies to large and small Vg subunits and protein A-colloidal gold of different sizes. Co-localization of the large and small Vg subunits in cisternae of the Golgi complex (G) and secretory granules (SG). The large Vg subunit is marked by 15 nm gold panicles (15) and the small Vg subunit by 10 nm gold particles (10). X70,000. (From 122, with permission from publisher.)

Biochemical and Molecular Aspects of Vitellogenin Synthesis

BIOCHEMICAL PROPERTIES OF MOSQUITO VITELLOGENIN AND VITELLIN

The major source of nutrients in an insect egg is the yolk protein, vitellin (Vn). The major yolk protein precursors, Vgs, are synthesized by the fat body as a single or multiple precursors, secreted into hemolymph and specifically accumulated by developing oocytes, where they are deposited as crystalline Vn. In most insects, Vgs are large oligomeric glycophospholipoproteins consisting of two or more subunits (87).

There have been discrepancies regarding the molecular weight and subunit composition of mosquito Vg/Vn. Hagedorn and Judson (65) first purified Vn from eggs of *A. aegypti* and reported that it was a glycolipoprotein with a molecular weight of 270,000 in its native form. Atlas et al. (2) estimated the molecular weight of *Culex* Vn to be 380,000. For *Aedes* Vn, Harnish and White (69) reported two values of 320,000 and 350,000 depending on the method utilized for molecular weight estimation. Using gel filtration chromatography, Borovsky and Whitney (18) found that Vg and Vn of *A. aegypti* had similar molecular weights (M_r 300,000). Utilizing non-denaturing polyacrylamide gel electrophoresis, Dhadialla and Raikhel (unpublished) have estimated molecular weights of *A. aegypti* Vg and Vn to be 380,000. The isoelectric points of Vg and Vn are 6.3 and 6.4, respectively (18).

In the first report on the subunit (apoprotein) composition, purified *Culex* Vn from newly-laid eggs was found to be composed of two subunits of 160 and 82 kDa (2). In a survey of insect vitellins, Harnish and White (69) reported that *Aedes* Vn consisted of two identical polypeptides of 170 kDa. Ma et al (97) showed that their monoclonal antibodies to Vn recognized two polypeptides (200 kDa and 66 kDa) on immunoblots after SDS PAGE of crude extracts of ovaries and the hemolymph of both *A. aegypti* and *A. atropalpus*. Based on SDS PAGE of *Aedes* Vn, Hagedorn (61) concluded that it was composed of two large subunits of 202 and 194 kDa and several small subunits around 56 kDa. In a more recent report, however, Vn of *A. aegypti* was reported to be composed of six subunits, while Vg consisted of three subunits (18).

In order to develop molecular probes which enable us to study a wide range of questions regarding mosquito vitellogenesis, and particularly biochemical properties of Vg and Vn, we have produced a library of monoclonal antibodies against Vn of the mosquito, *A. aegypti* (129, 132). These monoclonal antibodies recognize epitopes on both Vg and Vn (121, 129, 132). Immunoblot analysis have revealed that the monoclonal antibodies react only with either a 200 or 66 kDa polypeptide in extracts from vitellogenic fat bodies or ovaries (132, 128, 123).

Using these monoclonal antibodies in combination with radiolabeling and immunological techniques, Raikhel and Bose (123) have conclusively demonstrated that *A. aegypti* Vg is composed of only two subunits (apoproteins) with $M_r = 200,000$ and 66,000. In the oocyte, Vn is also composed of only these two subunits. No additional modification of Vn occurs after its internalization and crystallization in the oocyte prior to the cessation of vitellogenesis. The subunits of both Vg and Vn are not associated covalently, since they dissociate on SDS-PAGE under both reducing and non-reducing conditions. However, the stochiometric ratio of the two subunits in the native molecules of mosquito Vg and Vn is still not clear. The discrepancies concerning the structure of mosquito Vg and Vn could be attributed to instability of these molecules. A mixture of protease inhibitors with wide specificity is required to preserve their structural integrity (34, 123).

Early investigators have shown that mosquito Vn is glycosylated (2, 65). Labeling with radioactive carbohydrate precursors has revealed that both subunits of Vg and Vn are glycosylated (123). This situation is different from some other insects in which only the large Vg subunit appears to be glycosylated (109, 110). The carbohydrate moiety of mosquito Vg consists of mannose and N-acetylglucosamine and is sensitive to endo-beta-N-acetylglucosaminidase H (Endo-H). It remains Endo-H-sensitive after secretion from the fat body and crystallization inside the oocyte. This indicates that the high-mannose oligosaccharide moiety of Vg is not modified during processing in the fat body or oocytes. The carbohydrate moiety accounts for 10 kDa and 13 kDa of the molecular weight of the large and small subunits respectively (Figure 1.4B; 34).

Labeling with [^{32}P]-orthophosphate has shown that both subunits of Vg and Vn are phosphorylated and that the phosphorylated moieties are resistant to delipidation (123). Treatment of secreted Vg with Endo-H or alkaline phosphatase has revealed that most of the phosphorylated moieties are associated with the amino acid portion of the Vg molecule (34). Phosphorylation accounts for 7 kDa and 4 kDa of the large and small subunit molecular weights, respectively.

Mosquito Vg is also sulfated (34). Sulfation of both Vg subunits is most likely on the tyrosine residues, as evidenced by sensitivity of the [^{35}S]sulfate-labeled Vg to hydrolysis with HCI. Prior to this report, the only other insect yolk proteins known to be sulfated were those of *Drosophila melanogaster* (3).

BIOSYNTHESIS OF MOSQUITO VITELLOGENIN IN THE FAT BODY

Recently, the main biosynthetic steps for processing of the primary translation product into mature secreted Vg in the mosquito have been elucidated (19, 34, 124). The proposed pathway for Vg biosynthesis in the mosquito fat body is outlined in Figure 1.3.

Experiments using polyadenylated RNA from vitellogenic mosquitoes and in vitro cell-free translation assays have shown that both subunits of mosquito Vg originate from a common precursor and, therefore, are the products of the same gene (19, 34). The precursor polypeptide with $M_r = 224,000$ is a translation product specific to polyadenylated RNA from fat bodies of vitellogenic female mosquitoes. The identity of this Vg precursor has been confirmed by immunoprecipitation with monoclonal antibodies specific to the individual Vg subunits. Messenger RNA (mRNA) coding for Vg is about 6.5 kb long (Figure 1.5; 53). The size of the Vg precursor (pre-proVg), detected by in vitro translation assay of vitellogenic fat body mRNA, closely correlates with the size of a peptide product that can be derived from a 6.5 kb mRNA. Earlier, Harnish et al (70) reported incorrectly that translation of the mos-

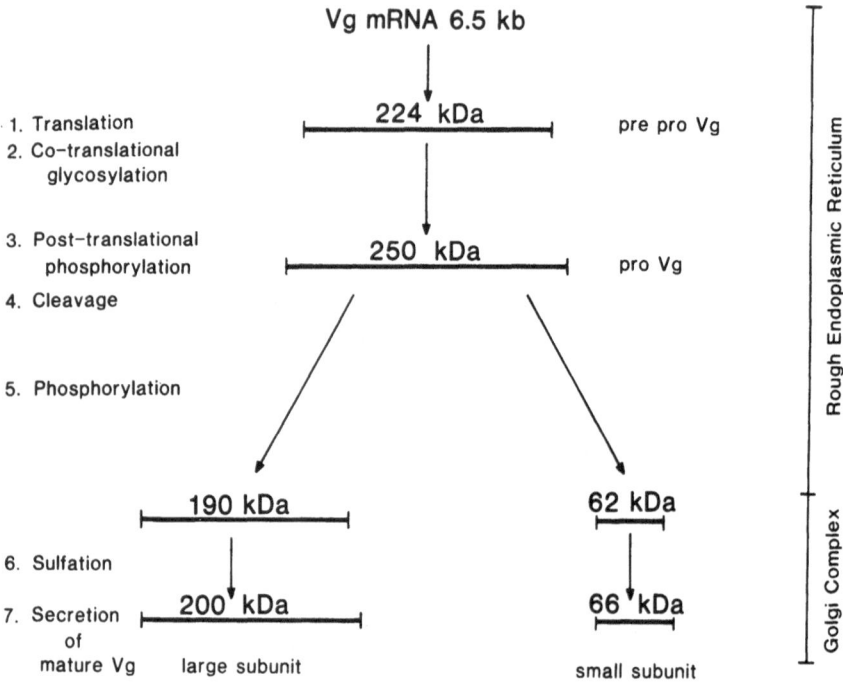

FIGURE 1.3. A schematic representation of the biosynthetic pathway for vitellogenin in the mosquito fat body. (From 34, with permission from publisher.)

quito Vg mRNA resulted into two polypetides of about identical size of 170 kDa.

In the fat body, a Vg precursor ($M_r = 250,000$) exists as a distinct interme-diate of the Vg biosynthetic pathway (Figure 1.4A) (34). The larger size of the cellular Vg precursor (proVg), as compared to pre-proVg observed in the cell-free translation, is due to its co-translational glycosylation and to a lesser degree to phosphorylation. After digestion with Endo-H, the molecular weights of the proVg and both the mature Vg subunits combined were reduced by 23 kDa each (Figure 1.4B). Furthermore, in vitro incubation of vitellogenic fat bodies with tunicamycin, an inhibitor of N-linked glycosyla-tion, has yielded the unglycosylated precursor with $M_r = 226,000$, which is close in size to the pre-proVg obtained from cell-free translation experiments. Thus, it appears that glycosylation of Vg occurs co-translationally and is completed at this stage of Vg biosynthesis.

Pulse-chase experiments in vitro have revealed rapid proteolytic cleavage of the 250-kDa proVg into two polypeptides with $M_r = 190,00$ and $62,000$ (34). Furthermore, transformation of the cleavage products into mature Vg subunits (200,000 and 66,000) occurs as one of the last steps prior to secretion of Vg. This increase in size of both the Vg subunits is sensitive to monensin,

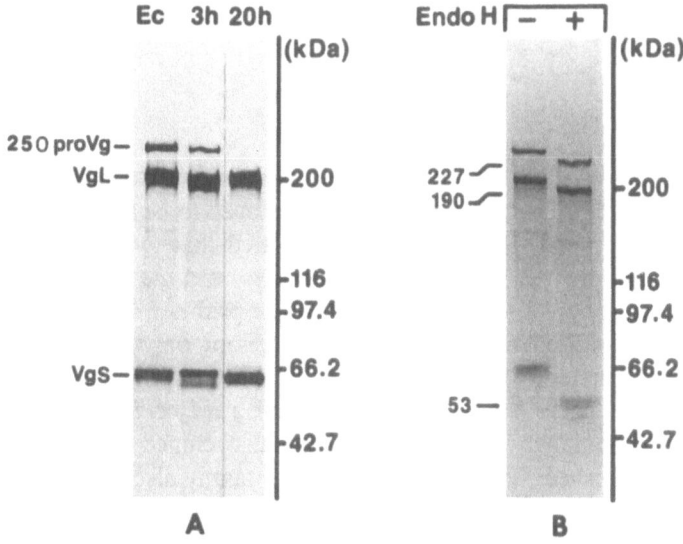

FIGURE 1.4 (A) In vitro pulse labeling of mosquito fat bodies of various physiological ages. Fat bodies, from previtellogenic females, after stimulation with 10^{-5} M 20-hydroxyecdysone for 6 hrs in vitro, and from vitellogenic females, 3 and 20 hrs after a blood meal were pulse labeled for two hr with [^{35}S]methionine and immunoprecipitated with a mixture of monoclonal antibodies against each Vg subunit. Immunoprecipitates of labeled fat body extracts were analyzed by SDS-PAGE and gel fluorography. 250 proVg, Vg precursor; VgL, large Vg subunit (200 kDa); VgS, small Vg subunit (66 kDa). (B) Effect of Endo-H on the mosquito Vg precursor and mature Vg subunits. Fat bodies, stimulated with 20-hydroxyecdysone as above, were pulse-labeled with [^{35}S]methionine. Immunoprecipitates of fat body homogenates were digested with Endo-H for 16 hr at 37°C. After digestion, the products were analyzed by SDS-PAGE. Presence and absence of Endo-H is indicated by + and −, respectively. The 227, 190, and 53 kDa molecules are the reduced M_r of the 250 proVg, VgL and VgS respectively, after Endo-H digestion (+). Molecular weights on the right in both figures, in order of decreasing M_r, are myosin, galactosidase, phosphorylase b, bovine serum albumin and ovalbumin (Bio-Rad). (From 124, with permission from publisher.)

an ionophore which disrupts function of the Golgi complex. Since sulfation, but not phosphorylation, has been predominantly blocked by monensin, the final maturation of Vg subunits in the Golgi complex is, at least in part, due to this modification (34).

Subunit-specific monoclonal antibodies and double gold immunolabelling have been used to investigate the exact intracellular location of two Vg subunits (200-kDa and 66-kDa) during their processing in the fat body cells. This study has shown that the same secretory pathway is employed by both the large and small subunits. The Vg subunits are located in all compartments of the secretory pathway, particularly in the Golgi complex and in secretory granules (Figure 1.2C; 122).

VITELLOGENIN GENES

In the mosquito, *A. aegypti* a multigene family consisting of five Vg genes was identified (53, 68). Four of these genes are believed to be actively transcribed (68). Restriction enzyme mapping demonstrated that two of the four genes (A1 and A2) have very similar restriction sites, and the other two have no sites in common with the first two genes. The probes from each of the four Vg genes hybridized to a 6.5 kb mRNA present only in the fat body of vitellogenic females (68).

Vitellogenin genes from *Anopheles gambiae*, a major vector of falciparum malaria in sub-Saharan Africa, have been also cloned (137, Romans and Miller, unpublished). The *Anopheles* Vg gene family also consists of four or five closely linked genes. Three of the genes are arranged in tandem with coding regions separated by approximately 3.0 kb. In addition to the three tandem genes, individual mosquitoes have one or two additional non-tandem Vg genes. Tandem and non-tandem genes are located together on the right arm of chromosome 2. In polytene chromosomes of ovarian nurse cells, the gene have been identified in position 18A in the dark band immediately adjacent to 18B. All genes apparently have extremely well conserved restriction sites within the 6.3 kb coding regions. This is in contrast to the divergent nature of *A. aegypti* Vg genes. Nucleotide sequence analysis of tandem genes has indicated that the genes are virtually identical in their protein coding regions. Changes are more frequent in introns or intergenic regions. So far, incomplete analysis of the coding regions shows that the amino termini encode a 16 amino acid signal peptide, which is interrupted by a 99 base pair intron. The sequence of the *Aedes* Vg A1 gene (Romans, Ke and Hagedorn, unpublished) predicts a very similar signal peptide of 16 amino acids, which is interrupted by a shorter (70 bp) intron in the eleventh codon. In *Anopheles*, the region upstream of the genes contain a comensus TATA box and two potential cap sites at a spacing normal for eukaryotic genes. Significantly, the conserved upstream regions have sequence homology to a known ovarian enhancer for *Drosophila* yolk protein genes. In addition, the upstream regions of vg2 and vg3 contain two element motifs similar to the fat body-specific enhancer of *Drosophila* yolk protein genes (52). These upstream regions also contain several sequence motifs similar to a *Drosophila* ecdysteroid response element (134). With the exception of the first fat body element and the ovarian enhancer sequence, regulatory elements found upstream of *Anopheles* Vg genes are also found upstream of *A. aegypti* Vg A1 gene (Romans, Ke and Hagedorn, unpublished). While at present, it is not known whether the ovaries of anophelines produced Vg, it is well established that *Aedes* ovaries do not (116, 122).

Identification of a Novel Protein with Serine Peptidase Activity Secreted by the Vitellogenic Fat Body and Accumulated by Oocytes

In order to fully understand the role of the fat body in vitellogenesis and oocyte maturation, activities, other than Vg synthesis, should also be elucidated. In several Lepidoptera, the fat body of adult females secretes a small 31-kDa protein, called microvitellogenin, which is accumulated in the oocytes (77, 78, 85, 86).

Analysis of proteins secreted by mosquito fat bodies has demonstrated that during vitellogenesis several proteins, in addition to Vg, are synthesized and secreted, and that one of these proteins is accumulated by the ovaries (71). This protein has an apparent molecular weight of 53,000, when separated under reducing conditions on SDS-PAGE. Glycosylation accounts for about 2–3 kDa in the size of this protein. It is not detectably phosphorylated or sulfated. In an in vitro cell-free translation assay, this protein is first synthesized as a 50-kDa unglycosylated precursor (26). Radioimmunoassay, using antibodies against the 53-kDa protein, has shown that this protein is synthesized only by the fat body of vitellogenic female mosquitoes. Its synthesis begins between 0–4 hrs PBM, peaks near 24 hrs PBM, and drops below detectable levels by 48 hrs (Figure 1.1B). Both the synthesis and secretion of the 53-kDa protein can be stimulated in previtellogenic female fat bodies cultured in vitro in the presence of physiological amounts of 20-HE (10^{-6}M).

Therefore, the properties of this additional sex-, stage- and tissue-specific protein, i.e., its size, kinetics and regulation of its synthesis, make it important for investigating hormonal control of the expression of fat body genes during vitellogenesis. Furthermore, the small size and simplicity of its posttranslational modifications make this vitellogenic protein an ideal marker for molecular studies of protein processing in the fat body and protein internalization by the oocyte.

The cDNA for the 53-kDa protein has been cloned by a combination of the lambda gt11 cDNA library screening and the polymerase chain reaction (26). The specificity of the cDNA clones has been confirmed by hybrid selection of mRNA and direct sequencing of the N-terminal of the 53-kDa protein. The cDNA clone recognizes a 1.5 kb mRNA present only in the fat body of vitellogenic females (Figure 1.5). The cDNA contains a single reading frame of 441 amino acids with characteristics of a secretory protein with M_r = 50,153. The predicted amino acid sequence of this protein does not have any similarity with lepidopteran microvitellogenin. Instead, it bears a significant similarity with several carboxypeptidases, in particular with human and mouse "protective proteins" and yeast carboxypeptidase KEX1, which are implicated in the processing and activation of a number of lysosomal enzymes (37, 50, 51,). At the onset of embryonic development, this protein, which we have named "vitellogenic carboxypeptidase" (VCP), undergoes processing as a result of which its size is reduced to 48-kDa, as identified by SDS-PAGE.

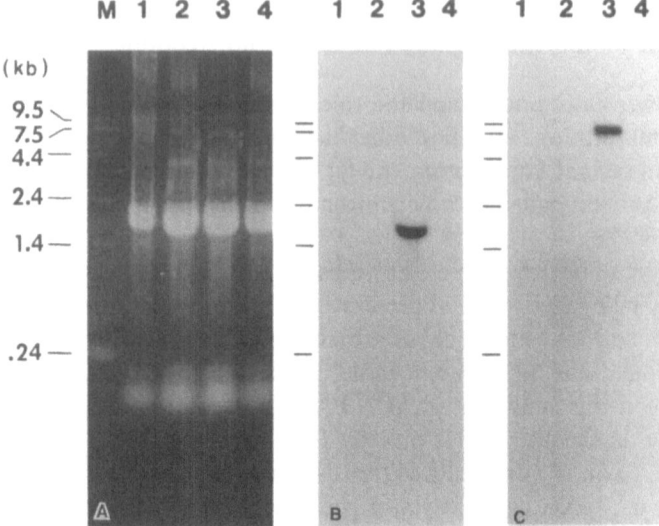

FIGURE 1.5. Northern blot analysis of stage and sex specificity of expression of vitellogenic carboxypeptidase and vitellogenin mRNAs. (A) agarose gel stained with ethidium bromide; (B) hybridization with 0.8 kb fragment of VCP; (C) hybridization with 1.8 kb fragment of mosquito Vg Al gene (gift of Dr. H. Hagedorn). Lanes: M-RNA standards; 1-total RNA from male mosquitoes; 2-total RNA from unfed, 4-day-old females; 3 and 4 total RNA from females 24 and 48 hrs after a blood meal respectively. (From 26, with permission from the publisher.)

Tests using serine protease inhibitor, diisopropyl fluorophosphate, have demostrated that only the 48-kDa polypeptide is an active enzyme (26). We suggest that VCP may play a role in activation of hydrolases associated with yolk bodies or other events at the onset of embryonic development. Serine proteases have been shown to participate in degradation of yolk during embryonic development in *Drosophila* (103), silkmoth (75) and ticks (40, 41). To our knowledge, however, this is the first example among oviparous animals of a processing enzyme, which is produced by an extraovarian tissue, internalized by developing oocytes, and apparently involved in embryonic development.

Regulation of Fat Body Activities During Vitellogenesis

ACTIVITIES OF THE ENDOCRINE SYSTEM AND HORMONE TITERS

For the purpose of the following discussion of the regulation of various aspects of vitellogenesis, I will briefly summarize the current state of knowledge concerning activity of the endocrine systems in the mosquito during vitellogenesis (Figure 1.1A). More detailed accounts of this subject can be found in the reviews listed in the Introduction.

The levels of juvenile hormone (JH), produced by the corpora allata (CA), in vitellogenic mosquitoes have recently been determined. Juvenile hormone III is the sole JH in adult mosquitoes (4, 133). The levels of JH III have been measured in whole body extracts of *A. aegypti* by coupled gas chromatography-mass spectrometry (141). The amount of JH rises over the first 2 days after emergence from 0.7 to 7.5 ng/g body weight, and then slowly declines over the next 5 days in females not given a blood meal (Figure 1.1A). In blood-fed females, JH levels decrease rapidly during the first 3 hrs to 2.3 ng/g body weight, followed by a slow decrease, such that by 24 hrs after the blood meal, JH levels has reach their lowest point (0.4 ng/g body weight). By 48 hrs, JH levels start to rise again until 96 hrs, when they are equivalent to pre-blood meal levels (Figure 1.1A). Readio et al. (133) have used in vitro incubation of CA and RIA to show that the glands from sugar-fed females of another anautogeneous mosquito *Culex pipiens*, synthesize JH III at a relatively constant rate of 271 fmol/hrs for 8 days after eclosion. Once again, blood-feeding suppressed the production of JH by CA. Corpora allata from blood-fed mosquitoes synthesize only 8.4 fmol/hrs at 4 hrs and 1.4 fmol/hrs at 12 hrs after the blood meal. In contrast to the above reports, Borovsky et al (15, 16) have indicated that there is an increase in synthesis of JH III by the female mosquito during the first several hours after a blood meal and that this synthesis occurs in the ovary. In two species of autogeneous mosquitoes, JH levels are reportedly low after eclosion and rise after a peak of ecdysteroids that correlates with the first cycle of egg maturation (58).

Ovaries of adult vitellogenic female mosquitoes synthesize and secrete a pro-hormone, ecdysone, which is converted to an active hormone, 20-hydroxyecdysone (20-HE), by the fat body (60, 61, 62, 66). In anautogeneous mosquitoes, *A. aegypti*, the levels of ecdysteroids are reportedly low during the first 8–10 hr after a blood meal with only a small peak at 4 hrs. Thereafter, the ecdysteroid level increases dramatically, reaching a maximum at 16–20 hrs PBM, and then rapidly declines to previtellogenic levels (Figure 1.1A). Although, these ecdysteroid levels were first determined in whole body preparations (66), they were later confirmed using hemolymph samples (60). During the first gonadotropic cycle of the autogeneous mosquito, *A. atropalpus*, the levels of ecdysteroids increase two-fold to 323 pg/female by 30 hrs after eclosion, which corresponds to the peak of Vg production (9, 100). The rates of ecdysone secretion by the ovary correlate to the kinetics of ecdysteroid levels in the mosquito bodies and hemolymph (9, 66).

In a series of microsurgical experiments, Lea has shown that a neurosecretory hormone, named "egg development neurosecretory hormone" (EDNH), is critical for development of eggs in mosquitoes (89, 91). It is produced by the medial neurosecretory cells of the brain and stored in the corpora cardiaca (CC), a neurohemal site for the neuroendocrine system (89, 91). In addition. in response to a blood meal, a neurosecretory hormone-releasing factor (NHRF) from ovaries stimulated the release of EDNH from the CC (94). Brain hormone, presumably EDNH, was shown to stimulate production of ecdysteroids by mosquito ovaries (67). Recently, ovarian ecdysteroidogenic hormones (OEH) with molecular weights between 6,500 and 13,000 have been isolated from the brain of the mosquito, *A. aegypti* (101). Since the site(s) of EDNH action has not been established in Lea's initial observations, it is not clear whether OEH is the full equivalent of EDNH, or whether there are more brain factors essential for vitellogenesis. According to Lea (91), EDNH is released during first 8 hrs after a blood meal (Figure 1.1A). Greenplate et al (56), however, have concluded that EDNH is released once before and once after 8 hrs following a blood meal.

Meola and Lea (105) have found that the mature ovary secretes a humoral factor, oostatic hormone (OSH), that inhibits yolk deposition in less developed follicles, if the female feeds on blood prior to egg deposition. Kelly et al. (79, 80) have suggested that OSH may act on the ovary affecting its production of ecdysone. A peptide factor, partially purified from ovaries, indeed arrested egg development (13). Later, Borovsky (14) has reported purification of a factor from mosquito ovaries which inhibits secretion of digestive enzymes in the midgut. Oostatic hormone appears to be secreted at the end of the vitellogenic cycle (Figure 1.1A), but it is not clear whether we are dealing with a single or multiple factors affecting one or several sites (14, 105).

In mosquito midgut epithelium, there are about 500 endocrine cells containing peptides, which cross-react with antisera to FMRF-amide, a molluscan cardioaccelerator, and to a number of vertebrate peptide hormones (23, 25). The function of these cells and their relation to events of vitellogenesis is not known at the present time.

REGULATION OF PREVITELLOGENIC DEVELOPMENT OF THE FAT BODY

Previtellogenic development is critical for the mosquito fat body to acquire competence for Vg synthesis and responsiveness to 20-HE. This period in the fat body development is controlled by JH, but the nature of this development has not been clear (46, 98). Recently, however, two cellular events, both controlled by JH, have been identified in trophocytes during previtellogenic development: 1) an increase in ploidy (36), and 2) the proliferation of ribosomes (130).

Using microspectrophotometric measurements of changes in nuclear DNA, Dittman et al., (36) have shown that the fat body cells in adult female *A. aegypti*, undergo two cycles of DNA replication. The first cycle begins after eclosion and resulted in 80% of diploid fat body cells becoming tetraploid and 20% becoming octoploid by the end of the third day (Figure 1.1B). (The second replication cycle occurred 48–72 h after a blood meal and resulted in an increase in octoploid nuclei to 67%). Topical application of JH or methoprene to abdomens*, isolated at emergence, stimulated an increase in ploidy levels above that normally seen in situ. Synthesis of DNA, estimated by incorporation of injected [³H]-thymidine, rose after emergence and remained high for 2 days (Figure 1.1B).

Juvenile hormone also regulates ribosomal production during previtellogenic development of the fat body (130). The fat body nucleoli enlarge three-fold, reaching maximal size between 2 and 3 days after eclosion Their granular component containing ribosomal precursors also increases considerably. The rate of RNA synthesis and the amount of ribosomal RNA in fat body cells rise during the first two days after eclosion and then decline gradually (Figure 1.1B). All characteristics of ribosomal proliferation, enlargement of nucleoli, accumulation of ribosomal RNA, and the increased rate of RNA synthesis, are blocked by removal of the CA in newly enclosed adult females and can be restored by either implantation of the CA or topical application of JH m or its analog, 7-S-methoprene, to allatectomized females.

DNA AND RIBOSOMAL SYNTHESIS IN VITELLOGENIC FAT BODIES

After a blood meal, synthesis of DNA increases again to a peak by 6 hrs and returns to a low level by 24 hrs PBM (Figure 1.1B; 36). Hormonal regulation of this DNA synthetic peak is not known, however, 20-HE has no effect (61).

The second peak in the activity of the nucleolus and the production of ribosomes occurs during the first 18 hrs after a blood meal (126). In a more recent report, Hotchkin and Fallon (73) have found that ribosome synthesis increases dramatically (4-fold compared to the previtellogenic levels) during the first part of the vitellogenic phase reaching its peak at about 18 hrs and by 48 hrs PBM has declined to levels below those found in previtellogenic fat bodies (Figure 1.1B). Significantly, the rate of ribosome production is higher at 10 hrs PBM than at 24 hrs, when the rate of Vg synthesis is at its peak (73). On the other hand, the amount of ribosomes in the trophocytes continues to increase until 24 hrs PBM, then it declines and by 48 hrs PBM has reached the levels found in previtellogenic fat bodies. Studies at the cellular level have demonstrated the participation of autolysosomes in destruction of ribosomes and RER during this time (118, 119). In females, the predominant ribonuclease activity has an acidic pH optimum, which indicates its lysosomal origin (47).

Thus, a cycle of build-up and degradation of biosynthetic organelles and ribosomes in particular, is coordinated with Vg synthesis. The regulation of ribosomal production in the trophocytes of the mosquito fat body during vitellogenic period, which is possibly different from that of Vg synthesis, remains to be elucidated.

In order to understand the mechanisms regulating the coordinated produc-

---

* isolation of abdomen removed the brain, corpora cardiaca and corpora allata.

tion of fat body ribosomes during mosquito vitellogenesis at the molecular level, genes coding for ribosomal RNAs (rRNAs) and proteins have been cloned (39, 49, 111). In the mosquito *Aedes albopictus*, the cDNAs, coding for L8 and L31 proteins of the large ribosomal subunit, hybridize to cytoplasmic mRNAs of 1.4 and 0.9 kb respectively (39). The structure of rRNA genes has been also reported for *A. aegypti* (49) and for *A. albopictus* (111). Mosquito rRNA genes are organized into units consisting of coding regions of 18S rRNA, 28Salpha and 28Sbeta rRNA and nontranslated spacer regions. As in all higher eukaryotes, the rDNA units are arranged in a head-to-tail, tandemly repeating manner (49). The length of the rDNA units is surprisingly different in the two closely related species being 9.0 kb in *A. aegypti* (49) and 15.6 kb in *A. albopictus* (111). The copy number of rRNA genes per haploid genome of *A. aegypti* increases from about 400 in larvae to about 1200 in adults (111). A similar increase in gene copy number was previously demonstrated for *Drosophila* rDNA (142). No significant changes in rRNA copy number in the fat body, however, have been observed between newly-emerged females, competent females prior to a blood meal, and females during active vitellogenin synthesis, 24 hrs PBM (111).

HORMONAL CONTROL OF VITELLOGENIN SYNTHESIS AND GENE EXPRESSION IN THE FAT BODY DURING VITELLOGENESIS

A great deal of information is available regarding the hormonal regulation of vitellogenin synthesis in mosquitoes. Although the importance of a neurosecretory hormone(s) from the brain, JH, and 20-HE for production of Vg by the fat body is well established, the precise role and interaction of each of these hormones in this process is still a matter of a considerable controversy (reviewed in 12, 20, 48, 60, 61, 62).

In the mosquito, *A. aegypti*, 20-HE was proposed as the primary factor initiating Vg synthesis in the fat body (43, 63, 64). The large doses of 20-HE needed to stimulate Vg synthesis in vivo in previtellogenic females of anautogeneous mosquitoes has been a major problem countering this hypothesis (48, 92). An attempt to explain this discrepancy by its rapid degradation in vivo (60) could not account for the fact that physiological doses of 20-HE are sufficient for initiation of Vg synthesis in autogeneous mosquito, *A. atropalpus* (48). Physiological doses of 20-HE are also effective in stimulation of Vg synthesis in the previtellogenic fat bodies in vitro as well as in decapitated females given blood by enema (48, 61, Raikhel, unpublished). It is possible that in previtellogenic anautogeneous mosquitoes, the response of the fat body to hormonal stimuli is suppressed by the state-of-arrest, the nature of which is poorly understood. High doses of 20-HE can override this arrest and stimulate Vg production in previtellogenic mosquitoes. The withdrawal of the fat body from this environment into an in vitro culture system possibly eliminates this arrest, and as a consequence, Vg production can be initiated by physiological doses of 20-HE in vitro. Factor(s) released after a blood meal could play a similar role in vivo. This arrest may be tissue-specific, since a physiological dose of 20-HE is effective in separating the secondary follicles in previtellogenic mosquitoes (5).

Another criticism has been directed at the low levels of Vg produced by the fat body after stimulation with 20-HE under permissive conditions such as incubations in vitro (48). This discrepancy was initially explained by a relatively poor nutritional status of previtellogenic females (60). Recent studies, however, have demonstrated a regulatory complexity of Vg production in the fat body at the cellular level (see above; 34, 126, 130). These studies suggest that the proliferation of the biosynthetic machinery preceeding massive production of Vg in the fat body is likely regulated independently of 20-HE. Therefore, it has been misleading to expect high levels of synthesis and secretion of Vg from a tissue lacking full biosynthetic capacity. When the effect of short exposures of 20-HE in vitro on fat bodies from previtellogenic females was tested, the rate of Vg synthesis was close to that in fat bodies of blood-fed mosquitoes (Table 1.1).

One of the most recent demonstrations of the regulatory complexity has been obtained in a study on Vg biosynthesis (34). As demonstrated by the analysis of microsomal fractions, the high-molecular weight Vg precursor (pro-Vg) exists as an intermediate in the biosynthesis of Vg. However, this precursor could be detected in whole fat body extracts only during the first few hours after the initiation of Vg synthesis by normal blood-feeding. Much greater accumulation of pro-Vg was found in fat bodies in which Vg synthesis was stimulated by 20-HE in vitro (Figure 1.4A). It is likely that 20-HE effects only transcription of Vg genes, which results in accumulation of pro-Vg. In normal blood-fed females, however, other factor(s) must affect production

TABLE 1.1. Effect of 20-hydroxyecdysone on vitellogenin synthesis by fat bodies from previtellogenic mosquitoes cultured in vitro.

Dose of 20-HE (M)	Percentage of control (6 hrs PBM)
0	9%
10^{-8}	16%
10^{-7}	47%
10^{-6}	60%
10^{-5}	73%
B 6h	100%

Fat bodies were dissected from previtellogenic females, 4 days after eclosion, and were incubated in the culture medium (81) for 6 hrs at 27°C in the absence or presence of 20-HE at the indicated concentration. Culture medium with 20-HE was replaced every 2 hrs. The fat bodies were then placed in the culture medium without the hormone but supplemented with [^{35}S]methionine for 1 hr. Fat bodies from blood-fed controls were placed in culture medium containing [^{35}S]methionine without prior incubation. Following several washings, the fat bodies were homogenized and the amount of newly synthesized Vg was determined by immunoprecipitation with monoclonal antibodies. Values are expressed as percentage of control which was fat bodies from females 6 hrs after blood feeding (B 6h). Three replicates, each containing 3 fat bodies, were used for every treatment. Raikhel, unpublished.

of the biosynthetic machinery, perhaps a cleavage protease required for the accelerated rate of post-translational cleavage of pro-Vg. In fact, recently, we have obtained experimental evidence showing the importance of head factor(s) for the cleavage of pro-Vg to occur at the normal rate (Raikhel and Hays, unpublished).

Several observations have indicated that low levels of Vg production can be initiated in *A. aegypti* by undetermined factor(s) associated with the blood meal. In the fat body of normal blood-fed females, Vg was detected immunocytochemically (126) and later at the level of mRNA (115) as early as 1 hr after a blood meal, prior to any increase in the 20-HE titer. Using RIA and anti-Vg monoclonal antibodies, Vg synthesis can be detected even at 30 mins PBM (Hays and Raikhel, unpublished). Low levels of Vg production in females can be stimulated by a blood meal in the absence of the head, thorax and ovaries (148). In decapitated females, a blood meal stimulates an increase in Vg mRNA that was 7.5% of the intact controls (115). These observations suggest a direct link between the gut and the fat body. Considering the rapid appearance of Vg in the fat body after a blood meal, it is possible that a hormonal, rather than nutritional factor, released by the midgut endocrine cells causes directly or indirectly the onset of Vg production at a low level.

The debate about the role of the corpora allata (CA) and JH in mosquito vitellogenesis has been going on since the work by Detinova (32). Lea (88, 90) has demonstrated that once the fat body is exposed to JH during its previtellogenic development, the CA is no longer required for the successful egg development after a blood meal; viable eggs can mature in female mosquitoes allatectomized several days before the blood meal. However, an application of small doses of JH analogue (25 pg) to isolated abdomens has reportedly reduced the amounts of exogenous 20-HE required for stimulation of egg development (17). These authors have suggested that JH may play a direct role in regulating Vg synthesis. A similar effect was recently reproduced by a small dose of 20-HE (99).

Bownes (20, 21) has hypothesized that JH along with products of sex determination genes plays a direct role in activation of yolk protein genes in *Drosophila*. She further suggested that in *Drosophila* 20-HE affects rates of expression of the yolk protein genes. It is possible that in the female mosquito the role of 20-HE is to enhance the expression of Vg genes, which has already been initiated by JH. Alternatively, 20-HE may act on Vg genes in the mosquito prior to a blood meal, but their actual expression remains arrested until a blood meal.

One of the most significant arguments for the involvement of 20-HE in the regulation of expression of Vg genes in mosquitoes has been the recent finding of hormone response elements in the regulatory sequences of Vg genes of *Anopheles* and *Aedes* (137, Romans and Miller, unpublished; Romans, Ke and Hagedorn, unpublished). These hormone response elements in mosquito Vg genes are similar to ecdysteroid response elements found in *Drosophila* genes (134). However, a comparative analysis of the genomic regulatory regions could be undecisive, since the regulatory sequences responsible for binding of

JH and 20-HE receptors may be very similar (112, 150). Future gene transformation experiments, utilizing modified promoter regions of mosquito Vg genes, should bring more definite answers on how 20-HE and JH affect the expression of Vg genes in mosquitoes.

Cessation of Vg synthesis by trophocytes is coordinated with termination of yolk accumulation by oocytes. At least two hypotheses have been introduced in order to explain the coordinated cessation of Vg synthesis: (1) the programmed termination after exposure of the fat body to 20-HE (10); (2) feedback control resulting from the rising Vg titer in the hemolymph (11). An ovarian inhibitory factor or oostatic hormone may also be involved directly or indirectly in regulation of the cessation of Vg synthesis in the mosquito fat body (13, 14, 80, 105).

Termination of expression of Vg genes may occur simultaneously or precedes the actual cessation of Vg synthesis in the mosquito fat body. This depends on the level of accumulation and the stability of Vg mRNA. In *Xenopus* liver, estrogens selectively stabilize Vg mRNA (22), but the effect of 20-HE on the stability of mosquito Vg mRNA is not known. While elucidation of this effect of 20-HE represents an important question for future studies, as a starting point, it is essential to characterize kinetics of Vg mRNA in the mosquito fat body during vitellogenesis. Although in all studies mentioned, it was shown that following a peak at 24–28 hrs PBM, Vg synthesis ceases rapidly, Racioppi et al. (115) reported that the amount of Vg mRNA reached its peak levels at 36 hrs PBM. However, in an attempt to explain this discrepancy, we have used the same methodology and have found that the amount of Vg mRNA has a peak at 24–28 hrs PMB which coincides with the maximal rate of Vg synthesis (Cho and Raikhel, unpublished). Further studies utilizing techniques such as nuclear run-on, which detects the rate of mRNA synthesis rather than the amount of mRNA, are required to determine the precise time and the mechanisms regulating the termination of Vg gene expression.

LYSOSOMAL REGULATION OF TROPHOCYTE SYNTHETIC AND
SECRETORY ACTIVITIES

Our studies have demonstrated that cellular events during the cessation of Vg synthesis are far more complex than just an interruption of Vg synthesis and that the lysosomal system is involved in the regulation of Vg secretion and trophocyte remodelling (118, 119, 120).

The rise in specific activities of several lysosomal enzymes coincides with a dramatic decline in Vg concentration in the fat body (Figure 1.1B; 118). Further analysis by video-enhanced fluorescent and electron microscopy has revealed two important physiological roles of lysosomes during termination of Vg production: (1) interruption of Vg secretion by degrading the Vg-containing secretory granules (Figure 1.2B), (2) destruction of the biosynthetic machinery, RER and Golgi complexes, and subsequent remodelling of trophocytes (118, 119). Thus, it appears that, in the mosquito fat body, two separate mechanisms are involved in regulating the cessation of Vg synthesis and secretion: one is responsible for turning off the expression of Vg genes, and another, the lysosomal activity, for interrupting the Vg secretion and organelle degradation.

Experiments with ovariectomized vitellogenic mosquitoes have suggested that a high concentration of Vg in the hemolymph provides a feedback stimulus in regulating this specific lysosomal activity in the fat body cells (120). 20-hydroxyecdysone may also be involved in the stimulation of lysosomal activity in the fat body of vitellogenic adult insects as was shown for *Calliphora* (147).

As an initial step towards a molecular approach to the study of this important cellular regulatory mechanism, a lysosomal aspartic protease from the mosquito *A. aegypti* was purified and characterized (27, 124). The structural characteristics of this enzyme are similar to those of mammalian cathepsin E, in that it has a native molecular weight of 80,000 and consists of two identical subunits with $M_r = 40,000$. The purified enzyme, however, exhibits properties characteristic of cathepsin D; it utilizes hemoglobin as a substrate, and its activity is completely inhibited by pepstatin-A and 6M urea but not by 10 mM KCN. This mosquito aspartic protease does not have isozymes, and its pI is 5.4. Density gradient centrifugation of organelles followed by enzymatic and immunoblot analyses demonstrated the lysosomal nature of the purified enzyme. A nineteen amino acid stretch from the N-terminus of the mosquito lysosomal protease was sequenced and found to have 74% identity with N-termini of human and porcine cathepsins D. The cDNA coding for this mosquito lysosomal protease was cloned by the polymerase chain reaction technique using two oligonucleotide probes; the first one based on the sequence of N-terminal of the mosquito lysosomal protease and the second one on the sequence of the pepstatin-binding site conserved among aspartic proteases. The cloned DNA fragment is being used for cloning a full-length cDNA corresponding to the mosquito lysosomal protease as well as for studying the expression of its gene (Cho and Raikhel, unpublished).

Selective Endocytosis of Yolk Protein Precursors in Mosquito Oocytes

Structure and Development of the Mosquito Ovary

The paired mosquito ovary consists of about 75 ovarioles of meroistic, polytrophic type (28). At eclosion, each ovariole contains a germarium and a single follicle. There are seven nurse cells and an oocyte surrounded by a monolayer of the follicular epithelium that comprise a follicle. When 1.5–2 μl blood is injested, follicles develop synchronously throughout all the stages of egg maturation.

At eclosion, *A. aegypti* oocytes have an undifferentiated cortex and are not competent for protein uptake. As the oocytes develop over next 48 hrs coated vesicles, microvilli and endosomes appear and render the oocyte competent to take up protein (Figure 1.1C; 127). The oocytes remain arrested until a blood meal initiates vitellogenesis. Accumulation of yolk proceeds during next 30–36 hrs. The formation of the first layer of the eggshell, the vitelline envelope, occurs during between 8 and 36 hrs PBM, and later, chorion is deposited (Figure 1.1C; 131). Egg maturation is completed in about 48–60 hrs PBM.

Accumulative Pathway of Vitellogenin in the Oocyte

Endocytosis of yolk proteins by developing oocytes is a central event in insect vitellogenesis. Selective uptake of extraovarian yolk protein precursors by *Hyalophora* oocytes suggested to Telfer a role for endocytosis in yolk accumulation (146). Coated vesicles, the cellular structure associated with selective endocytosis, were first observed in the mosquito oocyte (139). Selective or receptor-mediated endocytosis is now recognized as an ubiquitous mechanism for internalizing functionally important macromolecules in animal cells (55, 114). Molecules of a particular protein (ligand) are bound by specific receptors present in the cell membrane. Ligand-receptor complexes concentrate in coated pits that invaginate and pinch off to form intracellular coated vesicles (114). Coated vesicles carry the ligand-receptor complexes into the next cellular compartment, an endosome, which plays a key role in directing subsequent intracellular routes of both the ligand and receptor (72, 104). The ATP-dependent acidification of endosomes is the event that, in most cases, leads to the dissociation of ligand-receptor complexes followed by recycling of receptors back to the cell surface and delivery of free ligand into lysosomes (72, 104). Many pathogens utilize coated vesicles as a gate for cell entry, and the low pH of endosomes facilitates escape of pathogens from these organelles prior to lysosomal degradation (72)

Investigations of endocytotic organelles, stimulated by Roth and Porter's study of the mosquito oocyte, have led to the conclusion that the coated vesicle is universal organelle for the specific macromolecular transport in eukaryotic cells (114). The characteristic feature of a coated vesicle is an outer proteinaceous polyhedral cage enclosing its membrane (113, 114). Pearse (113) was the first to isolate a major protein of this cage. This protein, which she has d clathrin and is currently known as clathrin heavy chain, has a molecular weight of 180,000. Later, it has been found that three molecules of clathrin heavy chain together with three molecules of clathrin light chain (33 and 36 kDa) form a subunit or triskelion. Triskelions self-assemble into a polyhedral cage on the membrane of the coated pit/vesicle (114). In mosquito oocytes, clathrin and other protein components of coated vesicles are apparently synthesized in large amounts prior to onset of endocytotic activity. During previtellogenic development of the mosquito, *A. aegypti*, oocytes contain large aggregates of membrane-free lattices strikingly similar to poly-hedral cages assembled from purified clathrin (117).

Utilization of high-resolution immunocytochemistry gave important insight to the accumulative pathway of Vg in the mosquito oocyte (Figure 1.6) (116). Vitellogenin reaches the oocyte surface through interfollicular spaces. The oocyte membrane is differentiated into microdomains; Vg binds to receptors located at the base of and between the oocyte microvilli (Figures 1.6 and 1.7). After binding to its receptors, Vg is internalized by coated vesicles. The dissociation of Vg from its receptors occurs in the next compartment, the endosome. Endosomes coalesce into a transitional yolk body, a specific intermediate compartment, which plays an important role in Vg processing and presumably recycling of Vg receptors. In this compartment, Vg condensates and crystallizes. The transitional yolk body transforms into a mature yolk body, which is the final compartment containing crystalline Vn.

Our library of monoclonal antibodies against mosquito Vg/Vn was utilized to visualize the Vg accumulative pathway in the mosquito oocyte (121, 122). A method of double immunogold labeling was developed allowing simultaneous localization of both the Vg subunits (Figure 1.7B). This analysis has demonstrated that both Vg subunits are routed together through all steps of the accumulative pathway in the mosquito oocyte (122).

Fluorescent immunocytochemistry has demonstrated that distribution of the novel protein, vitellogenic carboxypeptidase (VCP), in the mosquito ovary is, for the most part, similar to that of Vg/Vn (71). Vitellogenic carboxypeptidase is present in interfollicular and perioocytic spaces of follicles and in yolk bodies of oocytes (Figure 1.7D). In the latter, however, it is located around the crystalline Vn (Figures 1.7C and 1.7D). Without immunocytochemical data at the ultrastructural level, it is not clear whether VCP is routed together with Vg or is delivered to yolk bodies by separate compartments.

One of the central problems of selective endocytosis is an understanding of the mechanism governing postendocytotic routing and sorting of macro-molecules. The sorting of the specific protein, Vg, and a non-specific protein, horseradish peroxidase, in the mosquito oocyte has been studied. These proteins, internalized by the oocyte, follow either a specific accumulative route or a lysosomal degradative route (128). Via coated vesicles, all proteins enter the same compartment, the endosome, where they dissociate from membrane-binding sites. The route to their final destination depends on the presence of the specific ligand, Vg. In its absence, the degradative route is followed and the endosome with a non-specific protein fuses with lysosomes. In the presence of the specific ligand, Vg, the accumulative route is followed, and both proteins are delivered into the transitional yolk body. We concluded that any protein bound to the membrane would be internalized by the oocyte, but only binding of Vg to its receptor served as a transmembrane signal stimulating the formation of accumulative compartments (128). The results of this study provide an important background for investigating the mechanism of trans-ovarial transmission of viruses and other pathogens in mosquitoes.

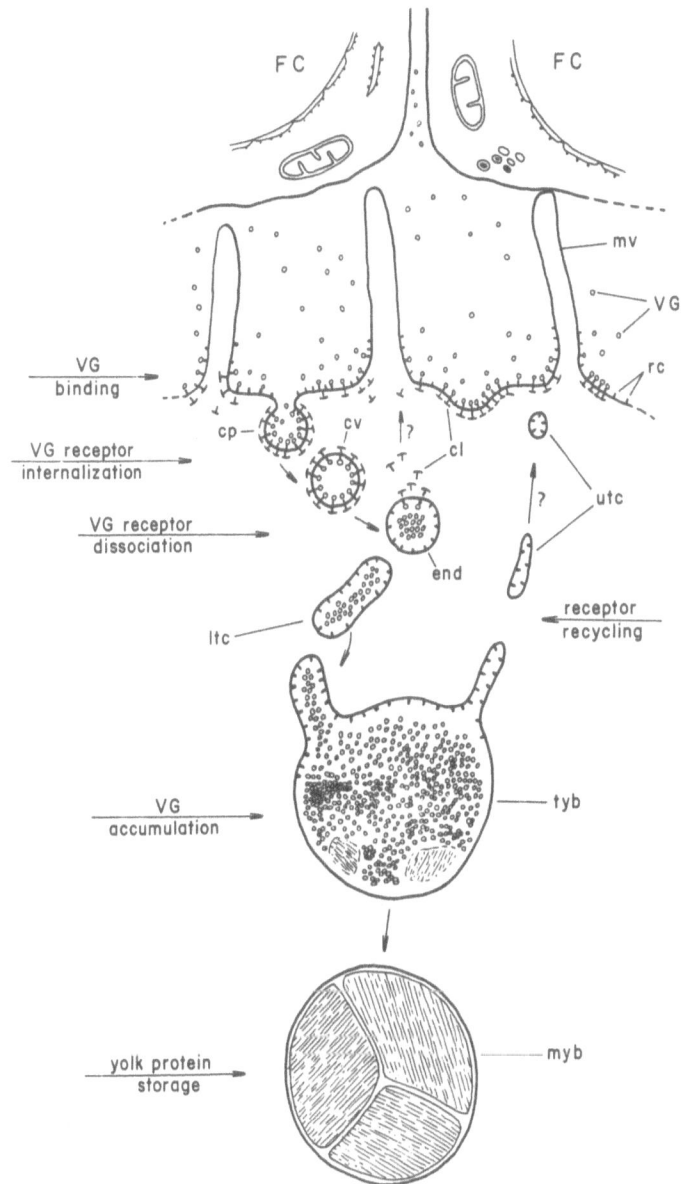

FIGURE 1.6. A schematic representation of the Vg-accumulative pathway in the mosquito oocyte. cl, clathrin; cp, coated pit; cv, coated vesicle; end, endosome (vesicular); FC, follicle cell; ltc, labeled (by Vg antibodies) tubular compartment (tubular endosome); mv, microvillus; myb, mature yolk body; rc, Vg receptor; tyb, transitional yolk body; utc unlabeled (with Vg antibodies) tubular compartment; VG, vitellogenin. (From 116, with permission from publisher.)

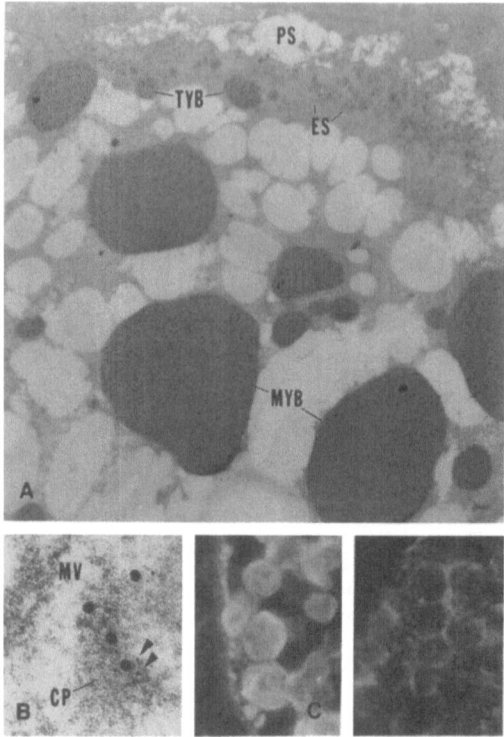

FIGURES 1.7. Internalization of Vg in the mosquito oocyte. Immunolabeling with anti-Vg monoclonal antibodies and colloidal gold probe. (A) Cross section of the vitellogenic oocyte. Immunolabeling was performed using monoclonal antibodies against the small Vg subunit, unconjugated rabbit anti-mouse IgG and protein A conjugated to colloidal gold, 15 nm particles. Note the labeling in endosomes, transitional and mature yolk bodies. × 12,000. (B) Double immunolabeling of the oocyte with monoclonal antibodies to large and small Vg subunits. 5 nm gold particles mark a small Vg subunit (arrows) and 15 nm gold particles a large Vg subunit. × 150,000. (C) and (D) Immunoflourescent localization of Vn and VCP in the yolk bodies of oocytes. Frozen sections of vitellogenic ovaries were labeled by polyclonal antibodies against Vg/Vn (C) or VCP (D) followed by rhodamine-conjugated goat anti-rabbit secondary antibodies. Note that, in contrast to Vn, which occupies the entire yolk body, VCP is located as a rim around Vn. CP, coated pit; ES, endosome; MV, microvillus; MYB, mature yolk body; PS, perioocytic space; TYB, transitional yolk body. (A and B are from 129; C and D are from 71, with permission from publisher.)

Selective Endocytosis of Vitellogenin by Mosquito Oocytes

The principal challenge facing successful studies of Vg endocytosis with mosquitoes was the isolation of sufficient quantities of purified Vg which could be radiolabeled to a high specific activity without loss of structural integrity. We have found that an in vitro system best satisfies these requirements. Vitellogenin is metabolically labeled to a high specific activity (3 to 6 $\times$ 10^5 CPM/μg) by incubating vitellogenic fat bodies in the presence of [^{35}S]methionine in vitro. Vitellogenin is then purified by a one step ion-exchange chromatography on DEAE-Sepharose CL-6B (81).

Using [^{35}S]-Vg as a probe, we have developed an in vitro assay to study the endocytosis of Vg by mosquito oocytes. In an optimized in vitro system, Vg endocytosis is maximal at pH 7.5. It declines at acid pH and ceases abruptly at pH 6.3, a pH value corresponding to the isoelectric point of mosquito Vg. Uptake of Vg by mosquito oocytes exhibits properties of receptor-mediated endocytosis; i.e. temperature dependence, saturability, selectivity, and tissue specificity. The uptake of Vg is inhibited at 4°C with only 6% of the uptake achieved at 27°C. At 24 hrs PBM, uptake of Vg by oocytes at 27°C, neares saturation with a Vg concentration in the medium of 8 μg/μl (Figure 1.8). Saturation kinetics generated for Vg by these oocytes has produced a $V_{max} = 3.2$ μg/μl/hrs and an apparent $K_{uptake} = 8.4 \times 10^{-6}$ M. Ovaries accumulate 10 times the amount of Vg compared to mouse IgG, while uptake of Vn has been 72% of that for Vg, indicating that oocytes are able to distinguish these proteins. The unbound fraction from ion-exchange chromatography of secreted fat body proteins, which does not contain Vg but included VCP, is taken up at 25% of the rate for Vg. This indicates the possibility that some proteins from this mixture are also taken up specifically. Non-ovarian tissues, i.e., fat body and Malpighian tubules, accumulate both Vg and IgG at equally low levels (81).

The endocytosis of Vg by oocytes in vitro has also been characterized for several other insect species (reviewed in 125). Interestingly, the concentration of Vg required for saturation of its uptake ranges from 80 μM for the cockroach *Nauphoeta* to 21 μM for the mosquito *A. aegypti*. The low Vg concentration saturating uptake by the mosquito oocyte indicates that the oocyte has receptors capable of a greater rate of turnover (recycling to the oocyte surface) than in the other insects. A higher rate of receptor turnover could be one of the adaptations essential for intense Vg accumulation characteristic of the very rapid development of the mosquito oocyte.

The rate at which Vg is internalized by oocytes changes during development. With the onset of vitellogenesis, the interfollicular and perioocytic spaces are formed (Figure 1.1C; 131), and that may facilitate the Vg uptake similarly to the well-established event, known as patency, in follicles of *Rhodnius prolixus* (31). The interfollicular spaces in *Aedes*, however, are not as large as those in *Rhodnius*. The endocytotic rate for Vg in mosquito oocytes increases in a bi-phasic manner. During the first 2.5 hrs PBM, the rate of Vg

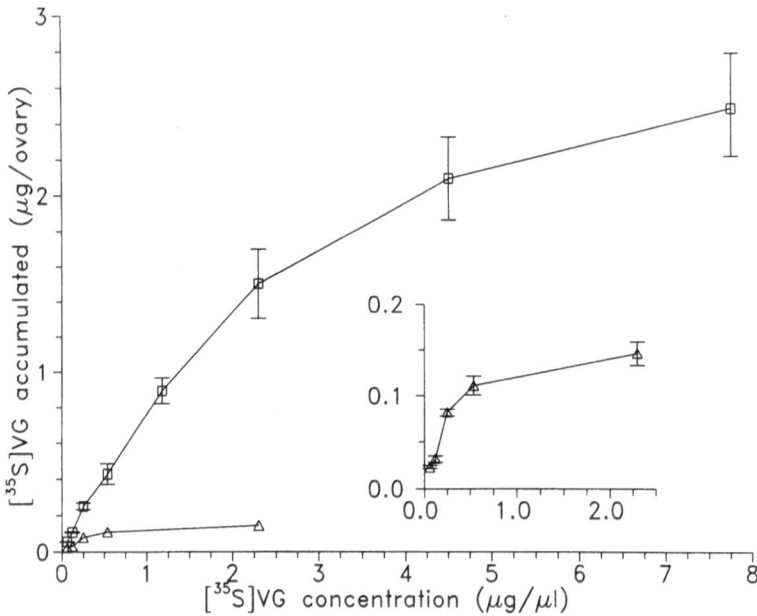

FIGURE 1.8. Saturation kinetics of Vg uptake by mosquito ovaries incubated in vitro. Ovaries with oocyte 160 μm long were selected 24 hr PBM. Ovaries were preincubated at 27°C ($\square$) and 4°C ($\triangle$) for 1 hr at an increasing concentration of Vg. Specific activity was 6.69 × 10^4 cpm/μg. Inset enlarges the uptake curve at 4°C. Points are for mean $\pm$ SE of 3 replicates. (From 81, with permission from publisher.)

endocytosis rises slowly and doubles by the end of the second hours. After this time, the rate of Vg endocytosis progresses more rapidly reaching a 10-fold increase during the next two hours (82). This rapid phase of endocytosis continues until 24 hrs PBM. The peak of Vg uptake occurs between 24 and 30 hrs PBM, followed by a precipitous decline and cessation of uptake by 36 hrs (Figure 1.1C; 81, 82). An interruption of Vg uptake is likely due to closing of interfollicular channels (Figure 1.1C). Tight junctions, which are formed between follicle cells, cease the permeability of interfollicular channels at about 32 hrs PBM (1, Raikhel, unpublished).

Regulation of Protein Accumulation in Mosquito Oocytes

HORMONE-MEDIATED DEVELOPMENT OF OOCYTE COMPETENCE OF PROTEIN UPTAKE

Gwadz and Spielman (59) have demonstrated that during the previtellogenic period primary follicles in mosquito ovaries grow from 45-50 um to 100 um in length and that this growth is under the control of JH. We have found that formation of endocytotic organelles in the oocyte cortex, which leads to oocyte competence to internalize proteins, is also controlled by this hormone (127). It appears that clathrin and possibly other proteins essential for endocytosis, such as Vg receptors, are synthesized in large amounts prior to the onset of the endocytic activity under the control of JH (117, 127). Recently we have demonstrated that differentiation of the follicular epithelium in primary follicles is also controlled by JH (131). In mosquitoes, therefore, the gonadotropic action of JH leads to the competence of cellular components of the primary follicles for their subsequent activities during vitellogenic period of oocyte maturation (Figure 1.1C).

INITIATION OF VITELLOGENIN UPTAKE IN THE MOSQUITO OVARY

Our knowledge about the regulation of protein uptake in oocytes of mosquitoes has been limited. Recently, however, utilizing an in vitro assay for monitoring Vg uptake, we have begun to investigate these events (82). As stated above, the rate of Vg endocytosis increases in two phases, the first, a slow increase, lasting 2.5 hrs, and a second rapid increase, which continues for the next 22–24 hrs. Decapitation, performed within 5 mins of the beginning of blood-feeding, eliminates the second, rapid phase, but not the first one. However, endocytosis is not stimulated in abdomens of previtellogenic females, which are isolated first and then given an enema of blood, and assayed 4 hrs later. Thus, in the intact animal endocytosis is first activated as a result of blood-feeding itself, which through nervous stimulation may result in a pulsed release of a factor possibly from the corpora cardiaca, as shown for diuretic hormone (8).

Since vitellogenic events can be activated by the enema of blood given to intact mosquitoes and viable eggs produced (148), it is clear that the activation of the first phase in the ovarian response through feeding may be bypassed. It is, however, an important physiological adaptation, which provides a rapid initiation of vitellogenic events under normal feeding conditions. Delays in decapitations longer than one hour result in only partial inhibition of the second, rapid phase of Vg endocytosis. This phase becomes fully independent from the head when decapitation is delayed for 16–20 hrs PBM. The second activation phase, therefore, is also dependent on a head factor, the continued release of which is essential for the maintenance of this ovarian function. Additional evidence in support of involvement of peptide neurohormones in the initiation of endocytosis has come from experiments using cAMP, a cyclic nucleotide which acts as second messenger in peptide hormone response. In vitro tests have shown that cAMP, but not cGMP, is effective in stimulating Vg uptake in previtellogenic ovaries (Koller and Raikhel, unpublished).

Injections of physiological doses of 20-HE into females in the state-of-arrest are ineffective in stimulating Vg accumulation in ovaries. In contrast, in females decapitated 5 mins after a blood meal, this hormone restores the second activation phase of Vg uptake (82). It is not clear, however, whether the effect of 20-HE on the rate of Vg uptake is due to an increase in the hemolymph titer of Vg or due to its direct action on the ovary. This hormone has been shown to regulate two other functions in the ovary; the separation of the secondary follicle from the germarium (5) and deposition of vitelline envelope (131).

Earlier studies indicated that JH may be involved in the regulation of protein accumulation in oocytes of several insects (7, 45). In *Rhodnius*, JH controls formation of spaces between follicle cells, patency (31). In mosquito follicles, perioocytic and interfollicular spaces are formed by 2 hrs PBM, but the control of their formation is not as clear as in *Rhodnius* (131). This event in mosquito follicle is clearly independent from 20-HE, since it was not induced when this hormone was injected into previtellogenic females. On the other hand, the separation was observed in previtellogenic decapitated females or isolated abdomens given an enema of blood. Surprisingly, a similar effect was obtained when previtellogenic females were implanted with a pair of active CA or treated with JH III. In the view of recent reports of the production of JH in the mosquito after a blood meal (15, 16), further careful investigation of the role of JH in the regulation of the initial stages of Vg uptake in mosquito ovary should be carried out.

Monitoring protein synthetic rates in the ovary provides an additional insight into the initiation of ovarian development after a blood meal. The rate of protein synthesis, measured by incubating ovaries in vitro in the presence of $[^{35}S]$methionine for 1 hr, increases dramatically during the first 30 mins PBM, after which it declines during the next hour and then rises gradually for the next 8 hrs. Like Vg uptake, activation of ovarian protein synthesis depends upon stimulation from blood-feeding and from head factors, although it becomes indepedent from the head factors 8 hrs PBM (82).

Properties of Mosquito Vitellogenin Receptor

Recently, the presence of specific receptors for Vg in ovarian membranes of *A. aegypti* has been confirmed by an in vitro binding assay (35). The binding reaction, which is dependent on pH and Ca^{2+}, uses Vn-free ovarian membranes, $[^{35}S]$-Vg labeled metabolically, and unlabeled Vn for competition. At pH 7.0 and in the presence of 5 mM Ca^{2+}, binding of Vg to its receptor reaches equilibrium within 60–90 mins at both 4° and 25°C. Binding is only specific for membranes prepared from ovaries. While both mosquito Vg and Vn bind with equal affinity to Vg receptors on ovarian membranes, neither locust Vg nor mouse IgG has any measurable affinity towards these sites. Analysis by non-linear least square fit of the saturation isotherms reveals a single class of Vg receptors on ovarian membranes with a dissociation constant (K_d) of 1.8×10^{-7} M (35). The K_d value for binding of Vg to its receptor in the mosquito is similar to K_d values obtained for *Locusta* and two species of cockroaches (83, 84, 135). The K_d value for *Manduca* Vg binding to its receptors on oocyte membrane was lower (1.3×10^{-8} M) (110).

In initial experiments, octyl-beta-D-glucoside, a non-ionic detergent, was found to be the most effective in solubilizing the Vg receptor from mosquito ovarian membranes (33) as had been previously shown for a number of other receptors (reviewed in 125). The properties of the solubilized receptors were tested using a solid-phase binding assay as introduced by Schneider et al. (140). Using this assay, we have demonstrated the tissue-specific nature and binding selectivity of solubilized Vg receptors. The sensitivity of Vg receptor to both heat and trypsin confirms its protein nature. Saturation analysis, performed for solubilized Vg receptors, has yielded the dissociation constant (K_d) of 2.8×10^{-8} M (35). Similarly, solubilized *Locusta* Vg receptors has had K_d 10-fold higher than that for membranes isolated from ovarian follicles (136).

A ligand blotting technique has successfully been utilized for visualization

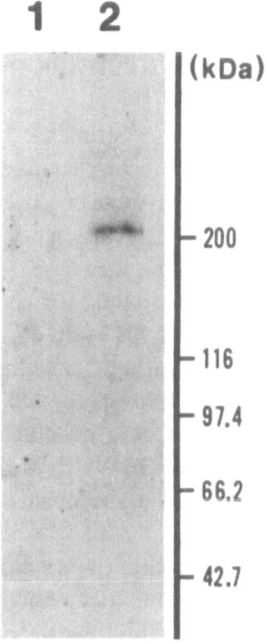

FIGURE 1.9. Visualization of the mosquito Vg receptor by the ligand-blotting technique. Membranes from ovaries 24 hrs PBM were solubilized with octyl-beta-D-glucoside and separated on 6% SDS polyacrylamide gel under reducing (lane 1) and non-reducing (lane 2) conditions followed by transfer to a nitrocellulose membrane. Ligand blotting was performed by incubating the nitrocellulose membrane with 150,000 cpm/ml of [^{35}S]Vg. Binding of radiolabeled Vg to its receptor was visulized by autoradiography of the washed and dried blot. The relative migration of standard molecular weight markers is shown by their weights on the right side of the autoradiogram. Note a band of a 205 kDa molecule, which is present in lane 2 (non-reducing conditions). (Dhadialla, Hays, and Raikhel, unpublished.)

of Vg receptors in the chicken (143) and in *Locusta* (136). We have used the ligand blotting protocol, initially developed for the low density lipoprotein receptor (30), in order to visualize the mosquito Vg receptor. Samples containing the solubilized putative Vg receptor were separated by SDS-PAGE under reducing and non-reducing conditions, electrophoretically transfered to a nitrocellulose membrane and probed with [^{35}S]-Vg. Radiolabeled Vg bound to a single polypeptide with $M_r = 205,000$ in an ovarian membrane preparation only under non-reducing conditions (Figure 1.9). This indicates that disulfate bonds are important in preserving a structural integrity of a functional Vg receptor. A polypeptide with $M_r = 156,000$ is identified as the Vg receptor in the *Locusta* ovary (136), while a 200 kDa protein is thought to be the Vg receptor in the cockroach, *Nauphtoeta cinerea* (74). More detailed account of the mechanisms of yolk accumulation in oocytes of insects, including mosquitoes can be found in a recent review on this subject by Raikhel and Dhadialla (125).

Conclusions and Future Directions

During the past decade significant progress has been achieved in our understanding of cellular, biochemical and molecular aspects of mosquito vitellogenesis. Studies at the cellular level have led to a realization of the complexity of events associated with Vg synthesis in the mosquito fat body and its uptake by the oocyte.

Regulation of Vg synthesis is not limited to just turning on and off its genes, but Vg synthesis is accompanied by a cycle of stringently controlled proliferation and degradation of biosynthetic cellular machinery (121, 124). The biosynthesis of Vg involves many steps of co- and posttranslational modifications (34). Furthermore, the vitellogenic mosquito fat body produces another protein, vitellogenic carboxypeptidase (VCP), which is also deposited in yolk bodies of developing oocytes and is involved in embryonic development (26, 71).

Elucidation at the molecular level of the developmental program of the fat body during vitellogenesis represents a challenging task for the future. The major genes involved in three sequential waves of activity of the vitellogenic fat body (proliferation of biosynthetic organelles, production of vitellogenic proteins and termination of their synthesis) have been identified and are currently under intensive investigation. The genes coding for Vg have been cloned and their structure partially elucidated for two mosquito species (68, 137). Importantly, putative ecdysteroid response elements in promoters of these genes have been found. A full-length cDNA coding for VCP has been cloned and characterized and cloning of its gene(s) is in progress (26). A number of genes for ribosomal RNAs and proteins have been characterized as well (44). Recently, a lysosomal protease has been purified (27), and its cDNA clone isolated (Cho and Raikhel, unpublished). Once these gene systems are structurally characterized, further progress in elucidating regulation of their expression, as well as their future biotechnological applications, will depend on the availability of gene transformation in mosquitoes. Germline gene transformation is still limited to *Drosophila*, because *Drosophila* transposable elements, which are critical for gene integration in the genome, are not fuctional in non-drosophilids and in which transposable elements have not yet been isolated (108). First reports of germline transformation in mosquitoes, however, provide a promise towards resolving this problem, which is the cornerstone of the molecular genetics of any organism (29, 102, 106, 107). Meanwhile, transfection of mosquito cell lines can provide an alternative to germline transformation for the promoter analysis of many important genes (38, 42, 54).

Considerable progress also has been made in our understanding of the mechanism of specific accumulation of yolk protein in mosquito oocytes. The pathway of Vg in the oocytes was mapped in detail by immunocytochemistry at the electron microscope level (116, 121). Properties of Vg endocytosis by developing oocytes and Vg binding to ovarian membranes have been investigated (35, 81). Finally, the Vg receptor, solubilized from ovarian membranes, has been visualized and characterized (33). This progress has paved the way for purification of the mosquito Vg receptor and further exploration of the molecular basis of interaction between Vg and its receptor. We have begun to elucidate regulatory aspects of biogenesis of endocytotic organelles and Vg endocytosis in the mosquito oocyte (82, 127). Recent progress which has been achieved in purification of mosquito neuropeptides (24, 93) will enhance further elucidation of the regulatory aspects of Vg synthesis and endocytosis.

In mosquitoes, vitellogenesis is a key event in oocyte maturation which is intimately tied to hematophagy and pathogen transmission. Elucidation of the molecular basis of vitellogenesis is critical, therefore, to the success in development of novel vector management strategies which will be based on genetic modifications of the vector capacity to transmit pathogens or vector's reproductive physiology. There are, however, many gaps in our understanding of mosquito vitellogenesis, and their solution will still require the efforts of physiologists and biochemists rather than of molecular biologists alone. The most important and least understood event is the nature of anautogeny, which, after all, is the very reason for a vector's ability to transmit pathogens.

Acknowlegments. I thank Drs. A. Fallon, H.H. Hagedorn, A.O. Lea, and P. Romans who kindly sent recent reprints and preprints of their unpublished work. Thanks to Drs. T.S. Dhadialla, A.O. Lea, and M.R. Brown for critical reading of the manuscript, and Mrs. R. Bickert for the excellent help in making the graphs and formatting the manuscript. This work was supported by grants AI-24716, HD-22958 and AI-32154 from the National Institutes of Health.

References

1. Anderson, W.A., and Spielman, A., 1971, Permeability of the ovarian follicle of *Aedes aegypti* mosquitoes., *J. Cell Biol.* **50**:201–221.
2. Atlas, S.J., Roth, T.F., and Falcone, A.J., 1978, Purification and partial characterization of *Culex fatigans* yolk protein, *Insect Biochem.* **8**:111–115.
3. Baeuerle, P.A., and Huttner, W.B., 1985, Tyrosine sulfation of yolk proteins 1, 2, and 3 in *Drosophila melanogaster, J. Biol. Chem.* **260**:6434–6439.
4. Baker, F.C., Hagedorn, H.H., Schooley, D.A., and Wheelock, G., 1983, Mosquito juvenile hormone: identification and bioassay activity, *J. Insect Physiol.* **29**:465–470.
5. Beckemeyer, E.F., and Lea, A.O., 1980, Induction of follicle separation in the mosquito by physiological amounts of ecdysterone, *Science* Washington, **209**:819–821.
6. Behan, M., and Hagedorn, H.H., 1978, Ultrastructural changes in the fat body of adult female *Aedes aegypti* in relationship to vitellogenin synthesis, *Cell Tiss. Res.* **186**:499–506.
7. Bell, W.J., and Barth, R.H., 1971, Initiation of yolk deposition by juvenile hormone, *Nature.* **230**:220–222.
8. Beyenbach, K.W., and Petzel, D.H., 1987, Diuresis in mosquito: role of a natriuretic factor, *New Physiol. Sci* **2**:171–175.
9. Birnbaum, M.J., Kelly. T.J., Woods. C.W., and Imberski R.B., 1984, Hormonal regulation of ovarian ecdysteroid production in the autogenous mosquito, *Aedes atropalpus, Gen. Comp. Endocrin.* **56**:9–18.
10. Bohm, M.K., Behan, M., and Hagedorn H.H., 1978., Termination of vitellogenin synthesis by mosquito fat body, a programmed response to ecdysterone, *Physiol. Ent.* **3**:17–25.
11. Borovsky, D., 1981, Feedback regulation of vitellogenin synthesis in *Aedes aegypti* and *Aedes atropalpus, Insect Physiol.* **11**:207–213.
12. Borovsky, D., 1984, Control mechanisms for vitellogenin synthesis in mosquitoes, *BioEssays.* **1**:254–267.
13. Borovsky, D., 1985, Isolation and characterization of highly purified mosquito oostatic hormone, *Arch. Insect Biochem. Physiol.* **2**:333–349.
14. Borovsky, D., 1988, Oostatic hormone inhibits biosynthesis of midgut proteolytic enzymes and egg development in mosquitoes, *Arch. Insect Biochem. Physiol.* **7**:187–210.
15. Borovsky, D., Carlson, D.A., Ujary, I., and Prestwich, G.D., 1990, Synthesis and degradation of JH III in the mosquito *Aedes aegypti*, in Hagedorn, H.H., Hildebrand, J.G., Kidwell, M.G., Law, J.H. (eds): Molecular Insect Science, Pergamon Press, New York, London, p. 281.
16. Borovsky, D., Carlson, D.A., Ujary, I., and Prestwich, G.D., 1990, Synthesis of (IOR) JH III by *Aedes aegypti* ovary, Annual Meeting of Entomological Society of America, p. 69.
17. Borovsky, D., Thomas, B.R., Carlson, D.A., Whisenton, L.R., and Fuchs, M.S., 1985, Juvenile hormone and 20-hydroxyecdysone as primary and secondary stimuli of vitellogenesis in *Aedes aegypti, Arch. Insect Biochem. Physiol* **2**:75–90.
18. Borovsky, D., and Whitney, P.L., 1987. Biosynthesis, purification, and characterization of *Aedes aegypti* vitellin and vitellogenin, *Arch. Insect Biochem. Physiol.* **4**:81–99.

19. Bose, S.G., and Raikhel, A.S., 1988., Mosquito vitellogenin subunits originate from a common precursor, *Biochem. Biophys. Res. Comm.* **155**:436–442.
20. Bownes, M., 1986, Expression of the genes coding for vitellogenin (yolk protein), *Ann. Rev. Entomol.* **31**:507–531.
21. Bownes, M., 1989, The roles of juvenile hormone, ecdysone and the ovary in the control of *Drosophila* vitellogenesis, *J. Insect Physiol.* **35**:409–413.
22. Brock, M.L., and Shapiro, D.J., 1983, Estrogen stabilizes vitellogenin mRNA against cytoplasmic degradation, *Cell.* **34**:207–214.
23. Brown, M.R., Crim, J.W., and Lea, A.O., 1986., FMRFamide- and pancreatic polypeptide-like immunoreactivity in midgut endocrine cells of a mosquito, *Tissue Cell* **18**:419–428.
24. Brown, M.R., and Lea, A.O., 1989, Neuroendocrine and midgut endocrine systems in the adult mosquito, *Adv. Disease Vector Res.* **6**:29–58.
25. Brown, M.R., Raikhel, A.S., and Lea, A.O., 1985, Ultrastructure of midgut endocrine cells in the mosquito, *Aedes aegypti, Tissue Cell* **17**:709–721.
26. Cho, W.-L., Deitsch, K.W., and Raikhel, A.S., 1991, Extraovarian protein accumulated in mosquito oocytes is a serine carboxypeptidase activated in embryos, *Proc. Natl. Acad. Sci. USA* **88**:10821–10824.
27. Cho, W.-L., Dhadialla, T.S., and Raikhel A.S., 1991, Purification and characterization of a lysosomal aspartic protease with cathepsin D activity from the mosquito, *Insect Biochem.* **21**:165–176.
28. Christophers, S.S.R., 1960, *Aedes aegypti* (L.), The Yellow Fever Mosquito, *Cambridge University Press*, England.
29. Crampton, J.M., Morris, A., Lycett, G., Warren, A., and Eggleston, P., 1990, Molecular characterization and genome manipulation of the mosquito, *Aedes aegypti*, in Hogedorn, H.H., Hildebrand, J.G., Kidwell, M.G., Law, J.H. (eds): Molecular Insect Science, Plenum Press, New York, London, pp. 1–11.
30. Daniel, T.O., Schneider, W.J., and Goldstein, J.L., 1983. Visualization of lipoprotein receptors by ligand blotting, *J. Biol. Chem.* **258**:4606–4611.
31. Davey, K.G., 1981, Hormonal control of vitellogenin uptake in *Rhodnius prolixus, Am. Zool.* **21**:701–705.
32. Detinova, T.S., 1945, On the influence of glands of internal secretion upon the ripening of the gonads and the imaginal diapause in *Anopheles maculipennis, Zoologitsheskij Journal* Moscow **24**:291–298.
33. Dhadialla, T.S., Hays, A.R., and Raikhel. A.S., 1992, Characterization of the solubilized mosquito vitellogenin receptor. *Insect Biochem. Mol. Biol.* (in press).
34. Dhadialla, T.S., and Raikhel, A.S., 1990, Biosynthesis of mosquito vitellogenin, *J. Biol. Chem.* **265**:9924–9933.
35. Dhadialla, T.S., and Raikhel, A.S., 1991, Binding of vitellogenin to membranes isolated from mosquito ovaries, *Arch Insect Biochem. Physiol.* **18**:55–70.
36. Dittmann, F., Kogan, P.H., and Hagedorn, H.H., 1989, Ploidy levels and DNA synthesis in the fat body cells of the adult mosquito, *Aedes aegypti*: The role of juvenile hormone, *Arch. Insect Biochem. Physiol.* **12**:133–143.
37. Dmochowska, A., Dignard, D., Henning, D., Thomas, D.Y., and Bussey, H., 1987, Yeast KEX1 gene encodes a putative protease with a carbosypeptidase B-like function involved in killer toxin and α-factor precursor processing, *Cell* **50**:573–584.
38. Durbin, J.E., and Fallon, A.M., 1985, Transient expression of the chloramphenicol acetyl transferase gene in cultured mosquito cells, *Gene* **36**:173–178.

39. Durbin, J.E., Swerdel, M.R., and Fallon, A.M., 1988, Identification of cDNAs corresponding to mosquito ribosomal protein genes, *Biochem. Biophys. Acta* **950**:182–192.

40. Fagotto, F., 1990, Yolk degradation in tick egg.: I. Occurrence of a cathepsin l-like acid proteinase in yolk spheres, *Arch Insect Biochem. Physiol.* **14**:217–235.

41. Fagotto, F., 1990, Yolk degradation in tick egg.: II. Evidence that cathepsin l-like proteinase is stored as a latent, acid-activable proenzyme, *Arch. Insect Biochem. Physiol.* **14**:237–252.

42. Fallon, A.M., 1986, Factors affecting polybrene-mediated transfection of cultured *Aedes aldopictus* (mosquito) cells, *Exper. Cell Res.* **166**:535–542.

43. Fallon, A.M., Hagedorn, H.H., Wyatt, G.R., and Laufer, H., 1974, Activation of vitellogenin synthesis in the mosquito *Aedes aegypti* by ecdysone, *J. Insect Physiol.* **20**:1815–1823.

44. Fallon, A.M., Park Y.-J., and Durbin, J.E., 1990. Analysis of ribosomal RNA in the mosquito, *Aedes aegypti*, Borovsky, D. and Spielman, A. (eds): *Host Regulated Developmental Mechanisms in Vector Arthropods*, University of Florida, IFAS, Florida Medical Entomology Laboratory, pp. 25–31.

45. Ferenz, H.-J., Lubzens, E.W., and Glass, H., 1981., Vitellin and vitellogenin incorporation by isolated oocytes of *Locusta migratoria migratorioides* (R.F.), *J. Insect Physiol.* **27**:869–875.

46. Flanagan, T.R., and Hagedorn, H.H., 1977, Vitellogenin synthesis in the mosquito: the role of juvenile hormone in the development of responsiveness to ecdysone, *Physiol. Entomol.* **2**:173–178.

47. Fritz, M.A., Hotchkin, P.G., and Fallon, A.M., 1986, Changes in ribonuclease activity during development of the mosquito, *Aedes aegypti*, *Comp. Biochem. Physiol.* **84B**:355–361.

48. Fuchs, M.S., and Kang, S-H., 1981, Ecdysone and mosquito vitellgenesis: a critical appraisal, *Insect Biochem.* **11**:627–633.

49. Gale, K, and Crampton, J., 1989, The ribosomal genes of the mosquito, *Aedes aegypti*, *Eur. J. Biochem.* **185**:311–317.

50. Galjart, N.J., Gillemans, N., Harris, A., van der Horst, G.T.J., Verheijen, F.W., Galjaard, H., and d'Azzo, A., 1988, Expression of cDNA encoding the human "protective protein" associated with lysosomal β-Galactosidase and neuraminidase: homology to yeast proteases, *Cell* **54**:755–764.

51. Galjart, N.J., Gillemans, N., Meijer, D., and d'Azzo, A., 1990, Mouse "protective protein", *J. Biol. Chem.* **265**:4678–4684.

52. Garabedian, M.J., Shepherd, B.M., and Wensink, P.C., 1986, A tissue-specific transcription enhancer from the *Drosophila* yolk protein 1 gene, *Cell* **45**:859–867.

53. Gemmill, R.M., Hamblin, M., Glaser, R.L., Racioppi, J.V., Marx, J.L., White, B.N., Calvo, J.M., Wolfner, M.F., and Hagedorn. H.H., 1986, Isolation of mosquito vitellogenin genes and induction of expression by 20-hydroxyecdysone, *Insect Biochem.* **16**:761–774.

54. Gerenday, A., Park, Y.J., Lan, Q., and A.M Fallon., 1989, Expression of a heat-inducible gene in transfected mosquito cells, *Insect Biochem.* **19**:679–686.

55. Goldstein, J.L., Brown, M.S., Anderson, R.G.W., Russell, D.W., and Schneider, W.J., 1985, Receptor-mediated endocytosis: concepts emerging from the LDL receptor system, *Ann. Rev. Cell Biol.* **1**:1–39.

56. Greenplate, J.T., Glaser, R.L., and Hagedorn, H.H., 1985., The role of factors from the head in the regulation of egg development in the mosquito *Aedes aegypti*, *J. Insect. Physiol.* **31**:323–329.

57. Gubler, D.J., 1987, Current research on dengue, *Current Topics Disease Vector Res.* **3**:37–54.
58. Guilvard, E., DeReggi, M., and Rioux, J.A., 1984, Changes in ecdysteroid and juvenile hormone titers correlated to the initiation of vitellogenesis in two *Aedes* species (Diptera, Culicidae), *Gen. Comp. Endocrinol.* **53**:218–223.
59. Gwardz, R.W., and Spielman, A., 1973., Corpus allatum control of ovarian development in *Aedes aegypti, J. Insect Physiol.* **19**:1441–1448.
60. Hagedorn, H.H., 1983, The role of ecdysteroids in the adult insect, in Downer, R.G.H. and Laufer, H. (eds): *Endocrinology of Insects*, Alan R. Liss, Inc., New York, pp. 271–304.
61. Hagedorn, H.H., 1985, The role of ecdysteroids in reproduction, in Kerkut, G.A., Gilbert, L.I. (eds): *Comprehensive Insect Physiology, Biochemistry, and Pharmacology*, Volume 8, Pergamon Press, Oxford, pp. 205–261.
62. Hagedorn, H.H., 1989, Physiological roles of hemolymph ecdysteroids in the adult insect, in Koolman, J. (ed.): *Ecdysone from Chemistry to Mode of Action*, Thieme Medical Publishers, Inc., New York, pp. 279–289.
63. Hagedorn, H.H., and Fallon, A.M., 1973., Ovarian control of vitellogenin synthesis by the fat body in *Aedes aegypti, Nature.* **244**:103–105.
64. Hagedorn, H.H., Fallon, A.M., and H. Laufer, 1973, Vitellogenin synthesis by the fat body of *Aedes aegypti, Develop. Biol.* **31**:285–294.
65. Hagedorn, H.H., and Judson., C.L., 1972, Purification and site of synthesis of *Aedes aegypti* yolk proteins, *J. Exp. Zool.* **182**:367–377.
66. Hagedorn, H.H., O'Connor, J.D., Fuchs, M.S., Sage, B., Schlaeger, D.A., and Bohm, M.K., 1975, The ovary as a source of α-ecdysone in an adult mosquito, *Proc. Natl. Acad. Sci. USA,* **72**:3255–3259.
67. Hagedorn, H.H., Shapiro, J.P., and Hanaoka, K., 1979, Ovarian ecdysone secretion is controlled by a brain hormone in an adult mosquito, *Nature* **282**:92–94.
68. Hamblin, M.T., Marx, J.L., Wolfner, M.F., and Hagedorn, H.H., 1987, The vitellogenin gene family of *Aedes aegypti, Mem. Inst. Oswaldo Cruz* **82**:109–114.
69. Harnish, D.G., and White, B.N., 1982, Insect vitellins: identification, purification and characterization from eight orders, *J. Exp. Zool.* **220**:1–10.
70. Harnish, D.G., Wyatt, G.R., and White, B.N., 1982, Insect vitellins: identification of primary products of translation, *J. Exp. Zool.* **220**:11–19.
71. Hays, A.R., and Raikhel, A.S., 1990, A novel protein produced by the vitellogenic fat body and accumulated in mosquito oocytes, *Roux's Arch Develop. Biol.* **199**:114–121.
72. Helenius, A., Mellman, I., Wall, D., and Hubbard, A., 1983, Endosomes, *TIBS.* **8**:245–250.
73. Hotchkin, P.G., and Fallon, A.M., 1987, Ribosome metabolism during the vitellogenin cycle of the mosquito *Aedes aegypti, Biochem. Biophys. Acta* **924**:352–359.
74. Indrasith, L.S., Kindle, H., and Lanzrein., B, 1990, Solubilization, identification, and localization of vitellogenin-binding sites in follicles of the cockroach, *Nauphoeta cinerea, Arch. Insect Biochem. Physiol.* **15**:1–16.
75. Indrasith, L.S., Sasaki, T., and Yamashita, O., 1988, A unique protease responsible for selective degradation of a yolk protein in *Bombyx mori, J. Biol. Chem.* **263**:1045–1051.
76. Kan, S.-P., and Ho, B.-C., 1972, Development of *Breinlia sergenti* (Dipetalonematidae) in the fat-body of mosquitoes. II. Ultrastructural changes in the fat-body, *J. Med. Entomol.* **9**:255–261.

77. Kawooya, J.K., and Law, J.H., 1983, Purification and properties of micro-vitellogenin of *Manduca sexta*. Role of juvenile hormone in appearance and uptake, *Biochem. Biophys. Res. Comm.* **117**:643–650.
78. Kawooya, J.K., Osir, E.O., and Law, J.H., 1986, Physical and chemical properties of microvitellogenin. Physical and chemical properties of microvitellogenin, *J. Biol. Chem.* **261**:10844–10849.
79. Kelly, T.J., Birnbaum, M.J., Woods, C.W., and Borkovec, A.B., 1984, Effects of housefly oostatic hormone on egg development neurosecretory hormone action in *Aedes atropalpus, J. Exp. Zool.* **229**:491.
80. Kelly, T.J., Masler, P.E., Schwartz, M.B., and Haught, S.B., 1986, Inhibitory effects of oostatic hormone on ovarian maturation and ecdysteroid production in diptera, *Insect Biochem.* **16**:273.
81. Koller, C.N., Dhadialla, T.S., and Raikhel, A.S., 1989, A study of receptor-mediated endocytosis of vitellogenin in mosquito oocytes, *Insect Biochem.* **19**:693–702.
82. Koller, C.N., and Raikhel, A.S., 1991, Initiation of vitellogenin uptake and protein synthesis in mosquito ovary, *J. Insect Physiol.* **37**:703–711.
83. Konig, R., and Lanzrein, B., 1985, Binding of vitellogenin to specific receptors in oocyte membrane preparations of the ovoviviparous cockroach *Nauphoeta cinerea, Insect Biochem.* **15**:735–747.
84. Konig, R., Nordin, J.H., Gochoco, C.H., and Kunkel, J.G., 1988, Studies on ligand recognition by vitellogenin receptors in follicle membrane preparations of the German cockroach, *Blattella germanica, Insect Biochem.* **18**:395–404.
85. Kulakosky, P.C., and Telfer, W.H., 1987, Selective endocytosis, *in vitro*, by ovarian follicles from *Hyalophora cecropia, Insect Biochem.* **17**:845–858.
86. Kulakosky, P.C. and Telfer, W.H., 1989, Kinetics of yolk precursor uptake in *Hyalophora cecropia*: stimulation of microvitellin endocytosis by vitellogenin, *Insect Biochem.* **19**:367–373.
87. Kunkel, J.G., and Nordin, J.H., 1985, Yolk proteins, in Kerkut, G.A., Gilbert, L.I. (eds.): *Comprehensive Insect Physiology Biochemistry and Pharmacology*, Volume 1, Plenum Publishing, New York, pp. 83–111.
88. Lea, A.O., 1963, Some relationships between environment, corpora allata, and egg maturation in Aedine mosquitoes, *J. Insect Physiol.* **9**:793–809.
89. Lea, A.O., 1967, The medial neurosecretory cells and egg maturation in mosquitoes, *J. Insect Physiol.* **13**:419–429.
90. Lea, A.O., 1969, Egg maturation in mosquitoes not regulated by the corpora allata, *J. Insect Physiol.* **15**:537–541.
91. Lea, A.O., 1972, Regulation of egg maturation in the mosquito by the neurosecretory system: the role of the corpus cardiacum, *Gen. Comp. Endocrinol. Supp.* **3**:602–608.
92. Lea, A.O., 1982, Artifactual stimulation of vitellogenesis in *Aedes aegypti* by 20-hydroxyecdysone, *J. Insect. Physiol.* **28**:173–176.
93. Lea, A.O., and Brown, M.R., 1990, Neuropeptides of mosquitoes, in Hagedorn, H.H., Hildebrand, J.G., Kidwell, M.G., Law, J.H. (eds): *Molecular Insect Science*, Plenum Press, New York, pp. 147–154.
94. Lea, A.O., and Van Handel, E, 1982, A neurosecretory hormone-releasing factor from ovaries of mosquitoes fed blood, *J. Insect Physiol.* **28**:503–508.
95. Ma, M., Gong, H., Newton, P.E., and Borovec, A.B., 1986, Monitoring *Aedes aegypti* vitellogenin production and uptake with hybridoma antibodies, *J. Insect Physiol.* **32**:207–213.

96. Ma, M., Gong, H., Zhang, J.-Z., and Gwadz, R., 1987, Response of cultured *Aedes aegypti* fat body to 20-hydroxyecdysone, *J. Insect Physiol.* **33**:89–93.

97. Ma, M., Newton, P.B., He, G., Kelly, T.J., Masler, E.P., and Borkovec, A.B., 1984, Development of monoclonal antibodies for monitoring *Aedes atropalpus* vitellogenesis, *J. Insect Physiol.* **30**:529–536.

98. Ma, M., Zhang, J.-Z., Gong, H., and Gwarz, R, 1988, Permissive action of juvenile hormone on vitellogenin production by the mosquito, *Aedes aegypti, J. Insect Physiol.* **34**:593–596.

99. Martinez, T., and Hagedorn, H.H., 1987, Development of responsiveness to hormones after a blood meal in the mosquito *Aedes aegypti, Insect Biochem.* **17**:1095–1098.

100. Masler, E.P., Fuchs, M.S., Sage, B., and O'Connor, J.D., 1981, A positive correlation between oocyte production and ecdysteroid levels in adult *Aedes. Physiol, Entomol.* **6**:45–49.

101. Matsumoro, S., Brown, M.R., Crim, J.W., Vigna, S.R., and Lea, A.O., 1989, Isolation and characterization of ovarian ecdysteroidogenic hormones from the mosquito, *Aedes aegypti, Insect Biochem.* **19**:651–656.

102. McGrane, V., Carlson, J.O., Miller, B.R., and Beaty, B.J., 1988, Microinjection of DNA into *Aedes triseriatis* ova and detection of integration, *Amer. J. Trop. Med. Hyg.* **39**:502–510.

103. Medina, M., and Vallejo, C.G., 1989, A serine proteinase in *Drosophila* embryos: yolk localization and developmental activation, *Insect Biochem.* **19**:687–691.

104. Mellman, I., Fuchs, R., and Helenius, A., 1986, Acidification of the endocytic and exocytic pathways, *Ann. Rev. Biochem.* **55**:663–700.

105. Meola, R., and Lea, A.O., 1972, Humoral inhibition of egg development of mosquitoes, *J. Med. Entomol.* **9**:99–103.

106. Miller, L.H., Sakai, R.K., Romans, P., Gwadz, R.W., Kantoff, P., and Conn, H.G., Stable integration and expression of a bacterial gene in the mosquito *Anopheles gambiae, Science* Washington, **237**:779–781.

107. Morris, A.C., Eggleston, P., and Crampton, J.M., 1989, Genetic transformation of the mosquito, *Aedes aegypti* by micro-injection of DNA, *Med. Vet. Ent.* **3**:1–7.

108. O'Brochta, D.A., and Handler, A.M., 1988, Mobility of P elements in drosophilids and nondrosophilids, *Proc. Natl. Acad. Sci. USA*, **85**:6052–6056.

109. Osir, E.O., Anderson, D.R., Grimes, W.J., and Law, J.H., 1986, Studies on the carbohydrate moiety of vitellogenin from the tobacco hornworm, *Manduca sexta, Insect Biochem.* **16**:471–478.

110. Osir, E.O., and Law, J.H., 1986, Studies on binding and uptake of vitellogenin by follicles of the tobacco hornworm, *Manduca sexta, Arch Insect Biochem. Physiol.* **3**:513–528.

111. Park, Y.J., and Fallon, A.M., 1990, Mosquito ribosomal RNA genes: Characterization of gene structure and evidence for changes in copy number during development, *Insect Biochem.* **20**:1–11.

112. Pau, R.N., Birnstingl, S., Edwards-Jones, K., Gillen, C.U., and Matsakis, E., 1987, The structure and organization of genes coding for juvenile hormone-regulated 16-kdalton oothecins in the cockroach, *Periplaneta americana.* in O'Connor, J.D. (ed): UCLA Symp. Mol. Cell. Biol. New Ser., *Molecular Biology of Invertebrate Development* **66**:265–277.

113. Pearse, B.M.F., 1987, Clathrin and coated vesicles, *EMBO J.* **6**:2507–2512.

114. Pearse, B.M.F., and Robinson, M.S., 1990, Clathrin, adaptors, and sorting, *Ann. Rev. Cell Biol.* **6**:151–171.

115. Racioppi, J.V., Gemmill, R.M., Kogan, P.H., Calvo, J.M., and Hagedorn, H.H., 1986, Expression and regulation of vitellogenin messenger RNA in the mosquito *Aedes aegypti., Insect Biochem.* **16**:255-262.

116. Raikhel, A.S., 1984, The accumulative pathway of vitellogenin in the mosquito oocyte; a high resolution immuno- and cytochemical study, *J. Ultrastr. Res.* **87**:285-302.

117. Raikhel, A.S., 1984, Accumulation of membrane-free clathrin-like lattices in the mosquito oocyte, *Euro. J. Cell Biol.* **35**:279-283.

118. Raikhel, A.S., 1986, Role of lysosomes in regulating of vitellogenin secretion in the mosquito fat body, *J. Insect Physiol.* **32**:597-604.

119. Raikhel, A.S., 1986, Lysosomes in the cessation of vitellogenin secretion by the mosquito fat body: selective degradation of Golgi complexes and secretory granules, *Tissue Cell* **18**:125-142.

120. Raikhel, A.S., 1986, Lysomal activity in the fat body—a novel mechanism involved in the termination of mosquito vitellogenesis, in Borovsky, D., and Spielman, A., (eds): *Host Regulated Developmental Mechanisms in Vector Arthropods*, University of Florida, IFAS, Florida Medical Entomology Laboratory, pp. 25-31.

121. Raikhel, A.S., 1987, The cell biology of mosquito vitellogenesis, *Memorias do Instituto Oswaldo Cruz* 82, suppl. **III**:93-101.

122. Raikhel, A.S., 1987, Monoclonal antibodies as probes for processing of yolk protein in the mosquito; a high-resolution immunolocalization of secretory and accumulative pathways, *Tissue Cell* **19**:515-529.

123. Raikhel, A.S., and Bose, S.G., 1988, Properties of mosquito yolk protein: A study using monoclonal antibodies, *Insect Biochem.* **18**:565-575.

124. Raikhel, A.S., Dhadialla, T.S., Cho, W.-L, Hays, A.R., and Koller, C.N., 1990, Biosynthesis and endocytosis of yolk proteins in the mosquito, in Hagedorn, H.H., Hildebrand, J.G., Kidwell, M.G., and Law, J.H. (eds): *Molecular Insect Science*, New York, Plenum Press, pp. 147-154.

125. Raikhel, A.S., and Dhadialla, T.S., 1992, Mechanism of yolk protein accumulation in insect oocytes, *Ann. Rev. Entomol.* **37**:217-251.

126. Raikhel, A.S., and Lea, A.O., 1983, Previtellogenic development and vitellogenin synthesis in the fat body of a mosquito: an ultrastructural and immunocytochemical study, *Tissue Cell* **15**:281-300.

127. Raikhel, A.S., and Lea, A.O., 1985, Hormone-mediated formation of the endocytic complex in mosquito oocytes, *Gen. Comp. Endocrinol.* **53**:424-435.

128. Raikhel, A.S., and Lea, A.O., 1986, The specific ligand, vitellogenin, directs internalized proteins into accumulative compartments of mosquito oocytes, *Tissue Cell* **18**:559-574.

129. Raikhel, A.S., and Lea, A.O., 1987, Analysis of mosquito yolk protein by monoclonal antibodies, in Law, J. (ed): *Molecular Entomology, UCLA Symposia on Molecular and Cellular Biology*, New Series, Volume 49, Alan R. Liss, Inc., New York, pp. 403-413.

130. Raikhel, A.S., and Lea, A.O., 1990, Juvenile hormone controls previtellogenic proliferation of ribosomal RNA in the mosquito fat body, *Gen. Comp. Endocrinol.* **77**:423-434.

131. Raikhel, A.S., and Lea, A.O., 1991, Control of follicular epithelium development and vitelline envelope formation in the mosquito; role of juvenile hormone and 20-hydroxyecdysone, *Tissue Cell.* **23**:577-591.

132. Raikhel, A.S., Pratt, L., and Lea, A.O., 1986, Monoclonal antibodies as probes for processing of yolk protein in the mosquito; production and characterization, *J. Insect Physiol.* **32**:879–890.

133. Readio, J., Peck, K., Meola, R. and Dahm, K.H., 1988, Corpus allatum activity (in vitro) in female *Culex pipiens* during adult life cycle, *J. Insect. Physiol.* **34**:131–135.

134. Riddihough, G., and Pelham, H.R.B., 1987, An ecdysone response element in the *Drosophila* hsp27 promoter, *EMBO J.* **6**:3729–3724.

135. Roehrkasten, A., and Ferenz, H.J., 1986, Properties of the vitellogenin receptor of isolated locust oocyte membranes, *Intern. J. Invert. Reprod. Develop.* **10**:133–142.

136. Roehrkasten, A., Ferenz, H.J, Bushchman-Gebhardt, B., and Hafer, J., 1989, Isolation of the vitellogenin-binding protein from locust ovaries, *Arch Insect Biochem. Physiol.* **10**:141–149.

137. Romans, P., 1990, The vitellogenin genes of the malaria vector *Anopheles gambiae,* Hagedorn, H.H., Hildebrand, J.G., Kidwell, M.G. and Law, J.H. (eds): *Molecular Insect Science,* Plenum Press, New York and London, p. 353.

138. Romer, F., Emmerich, H., and Nowock, J., 1975, Biosynthesis of ecdysones in isolated prothoracic glands and oenocytes of *Tenebrio molitor* in vitro, *J. Insect Physiol.* **20**:1975–1987.

139. Roth, T.F., and Porter, K.R., 1964, Yolk protein uptake in the oocyte of the mosquito *Aedes aegypti., J. Cell Biol.* **20**:313–332.

140. Schneider, W.J., Basu, S.K., McPhaul, M.J., Goldstein, J.L., and Brown, M.S., 1979, Solubilization of the low density lipoprotein receptor, *Proc. Natl. Acad. Sci. USA,* **76**:5577–5581.

141. Shapiro, A.B., Wheelock, G.D., Hagedorn, H.H., Baker, F.C., Tsai, L.W., and Schooley, D.A., 1986, Juvenile hormone and juvenile hormone esterase in adult females of the mosquito *Aedes aegypti, J. Insect Physiol.* **32**:867–877.

142. Sibatani, A., 1971, Difference in the proportion of DNA specific to ribosomal RNA between adults and larvae of *Drosophila melanogaster, Molec. Genetic.* **114**:117–180.

143. Stifani, S., George, R., and Schneider, W.J., 1988, Solubilization and characterization of the chicken oocyte vitellogenin receptor, *Biochem. J.* **250**:467–475.

144. Sturchler, D., 1989, How much malaria is there worlwide?, *Parasitol. Today* **5**:39–40.

145. Tadkowski, T.M., and Jones, J.C., 1979. Changes in the fat body and oocysts during starvation and vitellogenesis in a mosquito, *Aedes aegypti* (L.), *J. Morph.* **159**:185–204.

146. Telfer, W.H., 1960, The selective accumulation of blood proteins by the oocytes of Saturniid moths, *Biol. Bull. Mar Biol. Lab.* Woods Hole **118**:338–351.

147. Thomsen, E., and Thomsen, M., 1978, Production of specific-protein secretion granules by the fat body cells of blowfly *Calliphora erythrocephala,* Substitution of an ovarian key factor by, β-ecdysone, *Cell Tissue Res.* **193**:25–33.

148. Van Handel, E., and Lea, A.O., 1984, Vitellogenin synthesis in blood-fed *Aedes aegypti* in the absence of the head, thorax and ovaries, *J. Insect Physiol.* **30**:871–875.

149. Wyatt, G.R., 1980, The fat body as a protein factory, in Locke, M., and Smith, D.S. (eds): *Insect Biology in Future,* Academic Press, New York, pp. 201–225.

150. Wyatt, G.R., 1988, Vitellogenin synthesis and the analysis of juvenile hormone action in locust fat body, *Can J. Zool.* **66**:2600–2610.

2
Radioisotopes in Vector Research

Werner J. Kloft

Introduction

The physiology of the interaction of arthropods with their host-organisms has been studied by means of radioisotopes for a long time. As early as 1956, the author began to investigate these problems with the help of tracers. Very early in our research, our focus became the transmission of viral diseases by the saliva of plant-sucking aphids (23). In 1977 Harris and Maramorosch published and edited a comprehensive handbook entitled *Aphids as Virus Vectors*. The handbook included a chapter on "Radioisotopes in Aphid Research" (13) and Harris's (6) ingenious paper "Ingestion–Egestion Hypothesis of Noncirculative Virus Transmission." In the latter chapter, Harris presented evidence that even in aphids, which have very tiny food channels in their stylets, transmission of virus by regurgitation (egestion) occurs. Using tritiated water (THO) as a tracer, we verified this egestion phenomenon (13). Kloft started his research originally by using radioisotopes on plant-sucking insects. A review describing the method and results was published in 1968 (20). Due to Kloft's affiliation with the International Atomic Energy Agency (IAEA) as a teacher during the "International Training Courses on the Use of Radioisotopes and Radiation in Entomology," the results of many experiments have been published in the biannual *Proceedings of the FAO/IAEA Training Courses*. A summary of this research is given in reference 12.

The international courses from 1963 to 1986 for my part, in Gainesville, Florida, were continued in Nairobi, Kenya in 1987 and 1989, and again 1992 in Gainesville. They have always been broadly based to appeal to participants coming from very different fields of entomology. Therefore, we presented a wide variety of experiments and experiences with arthropods—mainly with insects, mites, ticks, and spiders. However our continuing interest centered on arthropods that are important in the transmission of disease agents. By

Werner J. Kloft, Institut für Angewandte Zoologie, Universität Bonn, Germany.
© 1992 by Springer-Verlag New York, Inc. *Advances in Disease Vector Research*, Volume 9.

analyzing whether or not haematophagous insects can transmit AIDS (15, 16), our research has been very intensively actualized.

Application of Suitable Radioisotopes

The selection of suitable radioisotopes depends on the purpose of the experiment and on the equipment available to detect the tracer (possible even with simple autoradiographic techniques) or to take reproducible measurements of its concentration. For work with aphids, ^{3}H, ^{14}C, ^{32}P, ^{86}Rb, ^{131}I, and ^{134}Cs have been used. Radiophosphate has been preferred, since this radioisotope is not too expensive, has a convenient physical half-life of 14.3 d, and has a β-disintegration energy of $E_{max} = 1.69$ MeV, which can easily be detected. For many experimental purposes, there is no need to work with the relatively expensive carrier-free radioisotopes. For special experiments radiolabeled biochemicals are available. The catalogs of international companies (e.g. Amersham-Buchler) provide a wide selection, and special substances, such as precursors that follow biochemical pathways, can be ordered.

L'Annunziata (27) has an interesting discussion of reaction mechanisms and pathways in biosynthesis. He also reports on double labels and demonstrates the use of stable isotopes (such as ^{13}C, ^{18}O, and ^{31}P) that can be detected by nuclear magnetic resonance (NMR) spectrometry. Application can be done via host organism (plant or animal in case of vectors for animal diseases). For aphids we presented (20) several proven methods of application. In Figure 2.1 a schematic summary of radioisotope methods is given. In parts B_1 and B_2 of this scheme, the transfer to other host plants or parts of them is shown. In principle the same line can be followed in case of zoophagous or zooparasitic arthropods. However here, due to the mobility of both partners in this system, we have additional complications, especially for proving transmission via saliva or egestion (regurgitation). Theoretically, the transfer of radioisotopes can also be arranged for radioactive host animals for ectoparasitic arthropods. However due to the amount and speed of metabolism, circulation, and excretion, especially in endothermic vertebrates (birds and mammals), applications via food and injections are difficult. We have found unpredictable changes in specific activities for the radioisotopes administered. A survey was done by l'Annunziata (26). Holleman et al. (8, 9) experimented with the transfer of radiocaesium from lichen to reindeer and caribous. The contamination resulted from fall-out after test explosions of atomic weapons.

In a very dramatic way, the reactor accident in Chernobyl on April 26, 1986 increased these problems (21). Neumann-Opitz showed (28) that the two-component model for the in vivo-kinetics of radiocaesium in herbivorous mammals (such as reindeer and caribou) should also apply to arthropods. Working with herbivorous and predacious beetles (*Epilachna varivestis* Mulsant or *Coccinella septempunctata* L., belonging to the family Coccinellidae) as well as with the aphid (*Acyrthosiphon pisum* Harris), the intake and retention of ^{134}Cs could be described in an exact mathematical way, using

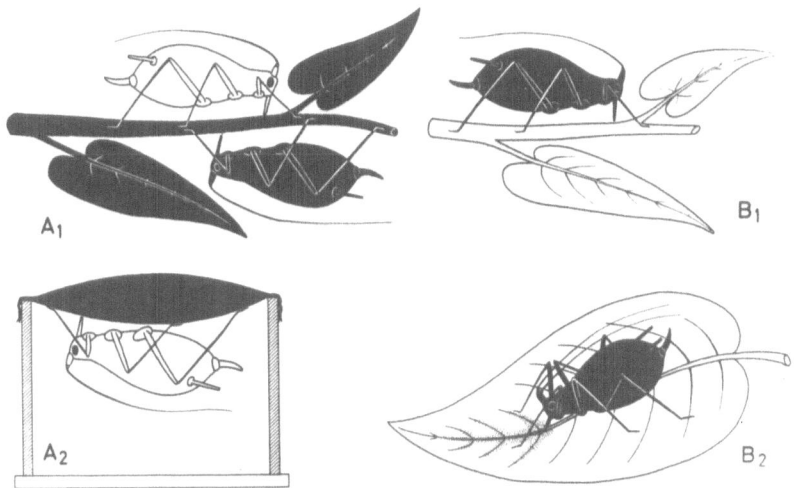

FIGURE 2.1. Schematic summary of radioisotope methods used: (A_1) Nonradioactive aphids white, transferred to a radioactive plant (black), ingest radioactive food and become radioactive (black). (A_2) When feeding aphids with artificial diet, the radioactively labeled diet (black) is enclosed between two membranes of Parafilm^R, stretched over a glass ring to form an aphid cage. (B_1) Labeled aphid (black) is transferred to a nonradioactive plant (white). (B_2) Radioactive aphid injects labeled saliva that disperses in the leaf (dotted area) primarily along veins. (After (20).)

a three-component model. Due to the other functional systems of insects (especially circulatory and excretory), three elimination fractions could be differentiated, which describe the radioactivity of the gut, hemolymph plasm, and cells and tissues. It was possible to quantify these fractions, thereby determining their biological half-life (12, 13). In this context, passage through the alimentary tract, transfer factors, and fast and slow elimination fractions out of the body pool (tissue compartment, or hemolymph-plasma compartment) could be calculated. Since these results are also of interest for more sophisticated tracer experiments with arthropod vectors, we present this model as Figure 2.2.

In special retention and excretion experiments with arthropod vectors, it is possible to inject the radioisotope directly into the haemolymph. 1 μl ^{134}CsCl, diluted in a sterile insect–ringer with an activity of 370 Bq/μl, has been injected (28). The composition of the Insect–Ringer's solution was 7.5 g NaCl, 35 g KCl, 21 g CaCl$_2$, and 1000 ml of distilled water.

Another method of applying radioisotopes to ectoparasitic, haematophagous arthropods is in vitro as shown in Figure 2.3, a design for flying stable flies, *Stomoxys calcitrans* L. However, special preconditions are necessary to prevent external contamination of the legs and wings. We designed two methods (Figure 2.4 and Figure 2.5) of avoiding such contaminations. Both are designed for experiments with individual insects—in this case, *Stomoxys calcitrans* L. For giving defibrinated blood to stable flies, microcapillaries

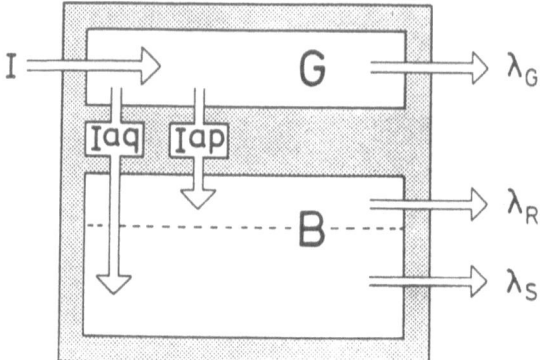

FIGURE 2.2. Model of the "in vivo-kinetics" of radiocaesium. I = Rate of Ingestion; G = Gut; B = Bodypool; a = factor of absorption; p = fraction of Rapid elimination ($p$lasma fraction); q = fraction of $s$low elimination (tissue fraction); λ_G, λ_R, λ_S, constants of elimination out of the Gut (λ_G), Plasma (λ_R) (rapid fraction), and tissue (λ_S) (slow fraction).

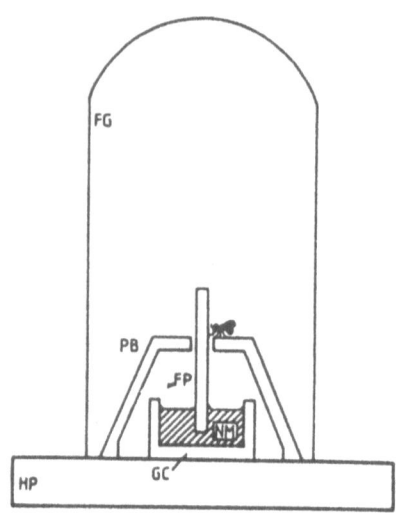

FIGURE 2.3. Arrangements for feeding S. calcitrans L.. The flies are enclosed in a flying glass (FG) in which they can move freely. However, they cannot get access to the labeled nourishing medium (NM)—in this case defibrinated human blood labeled with ^{32}P—which is in a glass container (GC), since this is covered with a plastic beaker (PB). The beaker has a slit through which a strip of filter paper (FP) is inserted into the medium, with the paper soaking up the medium. The arrangement is placed on a heating plate (HP), so that the blood is kept at 37°C. (From 22.)

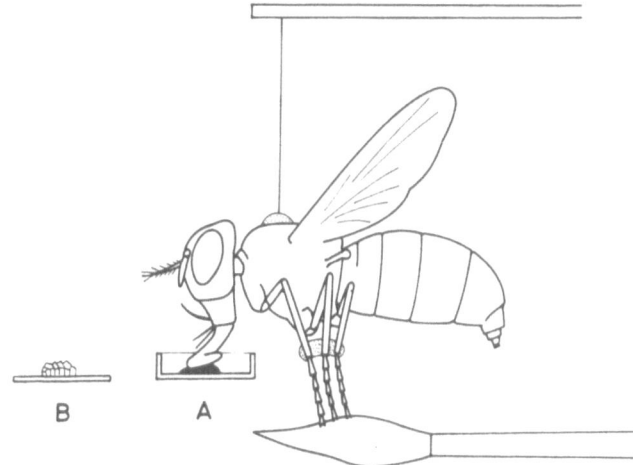

FIGURE 2.4. Experimental way to label flies (*musco domestica* L.) by offering them radioactive sugar solution (A). To prevent any contamination, the legs are glued together and, with a paint brush dipped in sucrose solution, the tarsal reflex (protrusion of the proboscis) must be released. After different time intervals nonradioactive sugar granules (B) are offered. To dilute these, the fly must secrete saliva. By withdrawing the sugar and checking for radioactivity, the time for resecretion of radioactive phosphorus with the saliva (λ_2, Fig. 2.7) can be obtained. (From 18.)

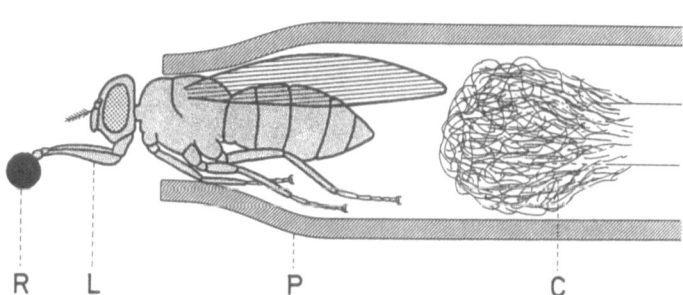

FIGURE 2.5. *Stomoxys calcitrans* enclosed in the tip of a glass pipet (P) for demonstration of regurgitation. Only the head with proboscis (L) are free, thus contamination of legs and wings is prevented. The ball of cotton (C) hinders the stable fly from moving back and can be used, by pushing forwards, to apply a slight pressure on the insect in order to study the provocation of regurgitation by pressure and stress. On the tip of the proboscis is regurgitated material (R). (Drawing by Dr. E. Wolfram, using an orginal photo (22).)

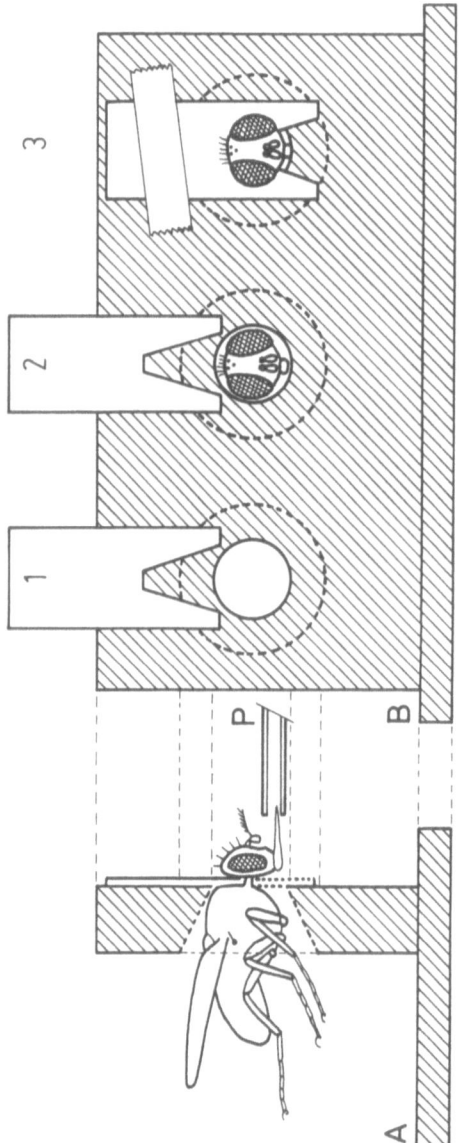

FIGURE 2.6. Arrangement to perform feeding and regurgitation experiments with the biting fly *Stomoxys calcitrans* L. in series. A vertical sheet of plexiglass (12 cm long, 2.5 cm thick) is fixed on a base sheet of same material. (A) The vertical sheet has conical bores (0.5–0.2 mm) through the smaller hole of which can be placed to the fly's head with proboscis. (B) Thin plastic sheets, notched in U-form (1), are then shoved behind the fly's head (2) and taped against the plexi glass sheet (3), thus preventing any withdrawing of the insects. Opposite the fly's head a second plexiglass strip is fixed, on which graduated microcapillaries (P in part A) can be fixed with plasticine, after being pushed over the proboscis. The microcapillaries (5 or 10 μl, graduated in steps of 1 μl) can be filled for feeding with marked blood or diets or they serve to collect the egested (regurgitated) material (with or without tracer and/or disease agents). (Drawing by Dr. E. Wolfram.)

having a diameter of 0,4 $\pm$ 0.1 mm can also be used. They should have a volume of 5 μl or 10 μl, with graduation marks in 1-μl steps. Thus, when offering radioactive food, a good combination of volumetric and radiometric monitoring is possible.

For serial experiments with the feeding (and regurgitation) of stable flies, we designed for Schlager–Vollmer (31) the feeding and regurgitation apparatus shown in Figure 2.6. This equipment allowed feeding (ingestion) or regurgitation (egestion) experiments with 10 insects at the same conditions at the same time.

Diets for blood-feeding arthropods can use defibrinated blood or FKS (fetales Kaelberserum = serum from fetal calves, SERVA Company) with the addition of 10^{-2} M ATP. The temperature is always important, as shown in Figure 2.3. The diet should be kept on a heating plate at 37°C. However remember that induction probing requires a positive difference of several degrees between the diet (surface temperature of the membrane) and the air temperature. In experiments with the bug *Rhodnius prolixus* Stal. (Family Triatomidae), we discovered that the difference is more important than the temperature itself. For the induction of feeding phagostimulants, such as ATP, nucleotides are generally necessary (5). To investigate the exact influence of phagostimulants, the diets should be marked with a radiotracer at the same specific activity.

Ingestion of Radiotracers with Food

If the radiolabeled food is ingested a detector system can determine whether or not and to what extent feeding has taken place in arthropods. A series of technical problems have to be taken into consideration. There are, of course, ways to avoid contamination of body and legs (Figures 2.4, 2.5, 2.7). However, parts of the mouth will be externally contaminated if inserted into capillaries that are filled with a radioactive medium or through parafilm membranes into a sachet (Figure 2.1, A_2) or a vial covered with stretched membrane (Figures 2.8, 2.9). As a result, different external decontamination techniques, using chasers, were developed for the mouth or for entire insects. These are nonradioactive forms of the liquids or media that are used as radiotracers that, after repeated insertions or washes can displace ("chase" away) the radioactivity. As an example, if $Na_2{}^{32}PO_4$ has been added as a tracer, a solution of inactive sodium phosphate is used as chaser. Further techniques for discrimination between external contamination ("mechanical transmission") will be discussed in later sections.

The amount of ingested food can often be measured with graduated capillaries. With radiolabeled food, it is difficult to check the ingested amount simply by measuring the arthropods entire body after feeding. It is, however, possible to determine if and to what degree an arthropod has been feed, if it is true that at certain developmental stages, ages, sexes, or functional phases,

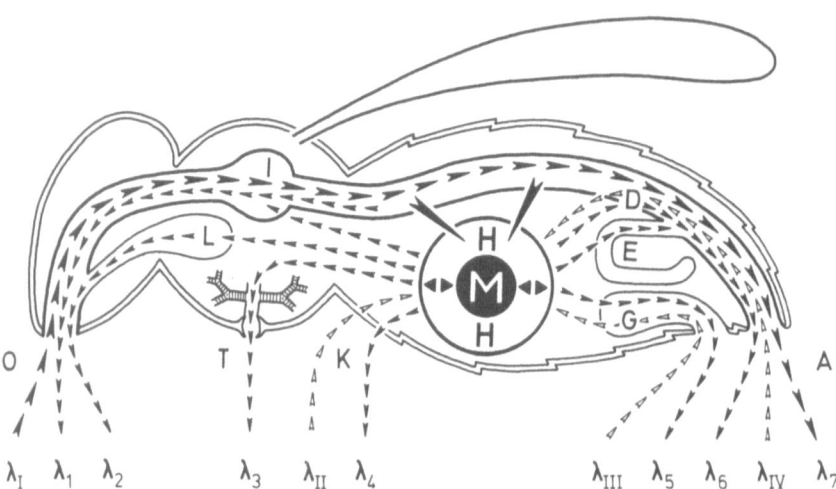

FIGURE 2.7. Pathways of radioisotope uptake (roman numerals) and excretion (arabic numerals) through an insect organism (E. Wolfram pinx.). Os = Os (mouth); L = Labial (salivary) glands; I = Ingluvies (crop); T = Tracheal system; K = Cuticle; H = Haemolymph; M = Metabolism; D = Gut; E = Excretion through Malpighian tubules; G = Genital organs; A = Anus. Possible pathways for uptake are: λ_I, via mouthparts and gut system; λ_{II}, through cuticle, including cuticular absorption of tritiated water: λ_{III}, transfer via sperm or auxiliary secretions during copulation with radioactively tagged males; λ_{IV}, via anus during rectal respiration and/or rectal osmo-regulation. Possible pathways for elimination are: λ_1, by regurgitation of the alimentary tract content; λ_2, tracer carried in the hemolymph enters the salivary glands (Labial glands, L) and is secreted with the saliva; λ_3 and λ_4, the tracer is excreted during transpiration (tritiated water) or respiration (output of $^{14}CO_2$) through the tracheal system (T), or through the cuticle (K), with cuticular secretion or pheromones; λ_5, tracer is eliminated by females with eggs, viviparous larvae or auxiliary secretions, and by males with their sperm and secretions; λ_6, via malpighian tubules (E) and Hindgut (A); λ_7, output with feces after normal passage through the alimentary tract. The haemolymph (H) and metabolism (M) are connected like a network with all these processes. λ_I = Ingestion λ_1 = Egestion. (From 12, 221a.)

they ingest different amounts. Also, researchers can check for the presence or absence of certain materials. The results can be compared with each other if all factors (e.g. detector type, distance between arthropod and detector, or the material used to hold or enclose the arthropod, especially important in the case of live specimen) are kept constant. Several problems are pointed out in the IAEA Manual (12). For example, β particle-emitting radioisotopes create special difficulties by the absorption and backscattering of the emitted β particles. Kloft (11) reported about the technical problems of radioisotope measurement in insect metabolism. In very small and thin arthropods, like aphids (12), mites, and small flies, these problems can be ignored. As shown in Figure 2.6, after a certain time, cuticular excretion begins for many ingested radioisotopes. The β-emitting tracer lying on the surface has no absorption,

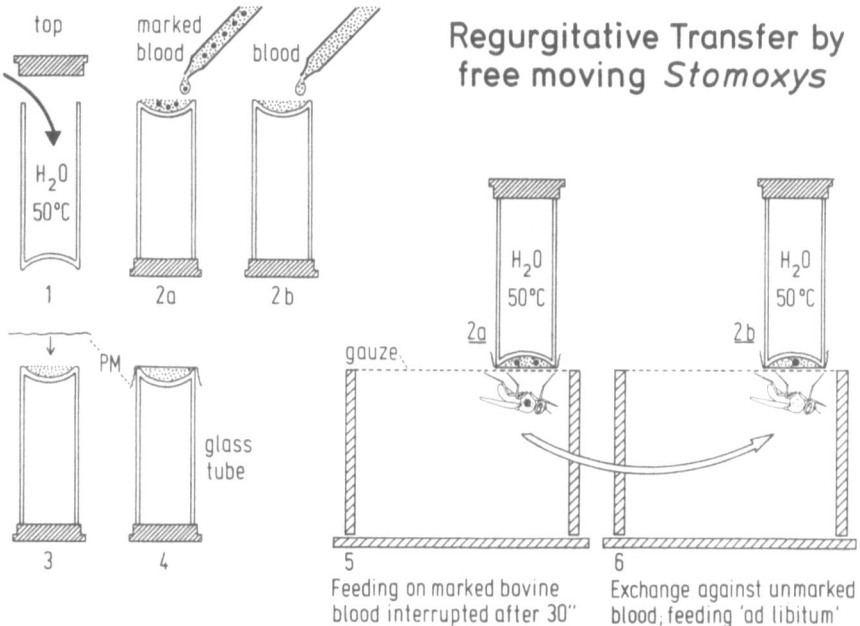

Regurgitative Transfer by free moving *Stomoxys*

FIGURE 2.8. Transfer of marked blood, ingested by *Stomoxys* on a simulated host (2a, 5) to unmarked blood (2b, 6). The blood is dropped on the concave bottom of a glass tube and enclosed by a Parafilm^R-membrane (PM) (1–4). (Drawing by Dr. E. Wolfram; experimental design by Dr. P. Neumann-Opitz.)

and whole body measurements give totally different results. Therefore, the elapsed time must always be taken into consideration. However using γ-emitting sources avoids these problems when the γ radiation is registered. This can be done by using solid Thallium-activated NaI-crystal detectors, connected with a counter-ratemeter system. By use of a well-type crystal, the efficiency of the count can be increased considerably. This is accomplished by drilling a hole in the center of a crystal to create a well for the sample. The radioactive source is, therefore, almost completely surrounded by the radiation-sensitive crystal. Radioiodine (^{131}I) has been frequently used to label blood. When ^{131}iodide is injected in the thyroid gland, the speed of concentration is a routine test in internal medicine. Radiocaesium (^{134}Cs) has a complex disintegration spectrum, with several β-particle energies and different γ energies. It is possible with a good crystal detector and spectrometer system to determine a convenient γ radiation for counting (12). Since caesium belongs to the monovalent alkali metals, it is very similar in its reactions to potassium, and it can therefore be used for many problems in plant as well animal physiology.

Liquid scintillation measurements avoid all problems for β-particle measurements, since the samples are totally diluted and mixed into a cocktail with sensitive ingredients. In the scintillation spectrometer, the light flashes regis-

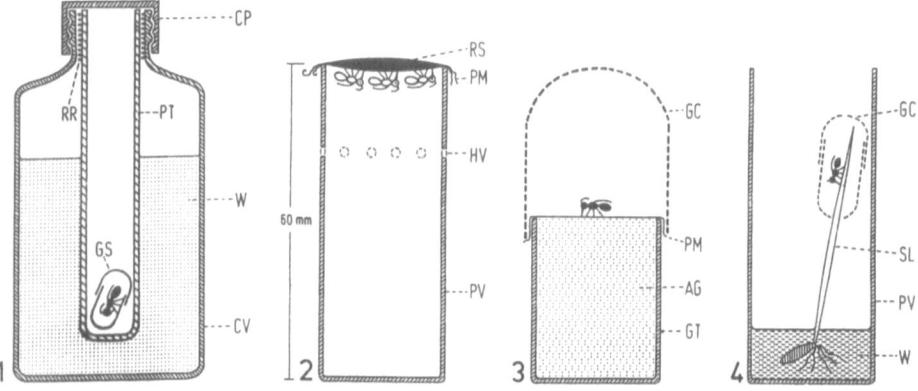

FIGURE 2.9. Arrangement for application of tracer (2), decontamination (3), transfer of egested material to a rice seedling (4) and measurement (1), working with green leafhopper (GLH), *Nephotettix cincticeps*. *Part 1*: Measurement of green leafhopper, marked with ^{32}P, by registration of Cerenkow radiation with liquid scintillation counter. The sample is enclosed in a gelatine capsule (GS), and put into a dry plastic tube (PT). This dry vial is fixed in a glass-counting vial (CV) that is filled with distilled water (W). The cap is tightened with a rubber ring (RR). *Part 2*: A parafilm sacchet (PM) that contains ^{32}P-radiolabel 10% sucrose solution (RS) has been fixed on top of a plastic vial (PV), thereby serving as feeding chamber for the enclosed leaf hoppers (GLH). It has holes for ventilation (HV). *Part 3*: A glass tube (GT), filled with sweetened agar (AG), is covered by a parafilm membrane (PM). A gelatin capsule (GC) on top, encloses the radioactive GLH as it decontaminates its stylets while feeding on a sucrose-sweetened agar medium. *Part 4*: Rice seedling (SL) with a feeding radioactive GLH enclosed in a gelatin capsule (GC). The seedling is in a plastic vial (PV) filled with some water (W).

ter the β-particle transformations as they react with scintillator substances. However, a secondary reaction, the so-called Cerenkow radiation, is created when β particles interact with a transparent media, such as water. This makes it possible to measure live insects (34). The first author worked with *Odonata*. He studied the food intake of larvae (L_1) of the dragonfly *Tetragoneuria cynosura* (Say, 1839). By designing a ^{32}P-labeled food chain (bacteria → paramecium → dragonfly larvae), the larvae could be maintained in water and put into water-filled liquid-scintillation vials for subsequent measurement. Based on this experience, we introduced Cerenkow radiation into vector research (25). Our methods and results are given later in this chapter.

Pathways of Radioisotope Uptake and Elimination

To use radioisotopes to elucidate ways of transmitting disease agents by arthropod vectors, we need to know the different pathways for the intake and elimination of tracers. After more than 20 years of working with arthropods, mainly insects of different taxonomic groups, to which a variety of radio-tracers were applied, the author and his group have outlined the possible intake pathways as shown in Figure 2.7 (12, 22a). In the figure, the greek Lambda (λ) is used to identify these processes. λ with a roman index number ($\lambda_I - \lambda_{IV}$) deals with intake; λ with an arabic index number ($\lambda_1 - \lambda_7$) refers to elimination. Note that λ_I equals intake via mouthparts and gut system (ingestion), and λ_1 equals elimination by regurgitation of the contents of crop or parts of the gut (egestion). These processes are well known and can be partly observed or indicated by use of dyes, especially fluorescent ones. But radioisotopes permit researchers to discriminate exactly between glandular secretion and egestion by regurgitation. For example, four species of flies were fed a sucrose solution, labeled with ^{32}P at a specific activity of 0.3–0.5 mC/ml (1.1–7.2·10^6Bq). After very careful decontamination of the mouthparts (see Figure 2.4 for M. domestica), secretion of the ingested ^{32}P can be checked. Table 2.1 shows the required time for this process. From feeding, reabsorption into the midgut, transport with the haemolymph to the labial salivary glands, incorporation of the tracer into the glandular cells, and secretion of the tracer with the saliva, a time lag of 6 to 8 hours is necessary (temperatures for the different species are given in Table 2.1 in (18)). In a further investigation with biting flies (1), we worked out the transfer of saliva by regurgitation by comparing the stable fly (Stomoxys calcitrans (L.)) with the hornfly (Haematobia irritans (L.)). We found that Stomoxys females resecreted radioactive saliva after 6 hours in 5 cases and after 8 hours 10 minutes in 8 cases. In one case, radioactive resecretion was detected in 3 hours and 21 minutes after food intake. For hornflies the required time was only 1 hour 57 minutes at 25–26.5°C. For decontamination and probing, the biting flies received agar blocks. Agar is cooked with serum (free of red blood cells = RBC) that is squeezed out of a thin rubber tube on a syringe. The agar is used as a column for probing and can be cut into pieces for radioactivity checks. In several cases, only 2 minutes after feeding and decontamination, we found radioactivity in the agar blocks that corresponded to about one tenth of the total amount of radioactivity ingested. These amounts were sufficient to produce a pink color in the blocks deriving from RBCs. In mass experiments in our training courses in Gainesville, we therefore checked the agar columns first to probe for red spots and then measured only these pieces for regurgitation. This was double checked by counting the radioactivity and by optical proofing for red spots. Other authors also used RBCs as markers for regurgitation. Straif et al. (32), in experiments with stable flies, used erythrocytes as markers for regurgitation. The same result was found with mosquitoes (Anopheles sp.) (3). We can now understand the real progress that radioisotopes brought into vector research. The selection of radioisotopes, like ^{32}P or tritium, that can be integrated into physiological and metabolic processes make it possible to bring into the discussion the component of time. This allows a clear discrimination between glandular secretion of saliva and direct delivery of gut content by regurgitation (egestion). Our functional scheme about pathways of radioisotopes gives additional insight into these physio-

TABLE 2.1. Time between peroral ingestion of ^{32}P and its resecretion with the saliva in different species of flies (18).

Species	Temp. in °C	Relative air humidity	Composition of liquid food	Amount of ingested food	First radioactive resecretion after	No. of measured flies	Last radioactive resecretion after	No. of measured flies
Musca domestica L.	20–23				7 hrs	98	21 hrs	78
Calliphora vicina (R.-D.)	20–23	70%	20% sucrose solution, labeled with ^{32}P at a total specific activity of 0.3–0.5 mCi/ml	per fly 0.01–0.03 ml	6 hrs	100	24 hrs	94
Callitroga macellaria (Fabr.)	28–30				8 hrs	96	20 hrs	81
Lucilia cuprina (Wiedemann)	28–30				7 hrs	93	21 hrs	79

logical processes. As early as 70 seconds after beginning feeding on a ^{32}P-labeled diet, larvae of the cabbage looper *Trichoplusia ni* had considerable amounts of ^{32}P (as phosphate) circulating in the haemolymph (30). As shown in Figure 2.7, the tracer is excreted via the malpighian tubules (E) and the hind gut (A). In our experiments, the first radioactive material was eliminated (λ_6) with fecal pellets as early as 6 to 7 minutes after beginning the feeding (for 1 minute only) of a radiolabeled diet. The normal passage time through the gut was 26 to 27 minutes. Therefore, the tracer bypasses the haemolymph and the excretion system within the digestive tract. This result is very important for the physiology of vector arthropods, especially haemophagous forms, whether insects or ticks. They all try to thicken the ingested blood meal by pushing out water, a phenomenon known as primary excretion. While insects use the pathway over the malpighian tubules (λ_6 in Figure 2.7), in Chelicerata (especially ticks), the primary excretion of water is through the coxal glands. In both tsetse flies (22) and ticks (7), droplets of this primary excretion contained the radiotracer. Therefore the tracer method assists in explaining the mechanisms of transmission of such pathogens as stercoraria among trypanosomes. In these species, infections found in feces travel through lesions into the host when the area of the bite is scratched. In all haematophagous arthropods, primary and normal excretion help to regulate the haemolymph volume, which was worked out during the oogenesis cycles of ticks (7).

The haemolymph volume of vector arthropods can be determined by isotope dilution technique. The vector is injected with an exactly measured volume (1 μl) of a solution of Inulin-(^{14}C)-carboxylic acid that has been diluted in insect-Ringer's solution to an exactly calculated specific activity. Inulin, a polysaccharide consisting of monomeric units of fructose, is used because it does not leave the extracellular space by an active transport process. After injection, it mixes with the circulating haemolymph. In 1 to 3 hours, 1 μl of haemolymph is removed. Its specific activity can be determined because it is reduced by isotope dilution. The first step is to determine the radioactivity (cpm) of 1 μl of the original Inulin (^{14}C-carboxylic acid) that has been diluted in Ringer's solution. We inject 1 μl of the original inulin into each of five liquid scintillation vials that were filled with a 10 ml "cocktail" and determine the cpm. The second step is to inject the vector arthropod. After the appropriate mixing time, units of 1 μl haemolymph are removed and put into the same type of liquid scintillation vials with a 10 ml cocktail. The cpm rate is then checked.

We are now able to calculate the haemolymph volume following the formula:

$$\text{haemolymph volume} = \frac{\text{cpm of } ^{14}\text{C-Inulin}}{\text{cpm of haemolymph}} - 1 \ \mu\text{l}$$

The volume unit used is 1 μl. This volume has to be substracted, since it was added to the vector haemolymph by injection. This procedure has been used in vector research with tsetse flies (changing volume during pregnancy cycles (35)), with mosquitoes (haemolymph volume of noninfected and Plasmodium-infected *Anopheles*, (27a)) and *Rhodnius prolixus* to follow haemolymph volume changes during flight (5a).

The Problem of Egestion by Regurgitation

Regurgitation is the functional reflux of gut content either out of the crop or the midgut itself; and therefore, it has different functions (22). In several papers (18a, 19, 1) we showed that radioisotopes were model substances that could be used to demonstrate the mode of pathogen transmission by arthropod vectors to animals and plants. We discussed four methods of transmission: a) via external contamination of the mouthparts, b) via secretion with saliva, c) via defecation after passage through the alimentary tract, and d) via regurgitation. The last is one of the more neglected methods, though in medical parasitology it has great importance in the transmission of plague by fleas and leishmaniasis by phlebotomine flies. In the mentioned paper (22) functional explanations and definitions are given. Systematic tracer studies have shown us the importance of egestion as a transfer mechanism in aphids (13), in stable flies (18, 18a, 19, 15, 16), and in ticks (7, 22). It was the extrusion of relatively large amounts of injected radiolabeled food that encouraged more intensive research into the study of egestion as a means of disease transmission. The egestion can be directly visible as a droplet (see Figure 2.10) or as an autoradiography of labeled blood. It can be measured directly by CPM counting or by volumetric techniques.

FIGURE 2.10. *Stomoxys calcitrans* with a droplet of regurgitated blood on the tip of the proboscis, immediately after having stopped feeding. (From Kloft, 7; photograph by Dr. E. Wolfram and E. Platten.)

Determination with Radiotracer Technique

Autoradiography of egested spots and the determination of the count rate (cpm) of egested material give qualitative—and under special conditions—semiquantitative results. However after collection of the regurgitant in microcapillaries, the regurgitated amount (volume) can be determined. As described earlier, under certain preconditions the volume can be calculated from the registered count rate.

Application of Volumetric Techniques for Determination of Egested Material

As discussed earlier, radiolabeled liquid food can be offered with microcapillaries. On the other hand, microcapillaries can also be used to collect regurgitated fluids. If the capillaries are graduated, even the volumes can be measured. For our experiments, we chose the stable fly, *Stomoxys calcitrans* L., as standard. This biting fly is a pool feeder and makes relatively large lesions in the skin of its host. Because its attacks are painful, feeding is frequently interrupted by the defense movements of the victims. All of these conditions favor transmission of disease agents from host to host. We developed a series of experiments (Figure 2.6) for feeding (application of radiotracers and/or disease agents) and collecting egested material (Figure 2.11).

The flies are immobilized by cooling them down for 2 minutes in a deep freezer at −4°C. This produces much better results than fixing the flies in suspended position (Figure 2.4) or in pipette tips (Figure 2.5). The results

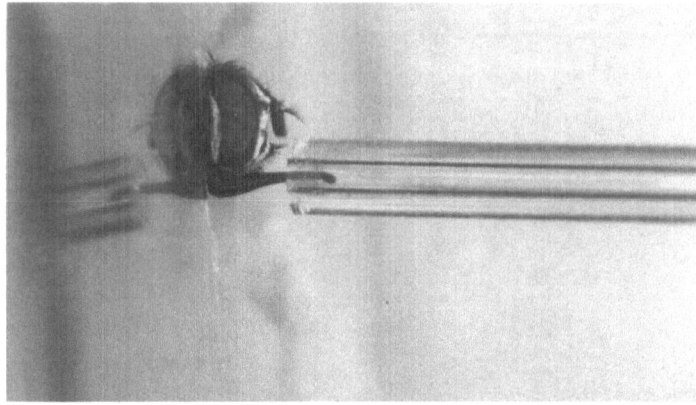

FIGURE 2.11. *Stomoxys calcitrans* has inserted its proboscis into microcapillary filled with transparent sucrose solution. It is obvious that this change of substrate induces regurgitation of some previously engorged blood—shown by reddish turbidity. (Photograph by C. Schlager-Vollmer (30).)

obtained with the feeding–regurgitation arrangement (Figure 2.6) were surprising. They support the ingestion–egestion hypothesis of noncirculative virus transmission, as published by K.F. Harris et al. (2a, 6, 6a, 6b). We obtained ingestion (feeding) volumes of between 6 to 10 μl and egestion (regurgitation) rates of up to 86.7%.

Factors Which Influence the Rate of Egestion by Regurgitation

The results discussed here were obtained with the arrangement shown in Figure 2.6. The stable flies ingested in the mean average 7 to 10 μl of defibrinated bovine blood in an imaginal age of 2 to 12 days. Only 1-, 13- and 14-day-old flies ingested 5 to 6 μl. The maximum of 22 μl was ingested by a female fly at the age of 6 days. Based on these results, we could check the extent to which the rate of regurgitation (egestion) depended on imaginal age. As can be seen from Figure 2.12, this rate is over 70% when the imaginal age is between 2 and 8 days (maximum 86.7% in 5-day-old stable flies). The rate drops to 28% in 14-day-old flies. The amount of ingested blood influences the rate (in %) as well as the volume of regurgitation (egestion). Stable flies, that were fed up to saturation, showed a 35% regurgitation rate. The average amount was 0.1 μl. Flies that were interrupted after ingesting 5 μl had a 75% regurgitation rate and egested 0.4 μl. In a third group, feeding was interrupted after intake of only 1 μl. There, the regurgitation rate was 45% with an average egestion of 0.05 μl.

However the most important factor was changing the food source after the interruption. Before starting the experiment, stable flies in an imaginal age of

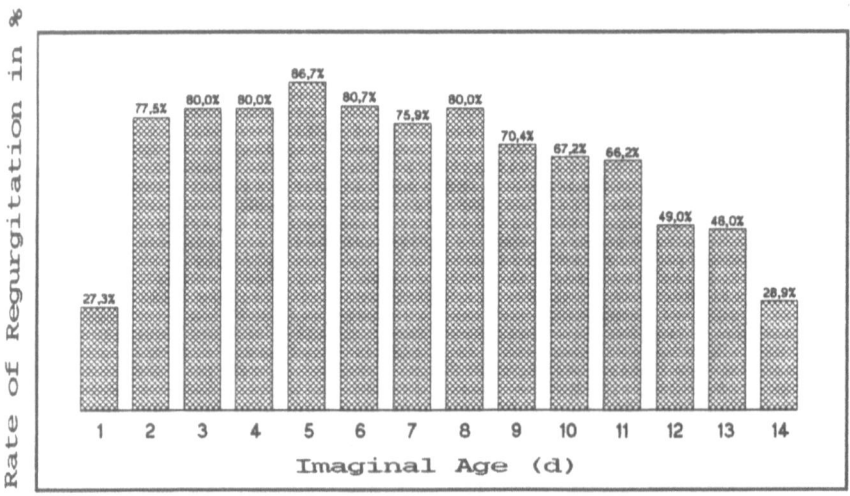

FIGURE 2.12. Rate of egestion by regurgitation (in %) of stable flies in different imaginal age. (From 31.)

TABLE 2.2. Capillary feeding with different food types. Changing the substrate after interrupted feeding of 3 μl increases the rate of regurgitation.

Capillary 1 offered with	Capillary 2 subsequently offered with	Rate of regurgitation (%)
Bovine blood	Pig blood	90
Horse blood	Human blood	93.3
Pig blood	Horse blood	90
Bovine blood	Sucrose (10%)	76.6
Sucrose (10%)	Bovine blood	70

5 days were starved for 24 hours. Then, they were fed out of capillaries with 3 μl of different blood types or a 10% sucrose solution. Immediately after ingestion of the 3 μl, feeding was interrupted; a second capillary, with another type of food, was shoved over the proboscis. This change of substrate led to increased rates of regurgitation, as shown in Table 2.2.

Stress conditions, such as slight pressure and narrow space, can also induce the rate of regurgitation, as shown in Table 2.3 (from 22). Slight pressure can be applied to the arthropods in pipettes (Figure 2.5). In tsetse flies, we only

TABLE 2.3. Demonstration of regurgitation in *Stomoxys calcitrans* L. by use of ^{32}P as a radiotracer. Through measurement of different volumes of labeled blood, a relationship between counts/min and volume (μl in the table) can be calculated (Feeding time is up to completion of meal, which takes on average 2.6 min) (22).

Total volume of blood taken in (μl)	Amount contaminating the mouthparts (μl)	Regurgitated volume (μl)	
		Undisturbed	Under stress conditions (narrow space, slight pressure)
9.5	0.03	—	0.18
9.3	0.02	0.07	0.09
8.5	0.02	0.05	—
7.5	0.04	—	0.11
13.0	0.03	—	0.09
8.3	0.04	0.08	0.04
5.3	0.03	0.06	0.05
6.4	0.06	0.01	0.11
7.9	0.04	—	—
6.8	0.03	—	0.17
6.3	0.03	0.14	—
9.0	0.03	—	0.12
9.3	0.02	—	0.07
7.5	0.04	—	0.16
8.9	0.03	—	—

found spontaneous regurgitation after slight pressure (21). The conclusion about the increased rate of regurgitation after a change of substrate is of enormous importance in studying the transmission of disease agents by egestion. Painful attacks by biting flies and other arthropod vectors induce, as consequence of defense movements of the host, the selection of another host. This can be compared with changing the substrate.

In the tick *Ornithodorus moubata* Murray (Ixodoidea: Argasidae), we also found evidence of regurgitation by using a radioactive tracer. To clarify this, the feeding solutions were labeled with the isotope ^{32}P. Regurgitation was shown to be part of the normal feeding behavior and as a stress reaction after slight pressing of the idiosoma. After feeding on bovine blood, female ticks showed a high rate of normal regurgitation (63.8%). After feeding on an artificial diet, no normal reflux occurred unless slight pressure was applied. The high rate of normal regurgitation after the blood meal might have the function of rinsing out the blocked food channel. Pool feeders ingest large portions of broken down tissue that, together with an extended feeding time or a lack of haemolysis in the pool, can result in the build-up of a plug in the narrow food channel.

Transfer of Egested (Regurgitated) Food to the Next Host

In experimental research using radioisotopes as models for the transmission of disease agents, simulated hosts are an important help. We have learned that even microcapillaries can play the role of the "next host." With tracers and optical methods and by using volumetric techniques, regurgitation of material can be shown.

As with aphids, we developed a combined system for the transfer of radioactive material (as model substance for disease agents) from a simulated host to the next one—either simulated or live (Figure 2.13). Stable flies feed on radioactive blood through artificial membranes. This feeding is interrupted after a short time; it can be continued either on a simulated second host, sacchet, filled with blood that does not contain any radioactivity, or on a baby mouse as live host.

In a similar arrangement (Figure 2.8) designed by Neumann-Opitz, the transfer of radiotracers from simulated host (a radioactive blood source) to a simulated second host by free moving stable flies (Figure 2.8) was studied. In the latter arrangement, the simulated hosts were made highly attractive by increased temperature.

Bovine blood has been diluted 1:1 with tritiated water (185 MBq/ml, Amersham, Code TRS.3) or ^{134}CsCl$_2$ in watery dilution (185 MBq/ml Cs, Amersham, Code CCS. 1). The resulting specific activity was 1.198×10^9 CPM/ml tritium-labeled blood or 1.3×10^9 CPM/ml blood in case of radiocaesium.

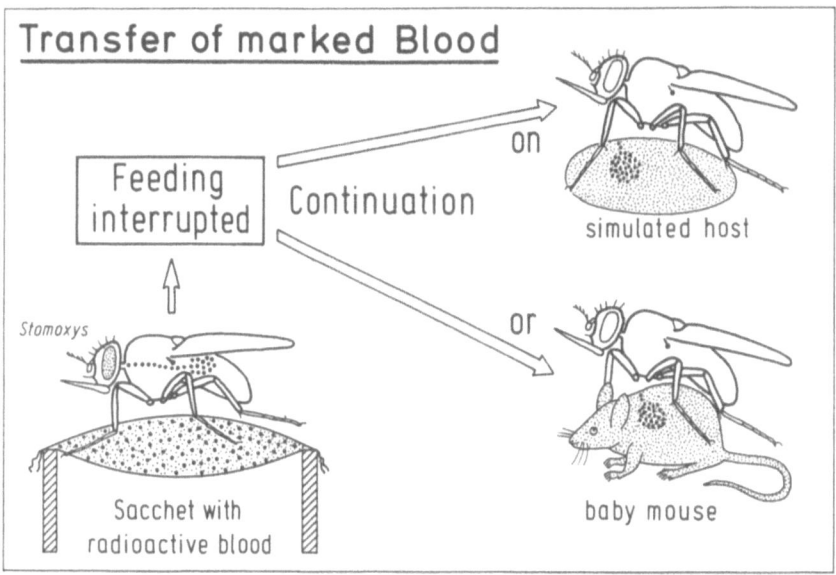

FIGURE 2.13. Regurgitative transfer of marked blood after feeding is interrupted for 30 seconds. Contamination on simulated host (sacchet with unlabeled blood) or on baby mouse. (Drawing by Dr. E. Wolfram.)

A direct transfer from mouse to mouse was not possible because an injection of THO did not result in a sufficient specific activity. Blood, ingested from such mice and regurgitated, was beyond the limit for determination. Therefore we did the application in both arrangements (Figure 2.8 and 2.13) over blood covered with membranes. The simulated host had been measured, together with the membrane, by a liquid scintillation counting system. To determine the volume of blood transferred to the second host, the radioactivity on the membrane (penetration area) and in the bloodpool was converted to volume after counting. The measurements were done with a PACKARD liquid scintillation counter in vials filled with Instant Scintillation Gel II (PACKARD). For reasons of standardization, the membranes were counted together with 50 μl of bovine blood. This meant that the quench factor could be kept nearly constant in all samples.

The baby mice were put as quickly as possible into liquid nitrogen. This form of fixation could prevent any "emigration" of transferred blood out of the area of insertion of the proboscis. Tissue probes (blocks of 2 × 2 mm) were carefully cut out, diluted in Soluene 350 (PACKARD), and measured in Hionic-Fluor (Packard) as the cocktail. In the mean average, a volume of 4.0×10^{-7} ml blood was transferred to the simulated host. Of this amount, 2.3×10^{-7} ml were on the outside of the membrane, whereas 1.7×10^{-7} ml were regurgitated into the bloodpool itself. In the live host—the baby mouse

—the transfer of radioactivity showed the same order of magnitude. Around the area of insertion/probing, we found a contamination of 2.5×10^{-7} ml; in deeper tissue, 1.6×10^{-7} ml could be detected. These amounts are rather small, especially if compared with spontaneous regurgitations (0.5×10^{-5} ml). It might be that during the contamination of feeding, the primary egested material can be partly reingested. Further experiments might give better results. The methods here described (Figures 2.8 and 2.13) were developed mainly in the context of our research about the possibility of transfer of retrovirus (HSV, HIV) by haematophagous insects. At the end of this section, there is an example of the use of the Cerenkow effect for measuring the transfer of radioactivity (as a model for virus-transmission by the rice green leafhopper *Nephotettix cincticeps*) (25). The possibility that the rice green leafhopper (GLH), *Nephotettix cincticeps*, regurgitates some of its food material from its gut into the plant when it feeds was observed by using ^{32}P as tracer. (Also see references 2a, 6a, and 6b.) The GLHs were fed on radioactive

TABLE 2.4. Labeling of rice green leafhopper (GLH) by feeding on ^{32}P-marked sucrose solution and acquired radioactivity (cpm) after different feeding times. Transfer to other media (agar or rice stem) induced regurgitation in several cases (25).

GLH No.	Feeding time[a] on radioactive food (min)	Radioactivity of GLH (cpm)	Medium of feeding site	Radioactivity of feeding site (cpm)	Remarks
1	60	409	agar	12	NR
2	60	5444	agar	5	NR
3	60	2750	agar	9	NR
4	60	1745	agar	10	NR
5	60	9128	agar	13	NR
6	60	2132	agar	11	NR
7	60	10458	agar	21[b]	NR
8	60	2823	agar	9	NR
9	60	1061	agar	7	NR
10	60	10985	agar	6	NR
11	30	8934	rice stem	5	NR
12	30	478	rice stem	104	R
13	30	2245	rice stem	5	NR
14	30	1747	rice stem	22[b]	NR
15	30	2339	rice stem	3	NR
16	15	625	rice stem	4	NR
17	15	853	rice stem	16[b]	NR
18	15	1379	rice stem	5	NR
19	15	809	rice stem	59	R
20	15	41	rice stem	8	NR

[a] Feeding time on medium was 30 minutes, GLH—green leafhopper
[b] Not significantly different from background
NR = No regurgitation
R = Regurgitation

sucrose solution for about 30 minutes. They were transferred immediately to a new feeding medium consisting of sweetened agar block or rice stem. About 20% of the GLHs regurgitated or transmitted a large portion of the radioactivity from the foregut into the medium. The sudden transfer was a disturbance that stimulated GLHs to regurgitate. Stylet decontamination and salivation, the well-known mechanisms of transmission, were also observed. The amounts of radioactivity, or similarly the disease agent, transmitted from the GLH's body into the medium through regurgitation is much greater than those transmitted through stylet decontamination or salivation (see Figure 2.9 and Table 2.4). Figure 2.9 (1–4) explains the procedure exactly.

Transmission of Disease Agents Via Egestion (Regurgitation)

The use of radioistopes as model substances made it possible to demonstrate regurgitation in aphids (13). The results offered possible support to Harris's (6a, 6b) ingestion-egestion hypothesis of noncirculative virus transmission. The transmission of disease agents by arthropods has received extensive coverage in these volumes. Therefore, only our own unpublished* results about regurgitative transfer of HSV and HIV will be briefly presented here. The basis of our research was the discovery of regular regurgitations (egestions) in the feeding behavior of our main model arthropod, the biting fly *Stomoxys calcitrans* L. With the presented methods of capillary feeding and collecting egested material (Figures 2.6 and 2.11), we also applied disease agents to the flies. In close cooperation with the virologist H. Brandner, we exposed *Stomoxys calictrans* to virus via capillary feeding culture of media with HSV as well as media containing human T-lymphocytes infected with HIV 1. The ingested material (MOLT $\frac{4}{8}$-HIV 1^+-cells) was collected after regurgitation with microcapillaries. When these cells were cultivated in special media, they showed positive immunofluorescence for HIV and also induced cytopathogenic effects (CPE). So far the infectiousness of regurgitated MOLT $\frac{4}{8}$-HIV 1^+-cells seems to be a proven fact. However further research is in progress and a summary of the present status together with future results will be published by G. Brandner et al. (37).

   The classical methods for pathogen transmission by bloodsucking arthropods are either "mechanical" or "biological." Both ways are rejected for retrovirus, the latter since no replication in the vector exists and its survival in the arthropod is very limited. However in *Stomoxys calcitrans*, the anterior part of the midgut has a storage function without any enzyme activity, as in Kirch's Ph.D. thesis (Bonn, 1990). Therefore surviving virus can be egested, and the hitherto neglected method of "regurgitative transmission" must be introduced as an additional term into parasitology (16).

---

* Meanwhile first results appeared (37).

Radioimmunoassay Techniques and Application in Disease Vector Research

The introduction of the radioimmunoassay (RIA) more than 25 years ago brought immense progress to the biological sciences, because it allows the identification of smallest amounts of species-specific proteins (10). It is a combination of immunological techniques with radiometric methods (2).

Collins et al. (4) worked with an immunoradiometric assay (IRMA) that attached a monoclonal antibody to the surface proteins of *Plasmodium falciparum* sporozoites. The goal was to determine the *P. falciparum* sporozoite rate in a West African population of the vector *Anopheles gambiae*. The sporozoite rate—the prevalence of female anopheline mosquitoes with sporozoites in their salivary glands—is the most sensitive parameter for describing the epidemiology of malaria in a given area. It is used to determine the entomological inoculation rate (the product of man-biting and sporozoite rates), and can be used to establish both vector identity and differences in transmission intensity over space and time.

Sporozoite rates in natural vector populations are determined conventionally by microscopic examination of the salivary glands of individual mosquitoes. This extremely labor-intensive procedure, which requires trained and dedicated personal, can only be performed on freshly captured mosquitoes. However it is practically useless in routine surveys of vector populations with sporozoite rates of less than 1%.

In many regions, endemic malaria is maintained by vector species with considerably lower sporozoite rates (29). The availability of monoclonal antibodies that are specific to the major surface-coat antigen of sporozoites of a number of different malaria species has enabled the development of immunological procedures for detecting sporozoites in infected mosquitoes (36). One such procedure, the immunoradiometric assay (IRMA), provides a rapid species-specific determination of both the presence and number of sporozoites in an infected mosquito. The IRMA can be performed on freshly caught or dead, dry mosquitoes, because the antigen appears to remain fully detectable for up to at least four months. Since the development and standardization of IRMA were done under laboratory conditions with heavily infected mosquitoes, it was necessary to evaluate the method with naturally infected specimens in which sporozoite rates and levels of infection might be lower and more variable. For this purpose, Collins et al. (4) performed a field trial in Gambia (West Africa). The field collections were done early in the morning by aspirating mosquitoes, one part of which was dissected in the classical way. The specimens in the second group were dried for 48 hours in a chemical desiccating chamber, and then stored individually in capped, polypropylene microfuge tubes. These mosquitoes were checked 3 to 5 weeks later at the United Kingdom Medical Research Council Laboratory in Fajara (West Africa). Researchers added 30 μl of phosphate buffered saline (PBS) containing 1% bovine serum albumin (BSA) to each microfuge tube containing a single dried mosquito. Protease inhibitors were also added (for the exact formulation, see (4)). This mixture is then incubated for one hour, triturated, and then diluted to 300 μl with PBS/BSA. A 30 μl aliquot is placed into the well of a polyvinyl microtiter plate previously coated with the anti—*P. falciparum* monoclonal antibody, incubated for 3 hours, and then washed thoroughly with PBS/BSA. The same monoclonal antibody labeled with [125]I (100000 cpm in 30 μl) in PBS/BSA with 10% human serum is then added and allowed to incubate for one hour. After several weeks the wells are cut and counted in a gamma counter. The results (4) offered rigorous criteria for distinguishing infected from uninfected mosquitoes among the unknowns. Using processed aliquots of uninfected mosquitoes to which are added known numbers of *P. falciparum* sporozoites, in dilutions from 12000 to less than 50, standard curves can be worked out. The number of sporozoites per infected mosquito (sporozoit load) can be estimated by graphic extrapolation to the standard curve.

As we see, a main advantage of RIA—or IRMA—techniques is the possibility of a separation in space and time between field collection and processing. From this perspective, RIA can be compared with the technique of neutron activation analysis. Here certain stable elements may be used to label insects. These are radioactivated by exposing them to a flux of neutrons from a reactor or neutron generator. In this neutron activation analysis, the induced radioactivity has a short half-life (seconds, minutes, or hours). While disintegrating to its stable form, the activated element can be identified by the characteristic gamma peak. Success in labeling insects has been achieved with certain rare earth or trace elements, such as dysprosium (Dy), europium (Eu), hafnium (Hf) and manganese (Mn). The labeling of tsetse flies (*Glossina spp.*) with rare earths and other tracer elements has proved to be successful. Neutron activation analysis has the advantage of not being restricted by time and place. Labeling and trapping of insects may be done in one part of the world, and the neutron activation and subsequent measurements may be carried out in an other part. The only problem is to assure that the chosen rare earth is not accumulated by chance in the respective experimental area (more details and literature in (14)). The possibility of separation in space and time is the only similarity between the techniques. In RIA, specifity and sensitivity for orders of magnitude are higher than in neutron activation analysis.

Acknowledgments. At first I want to thank K.F. Harris for inviting me to write this chapter. In this way I summarized past experiments and current research. Special thanks are also due to Professor Dr. H. Brandner, Professor Dr. W. Maier, Professor Dr. H.-M. Seitz, C. Schlager-Vollmer, Elisabeth Platten, and Ursula Wolf. Dr. P. Neumann-Opitz made many tracer applications and difficult measurements with calculations. He also designed the special feeding arrangement for free-moving *Stomoxys calcitrans*. But mainly I am obligated to Dr. E. Wolfram for drawing the self-explanatory illustrations that give a very special style to this chapter. For typing this manuscript, I thank my graduate student Burkhard Kape. I am also indebted to my wife and colleague Erika S. Kloft, who has cooperated with me on research in this field for more than 30 years.

References

1. Butler, J.F., Kloft, W.J., DuBose, L.A., and Kloft, E.S., 1977, Recontamination of food after feeding a ^{32}P food source to biting Muscidae, *J. Med. Ent.* **13**:567–571.
2. Chard, T., 1982, An Introduction to radioimmunoassay and related techniques, in Work, T.S., and Work, E. (eds): Laboratory Techniques in Biochemistry and Molecular Biology series, Elsevier Biochemical Press, Amsterdam, New York, Oxford, 284 pp.
2a. Childress, S.A., and Harris, K.F., 1989, Localization of virus-like particles in the foreguts of viruliferous *Graminella nigrifrons* leafhoppers carrying the semi-persistent maize chlorotic dwarf virus, *J. Gen. Virol.* **70**:247–251.
3. Chutmongkonkul, M., Maier, W.A., and Seitz, H.M., 1990, Regurgitation in Mosquitoes, *Z. angew. Zool.* **77**:367–373.
4. Collins, F.H. et al., 1984, First field trial of an immunoradiometric assay for detection of malaria sporozoites in mosquitoes, *Am. J. Trop. Med. Hyg.* **33**:538–543.
5. Galun, R., 1976, The physiology of haematophagous insect—animal relationships, *Proc. 15th Int. Congr. Entomol.*, Washington, D.C. pp. 257.
5a. Gringorten, J.L., and Friend, W.G., 1979, Haemolymph—volume changes in *Rhodnius prolixus* during flight, *J. Exp. Biol.* **83**:325.
6. Harris, K.F., 1977, An Ingestion—Egestion Hypothesis of Noncirculative Virus Transmission, in Harris, K.F., and Maramorosch, K., (eds): Aphids as Virus Vectors, Academic Press, New York, San Francisco, London, pp. 166–210.
6a. Harris, K.F., and Childress, S.A., 1981, Mechanism of maize chlorotic dwarf virus (MCDV) transmission by its leafhopper vector, *Graminella nigrifrons, Abst. 81st Ann. Mtg. Amer. Soc. Microbiol.*, 1–6 March, 1981, Daltas, Texas, p. 251.
6b. Harris, K.F. et al., 1981, Observations on leafhopper ingestion-egestion behavior: its likely role in the transmission of noncirculative viruses and other pathogens, *J. Econ. Entomol.* **74**:446–453.
7. Hesse, G.H., 1984, Haemolymph volume regulation during oogenesis of ornithodorus moubata (Murray, 1877) (*Ixodoiidea: Argasidae*), *Zbl. Bakt. Hyg.* **A 257**.
8. Holleman, D.F., Luick, J.R., and Whicker, F.W., 1971, Transfer of Radiocaesium from Lichen to Reindeer, *Health Physics* **21**:657–666.
9. Holleman, D.F., Luick, J.R., and White, R.G., 1979, Intake estimates for Reindeer and Caribou during winter, *J of Wildlife Management* **43**:192–201.
10. Keller, R., 1988, Radioimmunoassays and ELISA's: Peptides, in Gilbert, L.I., and Miller, T.A. (eds): Immunological Techniques in Insect Biology pp. 243–272.
11. Kloft, W.J., 1962, Technical problems of radioisotope measurement in insect metabolism—*Proc. Symp. Radioisotopes and Radiation Entomol.* IAEA Vienna, pp. 163–172.
12. Kloft, W.J., 1977, Part V in Laboratory Training Manual on the Use of Isotopes and Radiation in Entomology, Technical Reports Ser. 661, IAEA Vienna, pp. 141–214.

13. Kloft, W.J., 1977, Radioisotopes in Aphid Research, in Harris, K.F., and Maramorosch, K. (eds): Aphids as Virus Vectors, Academic Press, New York, San Francisco, London, pp. 292—310.

14. Kloft, W.J., 1984, Entomology, in L'Annunziata, M.F., and Legg, J.O. (eds): Isotopes and Radiation in Agricultural Sci., Volume 2, Academic Press, London, Orlando, pp. 51–103.

15. Kloft, W.J., 1988, Doch AIDS durch Insekten?, *Bild der Wissenschaft* 6:107.

16. Kloft, W.J., 1989, Besteht die Möglichkeit zur Übertragung von AIDS durch blutsaugende Insekten?, *Naturwissenschaften* 76:149–155.

17. Kloft, W.J., 1991, Entwicklung und Anwendung eines generalisierten Modells zur Aufnahme und Ausscheidung von Radioisotopen durch Insekten, Verh. Entomologentagung Wien (172.-6.4.1991), *Mitt. D.G.a.a.E.*, i. pr..

18. Kloft, W.J., and Bungard, K., 1975, Erfassung und Resekretion eines peroral aufgenommenen Tracers mit dem Speichel bei Fliegen, *Z. angew. Zool.* 62:83–87.

18a. Kloft, W.J., Butler, J.F., and Kloft, E.S., 1976, Radioaktive Isotope als Modell-substanzen zur Klärung des Uebertragungs-modus von Pathogenen durch Insekten bei Pflanzen und Tier, *Z. Pflanzenkr. Pflanzensch.* 83:80–86.

19. Kloft, W.J., Cromroy, H.L., and Kloft. E.S., 1980. Use of isotopes as model substance to elucidate the mode of transmission of pathogens to animals and plants by arthropod vectors, Isotope and Radiation Res. on Animal Diseases and their Vectors, *Proc. Int. Symp.* Vienna, 1979, pp. 299–312.

20. Kloft, W.J., Ehrhardt, P., and Kunkel, H., 1968, Radioisotopes in investigation of interrelationships between aphids and hostplants, in *Isotopes and Radiation in Entomology*, Int. Atomic Energy Agency Vienna, pp. 23–30.

21. Kloft, W.J., and Gruschwitz, M., 1988, "Oekologie der Tiere," 2nd. edit, UTB 729, Ulmer Verlag Stuttgart, 333 p.

22. Kloft, W.J., and Hesse, G.H., 1988, Use of radioisotopes to elucidate the role of regurgitation for direct transfer of parasites or disease agents between host organisms through arthropod vectors, in Modern Insect Control: Nuclear Techniques and Biotechnology, *Proc. Int. IAEA/FAO Symp.* Vienna, 1987, pp. 437–449.

22a. Kloft, W.J., and Kloft, E.S., 1978, Untersuchungen ueber Aufnahme und Ausscheidung von radioaktiven Tracern durch den Organismus von Insekten als Grundlage fuer radiooekologische Forschungen in der Entomologie, *Mitt. DGaaE* 1:6–10.

23. Kloft, W.J., and Kunkel, H., 1961, Einblicke in die Speichelausscheidung bei *Myzus ascalonicus* DONC. mit Hilfe der Tracermethodik, *Proc. XI Int. Congr. Entomol.* Vienna, 1960, 1:746–749.

24. Kloft, W.J., and Neumann-Opitz, P., 1989, Transfer of Radiocaesium (^{134}Cs) from contaminated food to insects of different nutritional types—Intake, Retention and Elimination, *Sec. Sci. Conf. Iraqui Atomic Energy Comm.* Baghdad, 1989, p. 115.

25. Kuswadi, A.N., and Kloft, W.J., 1988, Regurgitation as a possible mechanism of disease transmission by rice green leafhopper (GLH) *Nephotettix cincticeps* studied by using tracer techniques, *Ann. Entomol.* 6:15–24.

26. L'Annunziata, M.F., 1979, Radiotracers in Agricultural Chemistry, Academic Press, New York, London.

27. L'Annunziata, M.F., 1984, Agricultural Biochemistry: Reaction mechanisms and Pathways in biosynthesis, in L'Annunziata, M.F., and Legg, J.O. (eds): Isotopes

and Radiation in Agricultural Sciences, Volume 2, Academic Press, London, Orlando, pp. 108–182.

27a. Mack, S.R., Foley, D.A., and Vanderberg, J.P., 1979, Haemolymph volume of noninfected and *Plasmodium berghei*—infected *Anopheles stephensi, J. Invertebr. Pathol.* **34**:105–109.

28. Neumann-Opitz, P., 1990, "^{134}Cs–in -vivo–Kinetik" bei Insekten verschiedener ernährungsphysiologischer Typen (*Col.: Cocc.: Coccinella septempunctata* Linne', *Epilachna varvivestis* Mulsant; *Hom.: Aphidina: Acyrtosiphon pisum* Harris). Ph.D. dissertation, University Bonn, Germany.

29. Pampana, E.M., 1963, A Textbook of Malaria Eradication, Oxford University Press, London.

30. Ru Nguyen, and Kloft, W.J., 1976, Fast absorption and elimination of ^{32}P labelled food in larvae of the cabbage looper. *Trichoplusia ni (Lepidoptera: Noctuidae), Entomol. Germanica* **2**:242–248.

31. Schlager-Vollmer, C., 1991, Untersuchungen zur Frage einer regurgitativen Krankheitsübertragung durch *Stomoxys calcitrans* (L.), (*Diptera: Muscidae*). Ph.D. dissertation, University Bonn, Germany.

32. Straif, S., Maier, W., and Seitz, H.M., 1990. Regurgitation as a potential mechanism of patho-gen transmission in the biting fly *Stomoxys calcitrans, Z. angew. Zool.* **77**:357–373.

33. Subarao, S.K. et al., 1988, Subceptibility of *Anopheles culicifacies* species A and B to *Plasmodium vivax* and *Plasmodium falciparum* as determined by immuno-radiometric assay, *Transact. Royal Soc. Trop. Med. Hyg.* **82**:394–397.

34. Tennessen, K.J., and Kloft, W.J., 1979, Fluessigkeitsszin tillationsmessung lebender mit Radiophosphat markierter Erstlarven von *Tetragoneuria cynosura* (Say, 1839) zur Erfassung von Nahrungsaufnahme und Exkretion (*Anisoptera: Corduliidae*), *Odontologica* **1**:233–240.

35. Tobe, S.S., and Davey, K.G., 1979, Volume relationships during the pregnancy cycle of the TseTsefly *Glossina austeni, Can. J. Zool.* **30**:999–1010.

36. Zavala, F., Gwadz, R.W., Collins, F.H., Nusszweig, R.S., and Nusszweig, V., 1982, Monoclonal antibodies to circum-sporozoite proteins identify the species of malaria parasite in infected mosquitoes, *Nature* **299**:737–738.

3
Aegyptianella: An Appraisal of Species, Systematics, Avian Hosts, Distribution, and Developmental Biology in Vertebrates and Vectors and Epidemiology

Rainer Gothe

Introduction

The genus *Aegyptianella* (47) is comprised of intraerythrocytic, heteroxenous parasites of amphibians, reptiles and birds. On the basis of their morphological and biological characteristics, these parasites were assigned to the order Rickettsiales. They were thereby included in the family Anaplasmataceae, along with the genera *Anaplasma*, *Haemobartonella* and *Eperythrozoon*, with which they share analogous structure, place of residence in their vertebrate host, and developmental biology (111, 121, 122, 156–160, 210). Members of the Anaplasmataceae are small, cell-wall-less, obligatory, hemotrophic prokaryotes which, however, do not fit comfortably with the rickettsiale as classically defined and are, therefore, now considered as phylogenetically unaffiliated bacteria (158). Whether and to what extent other intraerythrocytic parasites of poikilotherme vertebrates, such as *Pirhemocyton*, *Toddia*, *Cytamoeba*, *Bertarellia*, *Immanoplasma*, *Haematractidium*, *Sauroplasma*, *Sauromella*, *Serpentoplasma*, *Tunetella*, *Haemohormidium* (150), and *Anaplasma*, or *Piroplasma*—like agents of birds are related or identical to *Aegyptianella* has not yet been clarified.

Aegyptianella spp. Described to Date

To date, nine species have been assigned to the genus *Aegyptianella* and named accordingly, including *A. pullorum*, *A. emydis* from a tortoise (44, 54), *A. moshkovskii* from various species of wild birds (168, 217), *A. botuliformis* from helmeted guinea fowl (*Numida meleagris*) (239), *A. carpani* from the

Rainer Gothe, Institute for Comparative Tropical Medicine and Parasitology, University of Munich, D-8000 Munich 40, Germany
© 1992 by Springer-Verlag New York, Inc. *Advances in Disease Vector Research*, Volume 9.

snake *Naia nigricollis* (30), *A. henryi* from ducks and geese (134, 170, 193), *A. anseris* from geese (132, 133, 193), *A. ranarum* from *Rana* spp. (76, 77), and *A. bacterifera* from *Rana esculenta* (27, 78).

Genus-typical characteristics within the order Rickettsiales specify that aegyptianellas parasitize exclusively erythrocytes, are not surrounded by a trilaminar cell wall (but by a double membrane), and cannot be cultivated in cell-free media (122, 156, 158, 232). As a result, *A. ranarum* and *A. bacterifera* may only be assigned to the genus *Aegyptianella* with some reservations, since both species are bounded by a trilaminar membrane (76, 78) and are therefore more similar to *Grahamella*. No cultivation experiments on these species have been performed yet. The same possibly applies to *A. botuliformis*, whose organisms are contained in a membrane-limited vacuole within the cytoplasm of the crythrocyte and are bound by a trilaminar membrane (239). For *A. emydis*, *A. moshkovskii*, *A. carpani*, *A. henryi* and *A. anseris*, neither ultra-structural nor cultivational investigations were carried out. Accordingly, their assignment to the genus *Aegyptianella* must likewise be viewed with some reservation. The species classified as *henryi* in ducks and geese, as well as *anseris* in geese, are certainly identical to *A. pullorum*. This can be determined because the light microscopical descriptions presented for "*A. henryi*" correspond to the morphological characteristics of *A. pullorum* in fowls. Ducks and geese are also extremely susceptible to aegyptianellas deriving from fowls, and develop high degrees of parasitemia as a result (111).

Therefore, in the context of host–vector relationships, only *A. pullorum* is unambiguously clarified with respect to its specific validity, and sufficiently dealt with experimentally. Therefore, in this chapter, exhaustive reference is limited to this species. Of other species assigned to the genus *Aegyptianella*, the host–parasite–vector link has only been studied experimentally for *A. ranarum*. It was determined that leeches of the species *Batracobdella picta* become infected and then transmit the agents, which multiply in cells of the esophagus and arrive in the lumen three weeks after the infectious blood meal, to susceptible frogs at their next intake of blood (76, 79).

On the Systematic Position of *A. pullorum*

As comprehensively reported (111), the systematic history of *A. pullorum* is an almost endless story that began with the classification of these intraery-throcytic parasites as piroplasms (7–10, 173)—unquestionably, however, as a protozoon (8–10). Subsequently, *A. pullorum* was determined to be an intra-erythrocytic stage of *Borrelia anserina* (11, 12, 40, 94, 214), and thereby either referred to as "after phase bodies" (11, 12) or "Balfour bodies" (40), which appear after surviving a borreliosis.

Since the "after phase bodies" were also seen before appearance of the borrelias (13, 14), the interpretation of *A. pullorum* was then changed (15–19). The borrelias no longer invade the erythrocytes, but emit numerous granules in internal organs that enter the erythrocytes, replicate within them, and leave the red blood cells as merozoites. Accordingly, these borrelias were now designated as *Spirochaeta granulosa penetrans* (15–17, 20). However, development of borrelias from these intraerythrocytic granules was not observed (20–24). Therefore, researchers could only conclude that the intraerythrocytic inclusions were possibly new blood parasites, probably a chlamydozooan, causing a new spirochaetosis-associated disease (20). A further possibility was that these granules might represent a part of the developmental cycle of a rickettsia (24).

Parallel to these interpretations concerning the identity of *A. pullorum* from 1907 to 1923, it was also proven that the corresponding intraerythrocytic granules did not appear in fowls infected with *Borrelia anserina* (82, 99, 100, 102–104, 141, 142, 174, 215), thus putting the intraerythrocytic stage of borrelias in question (154). On the other hand, intraerythrocytic settlement of *Borrelia anserina* was not excluded (88, 94, 151), and it was hypothesized that the spirochaetes subsequently degrade into granules in the blood. Following intake by the vector tick *Argas persicus*, these granules go through a cycle whose end forms, once again granules, reach the blood of the vertebrate host (88).

The Balfour bodies were, however, also interpreted as cocci (164), structures derived from erythrocytes (87) and nuclear degenerates of erythrocytes (140). They were also possibly *Anaplasma*-like bodies (198) or imagined to be the reaction products of toxically influencing spirochaetes (81, 103, 104, 152, 153) or erythrocytes damaged by chronic poisoning (153, 176). Later experiments, however, revealed that erythrocytic inclusions induced by toxic substances are not identical to *A. pullorum* (91). Furthermore, the possibility was discussed that these intraerythrocytic bodies are separate pathogens belonging to an infection that is independent of borreliosis (103, 104), and were relegated to the group of "structures of doubtful nature" (233).

A fundamental milestone on the road to species independence and classification of *A. pullorum* were investigations (47, 48) proving that these intraerythrocytic parasites go through a unique, closed developmental cycle and that only these parasites appear in the red blood cells of healthy fowls following the experimental transfer of infected blood. The intraerythrocytic replication was classified as schizogony, by which as many as 20 merozoites arise. From a systematic standpoint, these parasites were assigned to the haemosporidias and placed in the family Piroplasmidae, but were nomenclatorially demarcated from other piroplasmas as species of a new genus, and defined as *Aegyptianella pullorum* Carpano, 1928 (47). According to the definition of this genus, *Aegyptianella* includes small, round, oval, or piriform parasites of the red blood cells that replicate via schizogony to produce as many as 20 merozoites (47, 48). The species identity of *A. pullorum* was confirmed in nearly concomitant investigations in which it was proven that although borrelias disappeared in chickens after treatment with arsenic preparations, "Balfour bodies" were not affected (72). Parasites could be transmitted to susceptible chickens via inoculation of infected blood (41). The assignment to piroplasmas was not completely accepted, however; instead a relation to ana-plasmas was postulated (41, 42) and the species name altered to *Aegyptianella granulosa* (42).

Between 1930 and 1968, the independence of *A. pullorum* from borrelias was indeed always recognized, but its systematic classification was interpreted differently and the nomenclature also frequently changed. After an initial recommendation of *Aegyptianella granulosa penetrans* (178), *A. pullorum* was then referred to as *Babesia* (181, 182) and *Babesia pullorum* (183), or placed in the new genus *Balfouria*, Piroplasmidea, Sporozoa. It was subdivided into two species of parasites for geese and fowls, namely *Balfouria anserina* and *Balfouria gallinarum*, respectively (85, 86). Indeed, certain references were made concerning its relationship to species of the genera *Cytoecetes, Haemobartonella, Anaplasma, Eperythrozoon, Bartonella, Grahamella* and *Rickettsia* on the basis of similar morphological features (226, 227). Finally, however, *A. pullorum* was unanimously interpreted as a protozoon and assigned to babesias. The intraerythrocytic schizogony leading to numerous merozoites became a major criterion for the generic or subgeneric independence of *Aegyptianella* (1, 37, 43, 58, 66, 149, 163, 168, 169, 197, 206, 208, 209, 213, 218, 222). Furthermore it was noted, that *A. pullorum* parallel to the schizogony may also reproduce by binary fission in erythrocytes (229). Specifically, it was proposed that *Aegyptianella* be classified a subgenus of the genus *Babesia* (163) or that it be left as a genus assigned to the families Babesiidae, class Piroplasmasida (168), Piroplasmidae, Sporozoa (37), and Babesiidae, class Telosporidea (218) or to Sarcodina, and placed in the family Piroplasmidae (58), class Piroplasmasida (169) or the order Piroplasmida (66).

Finally, transmission electron microscopical investigations of erythrocytes from fowls parasitized by *A. pullorum* revealed that this parasite does not fulfill the morphological requirements of a protozoon (107, 111). Ultrathin sections of infected erythrocytes unambiguously documented that independent of form, size, and intracellular location, the individual parasites are bounded by a double membrane and that their interior structure consists of electron-dense aggregates of a finely granulated material embedded in a less dense substance. The parasites exhibited neither membrane-associated nuclei nor membrane-bounded organelles, were enclosed in a vacuole of the host cell, and did not replicate by schizogony as previously assumed, but via repeated binary fission (107, 109, 111). It was further determined that aegyptianellas of this species contain predominantly RNA; the amount of DNA is small, cannot be cultivated in cell-free media or in tissue cultures containing fowl fibroblasts, and only settle and multiply intraerythrocytically in embryonic chicken eggs (111). Furthermore aegyptianellas are not affected antibiotically by sulfonamides or drugs that are effective against babesias and plasmodias, but only by tetracyclines (111, 123–125, 161, 162) and dithiosemicarbazones (26, 111, 125) as has been confirmed in further investigations (185, 207). In consideration of all morphological and biological characteristics. *A. pullorum* was therefore placed in the family Anaplasmataceae of the order Rickettsiales, along with the genera *Anaplasma, Haemobartonella* and *Epery-*

throzoon (111). This systematic grouping of *A. pullorum* was then generally accepted (121, 122, 156–160, 210) and, with respect to ultrastructure, also extended to the intraerythrocytic parasites demonstrated in turkeys (56). On the basis of investigations with the transmission electron microscope, it was, however, also concluded that *A. pullorum* actually occupies a position between *Rickettsia* and the Psittacosis-, Lymphogranuloma-Trachoma-group of pathogens (38, 39). It was also proposed, to assign *Anaplasma* and *Aegyptianella* to the Rickettsiales; *Haemobartonella* and *Eperythrozoon*, however, to the Mycoplasmatales (89). Because aegyptianellas do not clearly fit with the rickettsiale, they are now placed into the group of phylogenetically unaffiliated bacteria (158). For further characterization of *A. pullorum*, it was determined that these parasites may be cryopreserved (28, 39, 143, 201, 202) and are sensitive to pleuromutilines (127).

On the Avian Host-Spectrum of *Aegyptianella*

The species validity of *A. pullorum* is unquestionably founded, and it is also indisputable. Transmission via blood from infected fowls has shown that ducks, geese, and quails are highly susceptible to this species (42, 111). It was also supposed that *Turtur erythrophrys*, *Balearica pavonina* (72), *Turtur senegalensis*, *Milvus aegyptiacus* and *Vidua principalis* (70) become infected. On the other hand, it has also been reported that parasites similar to *Aegyptianella* may be transmitted from species of wild birds to domestic poultry as from *Agapornis lilianae* to domestic poultry (145), from *Balearica pavonina* to a fowl and a pheasant (192), from wild to domesticated turkeys (56), and from *Amazona aestiva* to domestic poultry (195). It still remains unclear, however, whether and to what extent the etiology of natural infections demonstrated in wild birds, or even in domestic poultry, are to be associated with *A. pullorum* exclusively, or with other *Aegyptianella* spp. as well. Morphological characterizations of features on the basis of light and/or transmission electron microscopic investigations and experimental interspecific passages of aegyptianellas, are certainly not sufficient criteria for the differentation and classification of species. Accordingly, it is appropriate to recommend that aegyptianellas also be specifically differentiated, demarcated, and classified by means of DNA genomic probes and recombinant DNA probes, as has already been done for *Anaplasma* spp. (2, 105).

The species identities of *Aegyptianella*, therefore, still require definitive clarification. Accordingly, reservation is called for when referring to *A. pullorum* as a species. Subject to this reservation, *A. pullorum* can be classified as the causative agent of a noncontagious infectious disease of domestic poultry, which affects fowls primarily, but geese, ducks, and quails as well, parasitizes in erythrocytes exclusively, and is transmitted via argasid tick species cyclicalimentarily. The susceptibility of pigeons have been contradictorily stated, since occurrence of the infection has been reported on the one hand (212, 221) and denied on the other (13, 14, 17, 41, 42, 57, 111, 182, 188).

In all likelihood, pigeons are not susceptible, since, during an investigative period of 60 days, it was not possible to microscopically demonstrate *A. pullorum* in stained blood smears of ten 6-week-old pigeons that had received intramuscular applications of 400 million infected fowl erythrocytes per kilogram body weight each. In contrast, intramuscular inoculation of the same donor blood in fowls of like age resulted in infections with high parasitemia (111). Intravenous transfer of aegyptianellas from *Balearica pavonina* to a pigeon was likewise unsuccessful (192).

Whether *A. pullorum* and an additional species or another exclusive species occur in turkeys has not been definitively clarified either. The occurrence of aegyptianellas-infections is clearly demonstrated, however, inasmuch as babesia-like intraerythrocytic organisms were initially reported in turkey poults in California (177). Then blood parasites similar to *Aegyptianella* in their ultrastructure and referred to as *A. pullorum* were exhaustively described (56). When isolated from Rio Grande wild turkeys in Texas (56, 185), it could be transmitted to broadbreasted white turkey poults by means of infected blood (56). An occurrence of *A. pullorum* or similar parasites in turkeys was also reported in countries outside the Americas (33, 95, 166, 212). Transmission-experiments with *A. pullorum* from fowls to turkeys remained, however, unsuccessful (42, 111). Parasites were not demonstrable in the stained blood smears of two- and six-week-old turkeys during a period of 60 days following intramuscular application of 400 million infected fowl erythrocytes per kilogram of body weight. Fowls of like age, however, that were infected with blood from the same donor were clearly positive (111). Aegyptianellas were likewise unable to be passed to a turkey from guinea fowl (147) or *Balearica pavonina* (192).

The question as to whether guinea fowl can be infected with *A. pullorum* has also been answered ambiguously, since occurrence is described in Egypt (95) and Nigeria (6, 166) on the one hand, but infection-experiments by means of positive fowl blood turned out negative on the other (42). Intraerythrocytic parasites similar to *Aegyptianella* (148) or related ones referred to as *Karyonella cristata* (46) were also reported for guinea fowl from the Sudan (148) and the Philippines (46). Analogous to turkeys, it is likewise highly probable that there is at least one new *Aegyptianella* sp. in guinea fowl, since the aegyptianellas isolated from guinea fowl in South Africa differed from *A. pullorum* morphologically (147, 237, 241) and were not transmissible to fowls (147, 237). Nonetheless, guinea fowl also need to be considered as potential hosts for *A. pullorum*, but their susceptibility must be evaluated in light of the fact that infection indeed occurred following the transmission of a laboratory strain, but only a mild and transient parasitemia developed (147, 237).

Experiments are also necessary before ostriches can be added to the host-list for *A. pullorum*. However, in one epidemiological situation, two young birds in Chad became ill with aegyptianellas after being housed with fowls infected by *A. pullorum* in stalls infested with *Argas persicus* (205); this domestically held bird species also showed a susceptibility for such microscopically demonstrated intraerythrocytic parasites as *A. pullorum*.

Quails are highly susceptible to *A. pullorum* from infected fowl blood (42, 111) and should therefore be included in the spectrum of hosts and be considered epidemiologically. By inoculating 2- and 6-week-old *Coturnix coturnix japonica* with 400 million infected fowl erythrocytes per kilogram body weight, it was found that all the animals became infected and, analogous to fowls (111), developed varying degrees of maximum parasitemia depending on age, averaging 30.9% and 13.6% for 2- and 6-week-old birds, respectively (111). In Algeria, the natural infection of quails (*Coturnix coturnix coturnix*) with aegyptianellas was likewise demonstrated. The reported species classification as *A. pullorum* (220) is certainly correct, but should still be regarded with some reservation. In addition, frequent natural infections of quails with intraerythrocytic parasites referred to as *Babesia* sp., have also been reported in the United States (138).

It is currently impossible to make final judgments about the actual species identities of those intraerythrocytic parasites found in wild bird species that are described as being similar, identical, and/or related to *Aegyptianella, Anaplasma* and/or *Babesia*. This is because morphological descriptions, which rely on light microscopical results only, are insufficient for species differentation and classification. As summarized earlier (111, 113), corresponding intraerythrocytic parasites were described in Germany for *Athene noctua* and classified as *Anaplasma* (231). Blood parasites similar to Balfour bodies in *Hypoleis hypoleis* were found in Italy (96). *Gypaëtus barbatus* was listed in the Soviet Union as a further host species, with the responsible species being referred to as *Sogdianella moshkovskii* (217), which was then placed into the genus *Aegyptianella* (168, 197) and *Babesia* (163, 170–172). Furthermore, parasites of the red blood cells have been described as: *Babesia* (*Nicollia*) *ardeae* from *Ardea cinerae* in Indochina (224); *A. pullorum* from *Coturnix coturnix coturnix* in Algeria (220); *Aegyptianella* sp. from *Speniscus demursus* (64) and *Amadina erythrocephala* in South Africa (92), similar to *Aegyptianella* from *Colinus virginianus* in Mexico (32); *A. pullorum* from *Columba palumbus, Nyroca nyroca, Columba livia, Motacilla alba, Motacilla flava, Anas boschas* and *Anas crecca* as well as *Piroplasma avium* from partridges in Iran (212, 213); *A. pullorum* from *Struthio camelus* in Chad (205), similar to *Aegyptionella* from *Numida meleagris major* in Sudan (148), *Babesia moshkovskii* from *Corvus splendens* in Pakistan (163); *Karyonella cristata* from giunea fowl on the Philippines (46), *Nuttallia shortii* from *Falco tinnunculus rupicolaeformis* and *Falco naumanni naumanni*, as well as parasites similar to *Nuttallia shortii* from *Tyto alba alba* in Egypt (135, 136) and from *Falco tinnunculus tinnunculus* in Italy (67); *Nuttallia* sp. from *Bubulcus ibis ibis* probably in Egypt (151 a); *Babesia* sp. from *Corvus brachyrhynchos* and quails in the United States (138, 139); *Nuttallia emberizica* from *Emberiza bruniceps, Nuttallia frugilegica* from *Corvus frugilegus, Nuttallia kazachstanica* from *Galerida cristata, Nuttallia krylovi* from *Upupa epops, Nuttallia mujunkumica* from *Passer indicus* and *Nuttallia rustica* from *Hirundo rustica* in the Soviet Union (235); an agent similar to *Rickettsia* from *Turdus abyssinicus* in Kenya and from *Balearica pavonina* in Great Britain (192); *Nuttallia balearicae* from *Balearica pavonina*

pavonina and *Balearica pavonina gibbericeps* in Great Britain (193); *Babesia rustica* from *Hirundo rustica* in Kenya (194); *Babesia moshkovskii* from *Falco mexicanus* in the United States (69); *A. pullorum* from guinea fowl in Nigeria (166); *Aegyptianella* from *Amazona aestiva* in South America (195); *A. pullorum* from "desi birds" (scavengers) in Pakistan (200); *A. pullorum* from *Numida meleagris galeata* in Nigeria (6); *Aegyptianella* sp. from wild turkeys in the United States (185); *Aegyptianella* sp. from *Numida meleagris* in South Africa (147, 237, 238, 240, 241), *Aegyptianella botuliformis* from *Numida meleagris* in South Africa (239); and *A. pullorum* from *Meleagris gallopavo intermedia* in the USA (56). The parasites referred to as *Nuttallia* were then placed in the genus *Babesia* (194), and *Sogdianella moshkovskii, Babesia ardae, Nuttallia shortii* and *Piroplasma avium* combined in the species *Babesia moshkovskii* (170–172).

Occurrence and Distribution of *A. pullorum* in Domestic Poultry

To study the epidemiology of *A. pullorum* infections, the researcher has to consider not only their occurrence and distribution, but also the biotic factors that determine circulation of the agent in the context of the host-vector-relation. Thus, such studies must include the natural hosts of domestic poultry that function as carriers and donors of *A. pullorum*, as well as the biological vectors that act as reservoirs.

Although the literature primarily documents fowls, it also lists ducks, geese, and quails as natural hosts of *A. pullorum*. Natural infections in fowls have been demonstrated in Egypt (4, 5, 47, 48, 50–52, 90, 181–183), Algeria (83, 84, 219), Nigeria (3, 165, 166), Ethiopia (35), the French Sudan (40, 72), Senegal (167), Kenya (184), Sudan (7-10, 12, 17, 41, 42), Chad (204, 205), Zaire (101), the Republic of South Africa (28, 31, 59, 60, 151, 186, 211), Spain (65), Yugoslavia (85, 88, 179), Bulgaria (225), Albania (55), Greece (73, 74), Lebanon (57), the Soviet Union (85, 87, 187, 207, 230), Iran (75, 85, 212, 213), Pakistan (200), India (1, 153, 173, 180, 190), Sri Lanka (218), and, although a printing error is suspected, possibly in South America (63). An occurrence in fowls was also reported from Tahiti and the New Hebrides (203), as well as from Polynesia and Melanesia (45).

For geese, natural infections have been demonstrated in Egypt (49, 50, 90, 133, 134, 221), Sudan (12, 17), the Republic of South Africa (62, 186), Albania (55), Yugoslavia (85, 179), the Soviet Union (85, 87) and Iran (85), and for ducks in Egypt (90, 134, 181, 183), Kenya (184), the Republic of South Africa (60, 62, 186), Albania (55), Yugoslavia (179), Turkey (199), and Pakistan (200) as well as in Polynesia and Melanesia (45). Without citing the species of host animal, natural infections in domestic poultry were reported in Egypt (93), Ethiopia (36), French West Africa (70), Guinea (189), Cameroon (204), Senegal (189), Italy (223), Yugoslavia (86), Bulgaria (191), the Soviet Union (86),

Turkey (236), Iran (86), India (29, 228), Sri Lanka (80), and Kampuchea (34). Moreover, *A. pullorum* was isolated from wild argasid tick species, thereby proving its endemic occurrence in Israel (155), the Republic of South Africa (129, 216), Tunisia (97, 98), Burkina Faso (115) and the Soviet Union (234). On the assumption that parasites in fowls that are described only as aegyptianellas do indeed represent *A. pullorum*, infections have also been demonstrated in the United States (61), France (137), and the Philippines (175), and confirmed for India (1), Sri Lanka (68, 80), and the Republic of South Africa (61).

In addition to the documented geographic occurrence of *A. pullorum* infections in the scientific literature, the annual edition of the *Animal Health Yearbook* of the FAO-WHO-OIE reports aegyptianellosis since 1957 in Africa for Morocco (1962–1970), Tunisia (1958–1970), Lybia (1958–1970), Egypt (1957–1970), the Sudan (1959–1970). Mali (1963–1970), French West Africa (1957–1959), Cameroon (1959–1960), Tanzania (1957–1970), Zimbabwe (1957–1970), the Republic of South Africa (1957–1970), Zambia (1964, 1966–1970), Botswana (1964–1970), Swaziland (1964–1970), Lesotho (1969–1970), and Mauritius (1964–1969); in Europe for Yugoslavia (1957–1967) and the Soviet Union (1960–1970); in Asia for Jordan (1960–1968), Syria (1962–1970), Iraq (1967–1970), Iran (1957–1970), Pakistan (1957–1970), India (1957–1970), Burma (1957), Laos (1957–1960), Vietnam (1957–1970), Malaysia (1969–1970), and Taiwan (1961–1963), and in America for Mexico (1970).

Epidemiology of *A. pullorum*-Infection

It is certainly justified to assume that data on the distribution of *A. pullorum* are still incomplete and only reflect a fragmentary picture of the actual geographic extent of this agent. On the one hand, aegyptianellosis may be masked by the generally parallel and more severe infection with *Borrelia anserina*, which is transmitted by the same species of tick. On the other hand, *A. pullorum*-infections predominantly exhibit a latent course in older indigenous animals. Beyond this, it is quite likely that no investigations into the spread of these parasites have been carried out in many tropical and subtropical countries, despite known endemic occurrence of the vector ticks.

A. pullorum and Vertebrate Hosts Including Developmental Biology

Research with fowls (108, 109, 111) has shown that in vertebrate hosts only erythrocytes are parasitized by *A. pullorum* (Figure 3.1). Yet the processes of invasion, intraerythrocytic development, and expulsion or departure are very complex and many-sided. Transmission and scanning electron microscopic investigations revealed (107, 111, 116) that the 0.3–0.5 μm, round-to-oval parasites are bounded by a double membrane and are initially located epi-

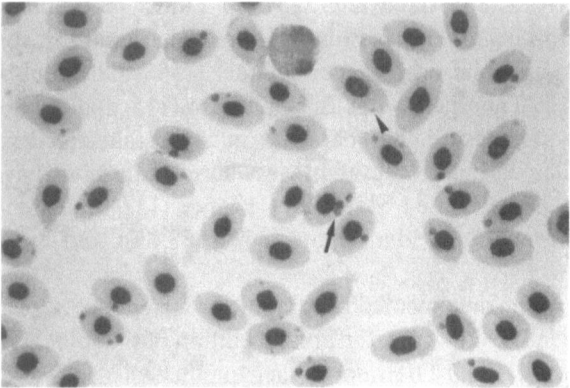

FIGURE 3.1. Initial (▲) and marginal bodies (↑) of *A. pullorum*, Giemsa-stained blood smear from fowl.

cellularly (Figure 3.2). The infective subunit later lies within a slight depression on the surface of the red blood cell. The penetration process is initiated via progressive development into an elongated cavity that follows the contours of the initial body (Figure 3.3). Final intracytoplasmic incorporation of initial bodies occurs with complete invagination of the cell membrane to form a vesicle that separates from the surface and is no longer attached to the cell membrane. The edges of the orifice fuse simultaneously, so that the initial body comes to lie inside a vacuole within the cytoplasm of the host cell.

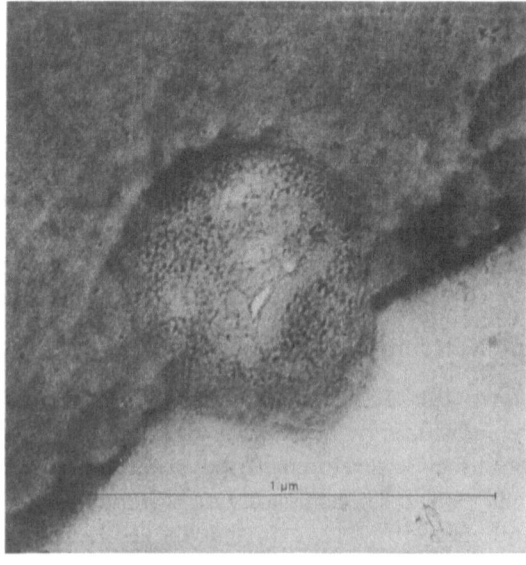

FIGURE 3.2. Initial body of *A. pullorum* invading a fowl-erythrocyte (TEM).

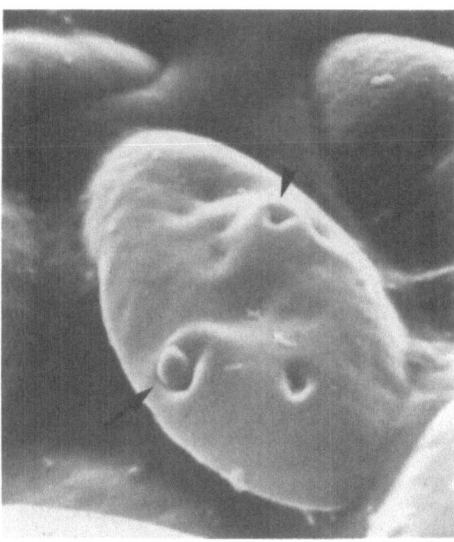

FIGURE 3.3. *A. pullorum*, epicellularly located initial body in a shallow indentation on the red cell surface (↑) and bulging of the erythrocyte after intracellular invasion (▲) (SEM).

The entrance mode of the initial bodies can therefore be classified as an endocytosis with subsequent erythrocytic vesiculation. Membranes of certain parasite-containing erythrocytic vacuoles that stained with ruthenium red retained a connection to the outer medium. Those membrances that did not stain were completely separated from the membrane during the fixation procedure. The presence of ruthenium red in the membranes of some, but not all, parasitophorous vacuoles indicates that entry of aegyptianellas is accomplished by invagination of the host cell membrane, and therefore is of host cell origin. Furthermore, these invaginated membranes must remain intact during *A. pullorum* invasion, because breakage of the erythrocytic membrane would otherwise permit entry of ruthenium red into the cytoplasm and lead to diffuse staining. At the end of the invasive process, the invaginated membranes fuse, resulting in vacuoles that are completely isolated from the outer medium and, therefore, cannot be stained with ruthenium red during the fixation procedure. This unequivocally proves intracellular parasitism of *A. pullorum* during its reproductive cycle (116).

In the vacuoles, which are generally submarginal and also paranuclear, the parasites enlarge to as much as 1 μm in diameter. This leads to occasional bulgings in the wall of the erythrocyte (Figure 3.3). The division process is initiated by slight one- or two-sided invaginations of the parasite's double membrane. These advance until the initial body of the intraerythrocytic development is completely tied off and two offsprings have been formed. Renewed invaginations of the double membrane and subsequent tying off leads to the formation of new parasites, which in turn replicate by binary

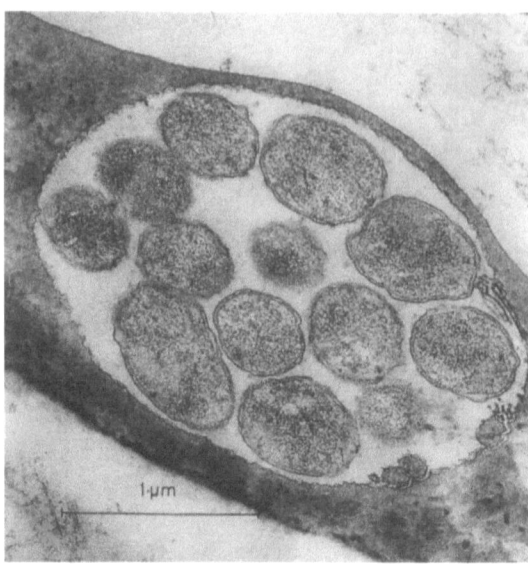

1·μm

FIGURE 3.4. Marginal body of *A. pullorum* showing 13 initial bodies in a fowl erythrocyte (TEM).

fission. This process of repeated binary fissions culminates in the formation of large round marginal bodies up to 4.1 μm in diameter, with numerous round-to-oval, 0.3–0.5 μm end forms. These forms fill the vacuole and terminate the intraerythrocytic development of *A. pullorum*. In electron micrographs, up to 13 of these end forms were counted within a vacuole (Figure 3.4), and as many as 26 could be seen during investigations with the light microscope (107, 111).

One mode of exit for these end forms, which actually represent new initial bodies, appeared to be the reverse of the invasive mechanism. This is indicated in several infected erythrocytes by the presence of evaginated margins from the orifice of the deep cavities which were distinctly above the surface of the red blood cell (Figure 3.5). By this process of expulsion or exocytosis without causing immediate lethal injury to the host cell, new initial bodies leave the red blood corpuscles via evagination. This avoids irreparable damage to the erythrocytic cell membrane. Exocytosis is, however, not the usual exit mechanism for aegyptianellas. In addition to the progressive parasite-induced anemia resulting from a reduction in the number of erythrocytes (111), scans of infected red blood cells revealed parasitogenic injury to the erythrocytic cell membrane and fragmentation of the inclusion membrane. This leads to the release of initial bodies into the plasma and subsequent lysis of the host cell (116). Using light- and scanning-electron microscopy in morphological and parasitological characterization of the same erythrocytes from fowls naturally infected with *A. pullorum*, it was further determined that pathomorphological changes in the red blood cells are only discernible when the marginal bodies formed by the aegyptianellas in the intraerythrocytic

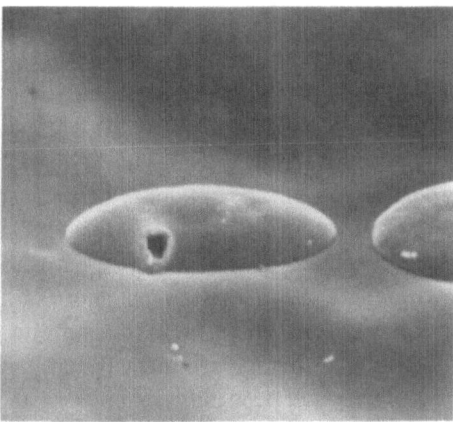

FIGURE 3.5. Fowl erythrocyte after expulsion of initial bodies of *A. pullorum*.

development cycle are at least 2 μm in diameter. They generally appeared as conical or crater-shaped protrusions and bulgings of the erythrocytic cell membrane (Figure 3.3), the extent, height, diameter, and location of which always corresponded to the size, position, and developmental phase of erythrocytes seen in concomitant observations with the light microscope (128).

A pre- and/or exo-erythrocytic phase of *A. pullorum* in the vertebrate host can be ruled out, since no parasites could be demonstrated outside of the erythrocytes during histological examinations of liver, spleen, bone marrow, kidney, brain, heart, and lung cells from fowls naturally infected through *Argas* (*Persicargas*) *walkerae*. Only during stages of high parasitemia can forms similar to *Aegyptianella* be observed in the Kupffer cells of the liver, in lymphocytes, neutrophils, eosinophils, monocytes, and in the blood plasma. Their occurrence, however, is isolated even in these cases. Exclusion of a pre-erythrocytic developmental phase is further substantiated by the fact that intravenous transfer of blood from fowls naturally infected with *A. walkerae* to fully susceptible, splenectomized chicks led to infection and the attainment of high degrees of parasitemia in the latter as early as 30 minutes after the tick's blood meal. The parasitemia continued to be seen at 1, 2, 4, and 6 hours, and in subsequent six hour intervals up to 120 hours. In addition, aegyptianellas were already demonstrated microscopically in the donor animals some 27 hours after attachment of the ticks. The entire intraerythrocytic development cycle of *A. pullorum* can therefore be completed within 36 hours. At the end of this time period, the first mature marginal bodies could be demonstrated in naturally infected chicks (108, 111).

As already substantiated (4, 59, 63, 73, 187, 188, 207), the course, maximal degree, and duration of parasitemia of *A. pullorum* depend on the age of the host animal (110, 111, 126, 216). With increasing age, microscopic demonstration of the aegyptianellas takes longer, but attainment of the first peak of parasitemia is shortened. The time required for the parasitemia to decline to a value of $>0.1\%$ parasitized erythrocytes following initial maximum is increased. The maximum rate of infected red blood cells is reduced from 63.1% to 30.3% and then to 0.3% when fowls at the ages of 1 day, 4 weeks, and 1 year, respectively, were infected with equivalent amounts of aegyptianellas based on 1 kilogram of body weight. After passing the first peak, the course of the parasitemia took on a wave form. The frequency of new rises diminished with increasing age under concomitant lengthening of the distance between the parasitemic peaks and reduction of their height. The parasitemia always fell to a microscopically undetectable level by six months post infection, but persistence of infection was more common in fowls that were younger at the time of transfer of the agent. After splenectomy, 100% of the fowls developed very high parasitemia, if they had been between one day and two weeks of age at the time of infection. The figures were reduced to 60, 40, and 10% when the aegyptianellas were transferred to animals aged 2, 6, or 8 weeks, 12 weeks and one year, respectively (111). The risk of infection for A. walkerae can certainly persist for a substantially longer period, as naturally infected fowls at 3 weeks of age can function as reservoirs for the pathogen for a period of at least 18 months. Blood transfer to splenectomized chicks further demonstrated that all 20 of the fowls included in this experiment remained carriers of the parasites at 18 months of age, despite the fact that aegyptianellas were not always microscopically demonstrable (111). Even clinically healthy fowls treated with the minimum therapeutic dose of tetracyclines and dithiosemicarbazones must be classified as potential sources of infection for ticks, since their therapeutic effectiveness was insufficient to sterilize *Aegyptianella* infections (124).

Hematological changes were analyzed with respect to the total number of erythrocytes, hemoglobin content, hematocrit, reticulocyte count, the ability to withstand hypotonic media, and sedimentation rate. They were scaled in direct correlation to the intensity of parasitemia and, thereby, to the ages of the fowls (111, 144) by distinct correlation between the degree of parasitemia and level of γ-globulins. Hypertrophies of the right ventricle were seen in chickens as well, possibly as a result of severe hypoxic reactions precipitated by the anemia (146).

Ducks, geese, and quails infected with the corresponding number of aegyptianellas at 2 and 6 weeks of age, but only followed up for a period of 30 days, attained maximum rates of parasitized erythrocytes of 9.3 and 28.0% in ducks, 47.0 and 64.5% in geese, and 30.9 and 13.6% in quails, respectively (111).

A. pullorum and Vectors Including Developmental Biology

To date, development of *A. pullorum* in the vector has only been evaluated for *Argas (Pesicargas) walkerae* (119, 120) using light-, immunofluorenscence-, and transmission electron microscopical investigations (106, 109, 111, 114). For larvae, I and II nymphs and adult females, it was concurrently demonstrated that the cycle in the ticks proceeds in three well-defined phases—development and reproduction within the gut epithelial cells, hemocytes, and cells of the salivary glands. In particular, it was concluded that aegyptianellas appear in cells of the gut epithelium as early as 24 hours after the infectious blood meal of the ticks. They begin to replicate by the 62nd hour and finally produce large conglomerates of parasites up to 75 μm in diameter that almost completely fill the host cell and contain numerous oval, compact, 0.4–0.6 μm large anaplasmoid bodies.

These anaplasmoid bodies, which represent the end forms of the intestinal development phase, penetrate the gut wall and appear in the hemolymph approximately 300–500 hours after the ticks became infected. Thereafter, aegyptianellas can no longer be demonstrated in the gut epithelial cells.

In the second developmental phase, these investigations revealed that the aegyptianellas initially swam freely in the hemolymph. They then adopt hemocytes as host cells and multiply intensively to form large conglomerates up to 36 μm in diameter approximately 600 hours after the infectious blood meal of the ticks. In turn, numerous oval anaplasmoid bodies of less than 1 μm are formed. These bodies are considered to be the end forms of the intrahemocytic development phase, which, having been liberated through the destruction of their host cells, proceed via the hemolymph to the salivary glands where they invade their cells. Some 620 hours following infection of the ticks, aegyptianellas invaded the cells of the salivary glands, replicated anew, and again grew to large conglomerates that filled their host cells almost completely. Numerous 0.3–0.5 μm oval bodies, probably the result of repeated binary fissions, were subsequently observed to appear. They were considered to be the forms infectious to vertebrate hosts that will be transmitted via the tick's saliva at the next blood meal.

In *A. walkerae*, the entire developmental cycle of *A. pullorum* therefore required about 30 days. It was synchronized with the oviposition and brooding behavior of the female ticks so that, as a rule, infection of the new host animal occurred at the next intake of blood. Subsequent stages of larvae as well as of I and II nymphs, on the other hand, appeared 9–10 and 11–14 days after the infectious blood meal. They did not transmit *A. pullorum* to susceptible chickens if infestation occurred shortly after the metamorphosis. If these subsequent stages were imposed on fowls after completion of the developmental cycle instead, i.e. approximately 30 days after the infectious blood meal, then these nymphs and adult ticks always proved to be effective vectors (111, 112),

To date, only argasid species of ticks, specifically *Argas reflexus* and *Argas africolumbae* of the subgenus *Argas* as well as *Argas walkerae, Argas persicus, Argas radiatus*, and *Argas sanchezi* of the subgenus *Persicargas*, have been proven to be potentially conceivable and/or natural vectors for *A. pullorum*. The spectrum of vector species is undoubtedly greater than hitherto determined, however, for *A. pullorum* infections are endemic in areas beyond the geographic range of these tick species.

Investigations of *A. walkerae* (106, 111, 112, 130), indicate that *Persicargas* and perhaps other species are particularly effective epidemiologically as vectors. Transfer experiments on fully susceptible fowls raised in a tick-free environment unequivocally demonstrated that all postembryonic stages become infected with *A. pullorum*. This occurs through the intake of blood containing parasites. The agent is transmitted to all subsequent stages or at the next blood meal. In instances of larval infection, both I, II, and III nymphs as well as male and female adult ticks were infectious. They remained so through at least their fourth blood meal. Corresponding horizontal passage and transmission of *A. pullorum* to or through the following stages/blood meals was also observed if I, II, or III nymphs or male and female adult ticks were infected. The vector function of adult female ticks actually persisted over eight ovipositions until the ninth blood meal. Furthermore, the pronounced reservoir capacity of *A. walkerae* for *A. pullorum* has been documented. These experiments revealed that infectiousness was retained over 730 days in I nymphs with no intermediate meal, and over 810 days in II nymphs and adult female ticks, which were infected in the corresponding previous stage. The chronological endpoint was not determined, but one can assume that the capability of infectivity persists in nymphs and adult ticks throughout the entire endurable fasting period for a given stage.

Vertical transmission of the agent via transovarian passage of *A. pullorum* to the next generation of ticks is possible in *A. walkerae* as well. It is, however, of only secondary importance from an epidemiological standpoint, since only a very small portion of the larval offspring that developed from the eggs of infected mother ticks were seen to be infectious. Of 40 chicks infested with 100 larvae derived from infected female ticks, 38 remained negative; aegyptianellas were found in only two animals. In other words, a maximum of only 200 of the 40,000 larvae studied had contracted an infection transovarially. But aegyptianellas were also transmitted if I and II nymphs, as well as male and female adult ticks, derived from the group of positive larvae, engorged on chickens.

A. walkerae ticks can actually infect one another, but this has only been demonstrated for female adult ticks. Infected female ticks and clean ticks were infested on noninfected chickens and removed after 15 minutes. After oviposition, they were again placed on fully susceptible chickens in order to suck blood. In these conditions, aegyptianellas were also passed by the originally negative ticks. Accordingly, it has to be concluded that intraerythrocytic parasites are not absolutely necessary for infecting the ticks, but that those forms from the salivary glands of ticks can also initiate colonization and development of *A. pullorum* in the vector (111).

The infectiousness of I nymphs that became infected as larvae, as well as the further nymphal stages and adult ticks that develop from them, are of particular epidemiological relevance. Based on the stationary–periodic nature of parasitism of larvae, which extends over a period of several days, the ticks are actually in a position to establish new endemic herds of aegyptianellas in regions previously devoid of both *A. pullorum* and ticks. This assumes that the occupied ecotope ensures the concomitant survival of ticks and cyclical propagation of the agent (111). The potential danger of an epidemic spread is certainly real, since *A. walkerae* possesses both a high natural infection potential (129) and a pronounced ecological tolerance capacity (196). With respect to natural infection potential, an experiment was conducted with 11 populations of wild strains collected from fowl coops and roosts in various regions of the Transvaal (Republic of South Africa). Of these, ten were infected with *A. pullorum* and five were additionally carriers of *Borrelia anserina*. Equally important, the rates of infection were very high (129). When 30 female ticks were randomly selected from these populations and allowed to individually infest fully susceptible chickens, 33.3% transmitted both agents simultaneously, 16.7% transmitted either *A. pullorum* or *Borrelia anserina* exclusively, and 33.3% were noninfectious (113). Either no or only very slight differences in virulence existed between the strains of *A. pullorum* isolated (216).

The particular susceptibility of *A. walkerae* for *A. pullorum* is also shown by experiments with virgin female adult ticks. Seventy-five were infested on highly positive, naturally infected fowls, and 98 were infested fowls that had been infected experimentally using cryopreserved aegyptianellas stored in liquid nitrogen for five years. Of the 75 and 98 exposed ticks, 89.3 and 72.7% became infected. Following copulation and oviposition, they transmitted the agent to susceptible fowls during the next blood meal. Cryopreserved aegyptianellas were actually passed to the larval offspring transovarially in the case of one tick (118). The high vectorial effectiveness in *A. walkerae* was also proven on the basis of successful intake and transmission of the agent through adult female ticks following infestation on fowls. These fowls had become naturally infected at the age of 3 weeks, but were first employed as blood donors when they were 18 months old, despite microscopically nondemonstrable parasitemia (111).

The transmission capability of *Argas (Persicargas) persicus* has also been proven. Its potential vector function has indeed been reported on frequently (1, 7–9, 13, 14, 17, 20–22, 25, 31, 43, 47, 48, 53, 55, 59, 60, 62, 85, 86, 88, 97, 98, 131, 155, 179, 180, 187, 188, 191, 207, 211, 230, 234), but no explanation or support was provided regarding the actual species identity of these ticks. At that time, these ticks were collectively grouped under this species classification, although they undoubtedly comprised multiple independent species (119, 120). Investigations with an unequivocally species-defined laboratory strain of *A. persicus* provided by Hoogstraal, demonstrated at first that *A. pullorum* is only transmitted by female adult ticks horizontally, and even then, by a relatively small proportion (117). The natural vector potential and agent reservoir of *A. persicus* must be seen in a broader scope, however, since *A. pullorum* was isolated in every case during infestation experiments with fully susceptible fowls and 19 wild-derived tick populations, from fowl coops, henhouses, and night roosts in various regions of Burkina Faso. The natural infection potential of the individual populations was consistently strong and very high, as evidenced by positive microscopic proof of *A. pullorum* and also of *Borrelia anserina* in all of the blood smears from fully susceptible chicks after they had been infested with 30–40 I and II nymphs as well as male and female adult ticks apiece. Infections of the ticks with both agents remained quantitatively unchanged and persisted over the entire period of investigation, for either nine months or until the third blood meal (i.e., until I and II nymphs had developed into adults or adult ticks had engorged three times). Transovarian passage to the next generation of ticks was also demonstrated, and was, when not cofrequent with transstadial transmission, likewise very common. Researchers studied the infestation of 30–40 larvae of the 200–300 total larvae populations investigated. After the first, second, and third meals of infected female adult ticks, 10–20% of the chicks always became infected with *A. pullorum* and 30–80% of the birds with *Borrelia anserina*. In contrast to transstadial transmission, only one species of these agents was passed on through transovarian transfer (115).

As with *A. walkerae*, we can justifiably conclude that all postembryonic stages of *A. persicus* can be infected with *A. pullorum*, and that this agent can be transmitted to subsequent stages transovarially or intrastadially at the next blood meal. This is in conjunction with the long concomitant infectiousness that persists over a period of at least 3 stages/blood meals. Compared with the species *A. walkerae*, the geographic distribution of which, on the basis of reports to date, appears limited to the southern portion of Africa, *A. persicus* must, due to its occurrence in the Ethiopian, palearctic, and possibly other fauna region(s), be classified as a significantly more important and epidemiologically more effective vector for *A. pullorum*. Furthermore, it also possesses an ecological tolerance capacity that is at least analogous to that of *A. walkerae* (196).

In addition to *A. persicus*, *Argas (Persicargas) radiatus*, and *Argas (Persicargas) sanchezi* can participate in an endemic manifestation of *A. pullorum* in North and Central America. Their vectorial function has been proven experimentally, but the investigations have thus far been limited to female adult ticks (117). The question as to whether and to what extent *A. pullorum* is already endemic to the regions inhabited by these 3 *Persicargas* spp. still presupposes an exact species characterization of aegyptianellas and *Aegyptianella*-like parasites that have been described within these regions. Such parasites have been isolated from wild birds, including *Corvus brachyrhynchos* in Michigan (138); frequently from Californian quails (138); from 6 of 30 young falcons investigated (*Falco mexicanus*) in Wyoming (69); from wild turkeys (185); from 24 of 300 birds (*Meleagris galloparvo intermedia*) in Texas (56), and from *Colinus virginianus* in Mexico (32). Among domestic poultry, occurrence of *Aegyptianella*-like parasites has been described for fowls in New York and Philadelphia (61); and for turkeys in California (177). *A. pullorum* is also reported for domestic poultry in Mexico in the 1970 edition of the *Animal Health Yearbook*. Regardless of these qualifications, the danger of avian aegyptianellosis becoming endemic and/or being introduced epidemically does exist in North and Central America due to the documented vectorial potential of *A. persicus, A. radiatus*, and *A. sanchezi*, coupled with their predilection for indigenous vertebrate hosts that are highly susceptible to *A. pullorum*. There is an obvious need to clarify whether or not development of the agent can take place in ticks under the ecological conditions prevailing in these regions. Yet, it should, at the same time, be noted, that the close association of this tick species to areas populated by humans and their domestic poultry certainly points to the existence of ecological niches that would allow for the cyclical propagation of *A. pullorum*. A corresponding risk must likewise be considered in the case of South America. There, *Persicargas* ticks associated with domestic poultry, whose vectorial function, incidentally, has not yet been investigated, and aegyptianellosis in *Amazona aestira* (195), and barring a suspected misprint, in fowls as well (63), have occurred.

Species from the subgenus *Argas* must also be tied into the epidemiological interplay of relationships pertaining to *A. pullorum* infections, although a vectorial function has only been proven for *Argas reflexus* and *Argas africolumbae*. In the case of *A. reflexus*, it was determined (111) that larvae, I and II nymphs, and male and female adult ticks become infected with *A. pullorum* and then transfer the agent to the corresponding subsequent stage or intrastadially at the next blood meal. Despite this high susceptibility, the natural vectorial potential for this species of tick needs to be viewed relatively, since all stages prefer pigeons which are most likely to be refractory for *A. pullorum*. Accordingly, the cyclical propagation of the pathogen cannot succeed in habitats otherwise typical for *A. reflexus*, although this would be possible if a location in fowl, goose, or duck pens is established.

As opposed to this, *A. africolumbae* enjoys a high degree of epidemiological effectiveness and may be significantly involved in the maintenance of the natural cycle of the pathogen—at least regionally. This species, which occurs in the Ethiopian fauna region and parasitizes wild birds predominantly (155a), also settles in primitive fowl pens and roosts. It thereby uses fowls as a blood donor while functioning as a very effective vector of *A. pullorum*. This vectorial significance of *A. africolumbae* is supported by investigations involving three wild populations collected from chicken coops, henhouses, and night roosts in Dedougou, Burkina Faso (115), which yielded high rates of infection. High degrees of parasitemia developed following the infestation of fully susceptible fowls with only 30-40 I and II nymphs, as well as male and female adult ticks. *Borrelia anserina* was transmitted as well. The vectorial potential of individual tick collectives from the three populations remained high and unchanged for both species of pathogens throughout the nine-month investigation period. Adult ticks that had reached their third engorgement transmitted both *A. pullorum* and *Borrelia anserina* to fully susceptible chicks during renewed infestations. This tick species actually passed aegyptianellas often vertically to its larval offspring, since up to 20% of the chicks became infected whenever the animals were infested with 200–300 larvae from the collective of female ticks subsequent to the latter's first, second, or third engorgement (115).

Conclusions

Thus, in considering the total framework of interrelationships between *A. pullorum* and its vertebrate hosts and vectors, it should first be emphasized that this heteroxenous species of pathogen occurs in Europe, Africa, Asia, and possibly in America. In other words, it is widely distributed and includes a broad host spectrum of domestic poultry. This is evidenced by the proven susceptibility of fowls, geese, ducks, and quails, as well as the concomittant and frequent documentation of natural infections. A similarly broad spectrum of vector species can be confirmed on the basis of its widespread geographic distribution, which was previously indicated by numerous isolates of the causative agent from wild strains of *A. africolumbae, A. persicus,* and *A. walkerae,* as well as by successful transfer tests with experimentally infected *A. radiatus, A. reflexus, A. sanchezi,* and *A. walkerae.*

The vectors, as proven for *A. walkerae,* not only become infected in every postembryonic stage and transmit *A. pullorum* transstadially, but they do not lose the pathogen in the same generation of ticks. They even allow it to be passed vertically and over lengthy fasting periods. This certainly represents the most important pathogen reservoir and, thereby, the epidemiologically most effective link in the vertebrate–vector–infection chain. Based on the adaption to fowls, one must further consider that in addition to the already proven vector–species, others may function to transmit *A. pullorum.* These include *A. abdussalami* in the Oriental region, *A. robertsi* in the Oriental and Australian region, *A. miniatus* in the neotropical and neoarctic region from the 15 *Persicargas* spp., and *A neghmei* and *A. magnus* in the neotropical region of the subgenus *Argas* (142a). The particular vectorial potential of the argasid species is further strengthened by their exceptional ecological plasticity to very effectively meet and resist critical abiotic conditions in their habitat over long periods of time. The potential consequence of this is the ability of the ticks to occupy additional settlement areas quickly and permanently, thereby establishing new endemic foci of *A. pullorum* infection. Of particular epidemiological relevance in this context is the susceptibility of larvae to the agent, since this stationary-periodic stage is the only one that remains permanently anchored to the host for several days. This allows it to accompany its host over long distances, thereby placing it in a position to set up new endemic foci in areas previously free of *A. pullorum* and/or ticks. Another fact is that argasid ticks become infected even at very low, microscopically elusive levels of parasitemia, or during concomitant infestation of infected and parasite-free ticks on "clean" hosts. Whether and to what extent transmission by other hematophage arthropods, as has been considered or suspected for ixodid ticks (47, 48), *Dermanyssus* (175, 207), *Ornithonyssus* (175), and other mites (47, 48), as well as for *Cimex* (9, 12, 47, 48, 207, 225), Nematocera (47, 48, 71, 175), and fleas (47, 48), might be possible still needs to be clarified. However, on the basis of simple transfer of infection through intravenous, intramuscular, and intraperitoneal inoculation of aegyptianellas and scarification of the skin, it cannot be ruled out.

Among vertebrates, fowls certainly represent the most important hosts, since these domestic animals are not only held more frequently in the endemic regions than geese, ducks, and quails, but are also the blood donor clearly preferred by all postembryonic stages of the primary vectors known to date—namely *A. persicus* and *A. walkerae*. These considerations are further substantiated by the published documentation or natural infections in fowls as compared with other avian species. Their epidemiological significance is further augmented by the potential longevity of infectiousness for the vector ticks, since fowls that become infected as chicks or young birds (as is the rule in endemic regions) can function as persistent pathogen reservoirs over a period of at least 18 months. Even clinically healthy fowls that have been treated with the minimum therapeutic dosage of tetracyclines and dithiosemicarbazones must be regarded as potential sources of infection for the vectors, since their therapeutic influence is not sufficient to completely eliminate the parasites from the host organism. Accordingly, it is appropriate to conclude that fowls, and perhaps other avian hosts, may be very effectively involved in the epidemic spread and establishment of endemic foci of avian aegyptianellosis. This is true even at great distances from the place of origin, provided, of course, that the birds' new habitats are concomitantly inhabited by vectors. In all probability, an infection reservoir of *A. pullorum* may also be epidemiologically operative through wild bird species, for aegyptianellas can be passed from wild to domestic birds and vice versa. However, the actual species identity and, therefore, the epidemiological relevance to domestic poultry of intraerythrocytic parasites found in wild birds and referred to as being similar or identical to *A. pullorum* as well as described as *Babesia* spp. or *Nuttallia* spp., is not yet clarified. Independent of this qualification, it should be emphasized that *A. pullorum* possesses an exceptional biological plasticity within the total framework of its relationships to vertebrate hosts and vectors. The important consequence epidemiologically is very effective infection reservoirs in the avian hosts and transmitter ticks, which complement and replenish one another continuously. This thereby ensures permanent natural circulation of the pathogen, as long a link in the infection chain does not break.

References

1. Abdussalam, M., 1945. Piroplasmosis of the domestic fowl in northern India, *Indian J. Vet. Sci.* **15**:17–21.
2. Aboytes-Torres, R., and Buening, G.M., 1990, Development of a recombinant *Anaplasma marginale* DNA probe, *Vet. Microbiol.* **24**:391–408.

3. Adene, D.F., and Dipeolu, O.O., 1975, Survey of blood and ectoparasites of domestic fowls in Ibadan, western state of Nigeria, *Bull. Anim. Health Prod. Afr.* **23**:333–335.

4. Ahmed, A.A.S., and Elsisi, M.A., 1965, Observations on aegyptianellosis and spirochaetosis of poultry in Egypt, *Vet. Med. J.* **11**:139–146.

5. Ahmed, A.A.S., and Soliman, M.K., 1966, Observations made during a natural outbreak of aegyptianellosis in chickens, *Avian Dis.* **10**:390–393.

6. Ayeni, J.S.O., Dipeolu, O.O., and Okaeme, A.N., 1983, Parasitic infections of the grey-breasted helmet guinea-fowl (*Numida meleagris galeata*) in Nigeria, *Vet. Parasitol.* **12**:59–63.

7. Balfour, A., 1907, A spirillosis and a haematozoal disease of domestic fowls in the Anglo-Egyptian Soudan. Preliminary note, *Br. Med. J.* **1**:744–745.

8. Balfour, A., 1907, A peculiar blood condition, probably parasitic, in Sudanese fowls, *J. Trop. Med. Hyg.* **10**:153–157, 322–323.

9. Balfour, A., 1907, A peculiar blood condition, probably parasitic, in Sudanese fowls, *Br. Med. J.,* **2445**:1330–1333.

10. Balfour, A., 1907, A peculiar blood condition, probably parasitic, in Soudanese fowls, *Lancet* **173**:708.

11. Balfour. A., 1908, Spirochaetosis of Sudanese fowls—an "after phase", *J. Trop. Med. Hyg.* **11**:37

12. Balfour, A., 1908, Spirochaetosis of Sudanese fowls, *Rep. Wellcome Trop. Res. Lab.* Karthoum, **3**:38–58.

13. Balfour, A., 1909, Further observations on fowl spirochaetosis, *J. Trop. Med. Hyg.* **12**:285–289

14. Balfour, A., 1910, Further observations on fowl spirochaetosis, *J. Trop. Vet. Sci.* **5**:309–322.

15. Balfour, A., 1911, The infective granule in certain protozoal infections, as illustrated by the spirochaetosis of Sudanese fowls. Preliminary note, *Br. Med. J.* **2654**:752

16. Balfour, A., 1911, The role of the infective granule in certain protozoal infections as illustrated by the spirochaetosis of Sudanese fowls. Preliminary note, *J. Trop. Med. Hyg.* **14**:113–114.

17. Balfour. A., 1911, Spirochaetosis of Sudanese fowls, *Rep. Wellcome Trop. Res. Lab.* Karthoum, **4A**:76–107.

18. Balfour, A., 1911, The role of the infective granule in certain protozoal diseases, *Br. Med. J.,* **2654**:1268–1269.

19. Balfour, A., 1912, The life-cycle of *Spirochaeta gallinarum*, *Parasitology* **5**:122–126.

20. Balfour, A., 1913, A contribution to the life-history of spirochaetes, *Zentralbl. Bakteriol. I. Abt. Orig.* **70**:182–185.

21. Balfour. A., 1914, Notes on the life-cycle of the Sudan fowl spirochaete, *Trans. 17th Int. Congr. Med. (London)*, sect. 21, Trop. Med. Hyg. 1914, part **2**:275–278.

22. Balfour, A., 1914, Note of the life-cycle of the Sudan fowl spirochaete, *Arch. Schiffs- und Tropenhyg.* **18**, Beihefte:844–845.

23. Balfour, A., 1922, Proceedings of a laboratory meeting, *Trans. R. Soc. Trop. Med. Hyg.* **16**:20–21.

24. Balfour. A., 1923, Note by Dr. Andrew Balfour on the above paper, *Trans. R. Soc. Trop. Med. Hyg.* **16**:484–485.

25. Barnes, H., 1975, Avian spirochaetosis in northern Nigeria, in Leeflang, P., and Ilemobade, A. A.: Tick-borne diseases of domestic animals in northern Nigeria. II. Research summary, 1966 to 1976, *Trop. Anim. Health Prod.* **9**:211–218.

26. Barrett. P.A., Beveridge, E., Bradley, P.L., Brown, C.G.D., Bushby, S.R.M., Clarke, M.L., Neal, R.A., Smith, R., and Wilde, J.K.H., 1965, Biological activities of some α-dithiosemicarbazones, *Nature (London)*, **206**:1340–1341.

27. Barta, J.R., Bouland, Y., and Desser, S.S., 1989, Blood parasites of *Rana esculenta* from Corsica: Comparison of its parasites with those of eastern North American ranids in the context of host phylogeny, *Trans. Am. Microsc. Soc.* **108**:6–20.

28. Bartkowiak, R.A., Huchzermeyer, F.W., Potgieter, F.T., van Rensburg, L., Labuschagne, F.J., and van Biljon, B.J., 1988, Freeze-drying of *Aegyptianella pullorum, Onderstepoort J. Vet. Res.* **55**:125–126.

29. Basu, B.C., 1944, A note on *Aegyptianella pullorum* in fowls in India, *Proc. 31th. Indian Sci. Congr.* **109**.

30. Battelli, C., 1947, Su di un piroplasma della *Naia nigricollis* (*Aegyptianella carpani* n. sp.), *Riv. Parassitol.* **8**:205–212.

31. Bedford, G.A.H., and Coles, J.D.W.A., 1933, The transmission of *Aegyptianella pullorum*, Carpano, to fowls by means of ticks belonging to the genus *Argas, Onderstepoort J. Vet. Res.* **1**:15–18.

32. Beltrán, E., 1944, Protozoarios sanguineos de las aves, *An. Escuela Nac. Cienc. Biol. Méx.* **3**:361–366.

33. Bennet, 1943, in Curasson. G. (ed): *Traité de protozoologie vétérinaire et comparée* Vigot Fréres, Paris, Tome **III**:242–249.

34. Bergeon. P., 1965., L'enseignement de la parasitologie des animaux domestiques et la production animale. Rapport au gouvernment du Cambodge, *Programme élargi d'assistance technique FAO* Rome, **2047**.

35. Bergeon, P., 1967, personal communication.

36. Bergeon. P., 1968, Report to the government of Ethiopia. Veterinary parasitology survey, *United Nations Dev. Program FAO* Rome, **2458**.

37. Berson. J.P., 1964, Les protozoaires parasites des hématies et du systéme histiocytaire des oiseaux. Essai de nomenclature. *Rev. Élev.* **17**:43–96.

38. Bird, R.G., and Garnham, P.C.C., 1967, *Aegyptianella pullorum* Carpano, 1928-fine structure and taxonomy, *J. Protozool.* Suppl. **14**:42.

39. Bird, R.G., and Garnham, P.C.C., 1969, *Aegyptianella pullorum* Carpano 1928-fine structure and taxonomy, *Parasitology* **59**:745–752.

40. Bouet, G., 1909, Spirillose des poules au Soudan Francais, *Bull. Soc. Path. Exot.* **2**:288–291.

41. Brumpt. E., 1929, No title, *Bull. Soc. Path. Exot.* **22**:318.

42. Brumpt. E., 1930, Rechutes parasitaires intenses, dues à la splénectomie, au cours d'infections latentes à *Argyptianella*, chez la poule, *C.R. Acad. Sci. (Paris)*, **191**:1028–1030.

43. Brumpt, E., 1949, *Precis de Parasitologie* Masson et Cie, Paris 6 edition.

44. Brumpt, E., and Lavier, G., 1935, Sur un piroplasmidé nouveau, parasite de tortue *Tunetella emydis* N.G.,N. Sp, *Ann. Parasit. Hum. Comp.* **13**:544–550.

45. Buxton, P.A., and Hopkins, G.H.E., 1927, Researches in Polynesia and Melanesia, *Sch. Hyg. Trop. Med. Mem.* London, **1**:52–53.

46. Cabrera, D.J., 1957, *Anaplasma*-like bodies in the guinea fowl, *J. Am. Vet. Med. Ass.* **130**:448–449.

47. Carpano, M., 1928, Piroplasmosis in Egyptian fowls (*Egyptianella pullorum*), *Bull. Minist. Agric. Egypt* Cairo, **86**:1–12.
48. Carpano, M., 1929, Su di un Piroplasma osservato nei polli in Egitto ("*Aegiptianella pullorum*"). Nota preventiva, *Clin. Vet.* (*Milano*) **52**:339–351.
49. Carpano, M., 1929, Sulla piroplasmosi dei volatili domestici determinata dalla "*Aegyptianella pullorum*," *Clin. Vet.* (*Milano*) **52**:475–476.
50. Carpano, M., 1929, Sur la piroplasmose des oiseaux domestiques déterminée par *Aegyptianella pullorum, Ann. Parasitol. Hum. Comp.* **7**:365–366.
51. Carpano. M., 1930, Sur la nature des anaplasmes et en particulier de l'*Anaplasma centrale, Ann. Parasitol. Hum. Comp.* **8**:231–240.
52. Carpano, M., 1930, Infections latentes a hémoprotozoaires. Maladies intercurrentes et récidives, *Ann. Parasitol. Hum. Comp.* **8**:638–658.
53. Carpano, M., 1936, Decouverte de corps intraleucocytaires chez les poulets d'Égypte et considérations sur leur nature, *Ann. Parasitol. Hum. Comp.* **14**:380–384.
54. Carpano, M., 1939, Sui piroplasmidi dei cheloni e su una nuova specie rinvenuta nelle tartarughe- *Nuttallia guglielmi, Riv. Parassitol.* **3**:267–276.
55. Carpano, M., 1939, Le infezione da "*Aegyptianella pullorum*" nei gallinacei e nei palmipedi in Albania, *Riv. Milit Med. Vet.* Roma, **2**:423–436.
56. Castle, M.D., and Christensen, B.M., 1985, Isolation and identification of *Aegyptianella pullorum* (Rickettsiales, Anaplasmataceae) in wild turkeys from North America, *Avian Dis.* **29**:437–445.
57. Chaillot, L., and Saunie, L. 1932. Contribution a l'étude de la spirillose aviaire dans les Etats du Levant, *Bull. Acad. Vét. Fr.* **5**:112–114.
58. Cheissin, E.M., 1965, Taxonomics of piroplasmae and some pecularities of their development in the vertebrate and invertebrate hosts, *Acta Protozool.* Warszawa **3**:103–109.
59. Coles, J.D.W.A., 1933, Mortality in fowls due to *Aegyptianella pullorum, Onderstepoort J. Vet. Sci.* **1**:9–14.
60. Coles, J.D.W.A., 1934, An outbreak of aegyptianellosis in Pekin ducks, *J.S. Afr. Vet. Med. Ass.* **5**:131.
61. Coles, J.D.W.A., 1937, A new blood parasite of the fowl, Onderstepoort *J. Vet. Sci.* **9**:301–305.
62. Coles, J.D.W.A., 1937, Aegyptianellosis and leg-weakness of the goose, *J.S. Afr. Vet. Med. Ass.* **8**:98–100.
63. Coles, J.D.W.A., 1939, Aegyptianellosis of poultry, *Proc. 7th World's Poult. Congr.* Washington D.C., U.S. Dep. Agric. pp. 261–265.
64. Coles, J.D.W.A., 1941, An epizootic in seabirds: A visit to Dassen and Malagas Islands, *J.S. Afr. Vet. Med. Ass.* **12**:23–30.
65. Cordero. M., 1967, personal communication.
66. Corliss, J.O., 1968, Definition and classification, in Weinmann, D., and Ristic, M. (eds): Infectious blood diseases of man and animals, Volume I, Academic Press, New York, London, pp. 139–147.
67. Corradetti, A., and Scanga, M., 1964, Segnalazione in Europa di *Babesia shortti* (Mohammed, 1958) in *Falco tinnunculus tinnunculus* e brevi note su questo parassita, *Parassitologia* **6**:77–80.
68. Crawford, M., 1947, *Diseases of poultry in Ceylon*, Ballière, Tindall, Cox, London.

69. Croft, R.E., and Kingston, N., 1975, *Babesia moshkowskii* (Schurenkova, 1938) Laird and Lari, 1957; from the prairie falcon, *Falco mexicanus*, in Wyoming, with comments on other parasites found in this host, *J. Wildl. Dis.* **11**:229–233.

70. Curasson, G., 1938, Notes sur la piroplasmose aviaire en A.O.F. *Bull. Serv. Zootech. Epizoot. A.O.F.* **1**:33–35.

71. Curasson, G., 1943, *Traité de protozoologie vétérinaire et comparée.* Vigot Frères, Paris, Tome **III**:242–249.

72. Curasson, G., and Andrejesky, P., 1929, Sur les «corps de Balfour» du sang de la poule, *Bull. Soc. Path. Exot.* **22**:316–317.

73. Debonera. G., 1933, La piroplasmose des poules en Grèce, *Bull. Soc. Path. Exot.* **26**:14–15.

74. Debonera, G., 1934, Unicité des agents des spirochétoses des volailles de Grèce. Relations entre spirochètes et *Aegyptianella, Rec. Méd. Vét.* **110**:467–471.

75. Delpy, L., 1946., Protozoaires observés en Iran dans le sang des animaux dometiques, *Bull. Soc. Path. Exot.* **39**:122–126.

76. Desser, S.S., 1987, *Aegyptianella ranarum* sp. n. (Rickettsiales, Anaplasmataceae): Ultrastructure and prevalence in frogs from Ontario, *J. Wildl. Dis.* **23**:52–59.

77. Desser, S.S., and Barta, J.R., 1984, An intraerythrocytic virus and rickettsia of frogs from Algonquin Park, Ontario, *Can. J. Zool.* **62**:1521–1524.

78. Desser, S.S., and Barta, J.R., 1989, The morphological features of *Aegyptianella bactifera*: An intraerythrocytic rickettsia of frogs from Corsica, *J. Wildl. Dis.* **25**:313–318.

79. Desser, S.S., Barta, J.R., Gruia-Gray, J., and Maclean, M., 1986, Pro- and eukaryotic parasites of the blood of frogs, in Howell, M.J. (ed): *Handbook 6th Int. Congr. Parasitol.*, Australian Academy of Science, Canberra, p. 121.

80. Dhanapala, S.B., 1962, Studies on some Sporozoa of *Gallus dometicus* and *Gallus lafayetti* of Ceylon, Thesis, London, p. 234. From: *Vet. Bull.* (*Weybridge*), 33 No. **1142**:182.

81. Dobell, C., 1912, Researches on the spirochaets and related organisms, *Arch. Protistenkd.* **26**:117–240.

82. Dodd, S., 1910, Spirochaetosis in fowls in Queensland, *J. Comp. Path.* **23**:1–17.

83. Donatien, A., and Lestoquard, F., 1931, Présence d'*Aegyptianella pullorum* chez les poules en Algérie, *Bull. Soc. Path. Exot.* **24**:371–372.

84. Donatien, A., and Lestoquard, F., 1934, Traitement curatif de la piroplasmose de la poule à *Aegyptianella pullorum, Bull. Soc. Path. Exot.* **27**:647–649.

85. Dschunkowsky, E., 1937, Pregled radova o Balfourovim granulama u vezi sa novim nazivima ovog parazita (*Aegyptianella* Carpano-*Balfouria* mihi), *Jugosl. Vet. Glasn.* **17**:315–321.

86. Dschunkowsky, E., 1937, Balfoursche Granula als echte Geflügelparasiten, ihre Natur und Stellung in der Systematik: *Aegyptianella pullorum* Carpano, *Balfouria* n. genus. *Balfouria anserina* n. sp. und *Balfouria gallinarum* n. sp., *Zentralbl. Bakteriol.* I. Abt. Orig. **140**:131–136.

87. Dschunkowsky, E., and Luhs, J., 1909, To the question of the study of protozoan illness in fowls in Transcaucasia, *9th Int. Vet. Congr. at the Hague*, pp. 18–19.

88. Dschunkowsky, E., and Urodschevitch, J., 1923, The spirochaetosis of fowls in Macedonia. (*Sp. anserina, Sp. gallinarum, Sp. granulosa penetrans.*), *Trans. R. Soc. Trop. Med. Hyg.* **16**:478–484.

89. Dyakonov, L.P., 1973, Structure, biology and systematic position of *Anaplasma* of ruminants, *Aegyptianella* of birds, *Haemobartonella* and *Eperythrozoon. Dokl. Vses. Akad. S-KH. Nauk.*, Moscow, **12**:22–24. From Vet. Bull. (Weybridge), 44 No. 1509 (1974).

90. Eddin Bey, S., 1952, Les piroplasmoses en Egypte, *Bull. Off. Int. Épizoot.* Paris **38**:593–603.

91. Eldin, M.S., 1932, The parasitic entity of *Aegyptianella pullorum* (Carpano, 1928), *Arch. Schiffs- Tropenhyg.* **36**:400–407.

92. Enigk, K., 1942., Blutparasiten bei südafrikanischen Vögeln, *Der Kolonialtierarzt, Dtsch. Tieraerztl. Wochenschr.* **50**:177–180.

93. Ezzat, A.M.E., 1963, Protozoal diseases of veterinary importance in Egypt, in Ludvik, J., Lom, L. and Vaura, J. (eds): Progress of Protozoology. *Proc. 1st Int. Congr. Protozool.*, Prague, 1961, Academic Press, New York, London.

94. Fantham, H.B., 1911, Some researches on the life-cycle of spirochaetes, *Ann. Trop. Med. Parasit.* **5**:479–496.

95. Farid, A., 1933, Veterinary service annual report 1930–1931, *Minist. Agric. Egypt*, Government Press: Cairo, **17**.

96. Franchini, G., 1924, Hématozoaires particuliers d'un oiseau (*Hypoleis hypoleis*), *Bull. Soc. Path. Exot.* **17**:884–885.

97. Galli-Valerio, B., 1909, Recherches sur la spirochétiase des poules de Tunisie et sur son agent de transmission: *Argas persicus* Fischer, *Zentralbl. Bakteriol.* I. Abt., Orig. **50**:189–202.

98. Galli-Valerio, B., 1913/1914, Recherches sur la spirochétiase des poules de Tunesie et sur son agent de transmission: *Argas persicus* Fischer, *Zentralbl. Baktriol.*, Abt. I. Orig. **72**:526–528.

99. Gerlach, F., 1924, Geflügelspirochätose in Oesterreich. *Zentralbl. Bakteriol.* I. Abt., Orig. **92**:84–96.

100. Gerlach. F., 1925, Geflügelspirochätose in Oesterreich, *Zentralbl. Bakteriol.* I. Abt., Orig. **94**:45–51.

101. Gillain. J., 1935, Note sur la présence d'*Aegyptianella pullorum* chez les poules au Congo Belge. *Ann. Soc. Belg. Méd. Trop.* **15**:299–300.

102. Gilruth. J.A., 1910, Note on the existence of spirochaetosis affecting fowls in Victoria, *Proc. R. Soc.* Victoria, **23**.

103. Gleitsmann, 1913, Beitrag zur Entwicklungsgeschichte der Spirochäten (Borrelien), *Zentralbl. Bakteriol.* I. Abt., Orig. **68**:31–49.

104. Gleitsmann, 1913, Beitrag zur Entwicklungsgeschichte der Spirochäten (Borrelien), *Zentralbl. Bakteriol.* I. Abt., Orig. **70**:186–187.

105. Goff, W.L., Stiller, D., Roeder, R.A., Johnston, L.W., Falk, D., Gorham, J.R., and McGuire, T.C., 1990, Comparison of a DNA probe, complement-fixation and indirect immunofluorescense tests for diagnosing *Anaplasma marginale* in suspected carrier cattle, *Vet. Microbiol.* **24**:381–390.

106. Gothe, R., 1967, Zur Entwicklung von *Aeyyptianella pullorum* Carpano, 1928, in der Lederzecke *Argas (Persicargas) persicus* (Oken, 1818) und Übertragung, *Z. Parasitenkd* **29**:103–118.

107. Gothe, R., 1967, Ein Beitrag zur systematischen Stellung von *Aegyptianella pullorum* Carpano, 1928. *Z. Parasitenkd* **29**:119–129.

108. Gothe, R., 1967, Untersuchungen über die Entwicklung und den Infektionsverlauf von *Aegyptianella pullorum* Carpano, 1928, im Huhn, *Z. Parasitenkd.* **29**: 149–158.

109. Gothe, R., 1967, Die Entwicklung von *Aegyptianella pullorum* im Wirbeltierwirt und in der Überträgerzecke *Argas* (*Persicargas*) *persicus*. *Proc. 3rd Conf. Dtsch. Tropenmed. Ges.* Hamburg, pp. 22–24.

110. Gothe, R., 1969, Zur Pathogenese der *Aegyptianella pullorum*-Infektion beim Huhn. *Z. Parasitenkd.* **31**:3.

111. Gothe, R., 1971, Ein Beitrag zum Wirt-Parasit-Verhältnis von *Aegyptianella pullorum* Carpano, 1928, im biologischen Übertrager *Argas* (*Persicargas*) *persicus* (Oken, 1818) und im Wirbeltierwirt *Gallus gallus domesticus* L., *Adv. Vet. Med.*, Suppl. to *Zentralbl. Veterinaermed.* Verlag Paul Parey, Berlin, Hamburg **16**:14 pp.

112. Gothe, R., 1977, Zur Entwicklung und Übertragung von *Aegyptianella pullorum, Tropenmed. Parasitol.* **28**:283–284.

113. Gothe, R., 1978, New aspects of the epizootiology of aegyptianellosis in poultry, in Wilde, J.K.H. (ed): *Tick-borne diseases and their vectors* University of Edinburgh, Center for Tropical Veterinary Medicine, Lewis Reprints Ltd., Tonbridge, pp. 201–204.

114. Gothe, R., and Becht, H., 1969, Untersuchungen über die Entwicklung von *Aegyptlanella pullorum* Carpano, 1928, in der Lederzecke *Argas* (*Persicargas*) *persicus* (Oken, 1818) mit Hilfe fluoreszierender Antikörper *Z. Parasitenkd.* **31**:315–325.

115. Gothe, R., Buchheim, C., and Schrecke, W., 1981, *Argas* (*Persicargas*) *persicus* und *Argas* (*Argas*) *africolumbae* als natürliche biologische Überträger von *Borrelia anserina* und *Aegyptianella pullorum* in Obervolta, *Berl. Muench. Tieraerztl. Wochenschr.* **94**:280–285.

116. Gothe, R., and Burkhardt, E., 1979, The erythrocytic entry- and exit-mechanism of *Aegyptianella pullorum* Carpano, 1928, *Z. Parasitenkd.* **60**:221–227.

117. Gothe, R., and Englert, R., 1978, Quantitative Untersuchungen zur Toxinwirkung von Larven neoarktischer *Persicargas* spp. bei Hühnern, *Zentralbl. Veterinaermed*, **B25**:122–133.

118. Gothe, R., and Hartmann, S., 1979, The viability of cryopreserved *Aegyptianella pullorum* Carpano, 1928 in the vector *Argas* (*Persicargas*) *walkerae* Kaiser and Hoogstraal, 1969, *Z. Parasitenkd.* **58**:189–190.

119. Gothe, R., and Koop, E., 1974, Zur biologischen Bewertung der Validität von *Argas* (*Persicargas*) *persicus* (Oken, 1818), *Argas* (*Persicargas*) *arboreus* Kaiser, Hoogstraal und Kohls, 1964 und *Argas* (*Persicargas*) *walkerae* Kaiser und Hoogstraal, 1969. I. Untersuchungen zur Entwicklungsbiologie, *Z. Parasitenkd.* **44**:299–317.

120. Gothe, R., and Koop, E., 1974, Zur biologischen Bewertung der Validität von *Argas* (*Persicargas*) *persicus* (Oken, 1818), *Argas* (*Persicargas*) *arboreus* Kaiser, Hoogstraal und Kohls, 1964 und *Argas* (*Persicargas*) *walkerae* Kaiser und Hoogstraal, 1969. II. Kreuzungsversuche, *Z. Parasitenkd.* **44**:319–328.

121. Gothe, R., and Kreier, J.P., 1977, Aegyptianella, Eperythrozoon, and Haemobartonella, in Kreier, J.P. (ed): *Parasitic Protozoa* Volume IV, Academic Press. New York, San Francisco, London, pp. 251–294.

122. Gothe, R., and Kreier, J.P., 1984, Genus II. *Aegyptianella* Carpano, 1929, in Krieg, N.R., and Holt, J.G. (eds): *Bergey's Manual of Systematic Bacteriology* Volume 1, 9th. edition Williams & Wilkens. Baltimore, London, pp. 722–723.

123. Gothe, R., and Lämmler, G., 1970, Die *Aegyptianella pullorum*-Infektion des Huhnes, ein Modell zur quantitativen Auswertung der therapeutischen und

prophylaktischen Wirksamkeit von Tetracyclinen, *Arzneim. Forsch.* (*Drug Res.*) **20**:92–95.

124. Gothe, R., and Lämmler, G., 1970, Über die Persistenz von *Aegyptianella pullorum* Carpano, 1928, in Küken nach chemotherapeutischer Behandlung. *Zentralbl. Veterinaermed.* **B17**:806–812.

125. Gothe, R., and Lämmler, G., 1971, Zur Chemotherapie und Chemoprophylaxe der *Aegyptianella pullorum*-Infektion des Huhnes. *Proc. 2nd Conf. Oester. Ges. Tropenmed., 4th Conf. Dtsch. Tropenmed. Ges., Salzburg and Bad Reichenhall.* Hanseatisches Verlagskontor H. Scheffler, Lübeck, 1969, p. 212.

126. Gothe, R., and Lämmler, G., 1971, Über das Verhalten der Serumproteine bei Hühnern im Verlauf der *Aegyptianella pullorum*-Infektion. *Zentralbl. Veterinaermed*, **B18**:162–169.

127. Gothe, R., and Mieth, H., 1979, Zur Wirksamkeit von Pleuromutilinen bei *Aegyptianella pullorum*-Infektionen der Küken. *Tropenmed. Parasitol*, **30**:323–327.

128. Gothe, R., and Nolte, I., 1981, Kombinierte licht- und rasterelektronenmikroskopische Betrachtungen der Wechselwirkungen im intrazellulären Kontaktbereich von *Aegyptianella pullorum* Carpano, 1928, und Erythrozyten, *Berl. Muench. Tieraerztl. Wochenschr.* **94**:261–263.

129. Gothe, R., and Schrecke, W., 1972, Zur epizootiologischen Bedeutung von *Persicargas*-Zecken der Hühner in Transvaal, *Berl. Muench. Tieraerztl. Wochenschr.* **85**:9–11.

130. Gothe, R., and Schrecke, W., 1972, Zur Epizootiologie der Aegyptianellose des Geflügels. *Z. Parasitenkd.* **39**:64–65.

131. Hadani, A., and Dinur, Y., 1968, Studies on the transmission of *Aegyptianella pullorum, J. Protozool.* Suppl. **15**:45.

132. Haiba, M.H., and El-Shabrawy, M.N., 1967, Some haematological studies on *Babesiosoma anseris* (Haiba and El-Shabrawy, 1967) in the goose, *Cygnopsis cygnoides* in Egypt, *J. Vet. Sci. U.A.R.* **4**:177–187.

133. Haiba, M.H., and El-Shabrawy, M.N., 1967, *Babesiosoma anseris* (n. sp.) investigated in the goose, *Cygnopsis cygnoides* in Egypt. *J. Vet. Sci. U.A.R.* **4**:189–194.

134. Haiba, M.H., and El-Shabrawy, M.N., 1967, On *Nuttallia henryi* n. sp., from the blood of ducks and geese in Egypt, *J. Helmithol., Protozoology Suppl.* **2**:215–221.

135. Helmy Mohammed, A.H., 1952, Protozoal blood parasites of Egyptian birds. Thesis from Dept. Parasitol., Sch. Hyg. Trop. Med. London.

136. Helmy Mohammed, A.H., 1958, Systematic and experimental studies on protozoal blood parasites of Egyptian birds. Volume I, and II, University Press, Cairo.

137. Henry, C., 1939, Présence dans les hématies de poulets d'éléments rappelant les corps de Balfour, *Bull. Soc. Path. Exot.* **32**:145–149.

138. Herman, C.M., 1968, Blood protozoa of free-living birds, *Symp. Zool. Soc. London* **24**:177–195.

139. Herman, C.M., 1969, in Diarmid, A. (ed): Diseases in free-living wild animals, *Smyp. Zool. Soc. London*, **24**:177–193, Academic Press, London. From: Peirce, M.A. 1975. *Nuttallia* Franca, 1909 (Babesiidae) preoccupied by *Nuttallia* Dall, 1898 (Psammobidae): A reappraisal of the taxonomic position of the avian piroplasms. *Int. J. Parasitol.* **5**:285–287.

140. Hindle, E., 1911, On the life-cycle of *Spirochaeta gallinarum*, Preliminary note, *Parasitology* **4**:463–477.

141. Hindle, E., 1912, Note on the foregoing communication by Dr. Andrew Balfour, *Parasitology* **5**:127.

142. Hindle, E., 1912, The inheritance of spirochaetal infection in *Argas persicus*. *Proc. Cambridge Philos. Soc.* **16**:457–459.

142a. Hoogstraal, H., Clifford, C.M., Keirans. J.E. and Wassef, H.Y., 1979, Recent developments in biomedical knowledge of *Argas* ticks (Ixodoidea: Argasidae), in Rodriguez, J.G. (ed): Recent Advances in Acarology, Volume II, Academic Press, New York, San Francisco, London, pp. 269–278.

143. Huchzermeyer, F.W., 1965, Das Tiefgefrieren von *Aegyptianella pullorum* in flüssigem Stickstoff mit einigen Bemerkungen über die künstliche Infektion beim Huhn, *Berl. Muench. Tieraerztl. Wochenschr.* **78**:433–435.

144. Huchzermeyer, F.W., 1967, Die durch künstliche *Aegyptianella pullorum*-Infektion beim Haushuhn hervorgerufene Anämie. *Dtsch. Tieraerztl. Wochenschr.* **74**:437–439.

145. Huchzermeyer, F.W., 1969, Personal communication to Keymer, I.F., in Petrak, M.L. (ed): 1969, Diseases of cage and aviary birds, Lea and Febiger, Philadelphia.

146. Huchzermeyer, F.W., Cilliers, J.A., Diaz Lavigne, C.D., and Bartkowiak, R.A., 1987, Broiler pulmonary hypertension syndrome. I. Increased right ventricular mass in broilers experimentally infected with *Aegyptianella pullorum, Onderstepoort J. Vet. Res.* **54**:113–114.

147. Huchzermeyer, F.W., Horak, I.G., and Braak, L.E.O., 1989, Bloodparasites and argasid ticks from guineafowl (*Numida meleagris*) from the Skukuza area of the Krüger National Park with a note on the suitability of the guineafowl as experimental host for *Aegyptianella pullorum, 18th Ann. Symp. Parasitol. Soc. S. Afr.* Johannesburg p. 11.

148. Huff, C.G., Marchbank, D.F., Saroff, A.H., Scrimshaw, P.W., and Shiroishi, T., 1950, Experimental infections with *Plasmodium fallax* Schwetz isolated from the Uganda tufted guinea fowl *Numida meleagris major* Hartlaub, *J. National Malaria Soc.* **9**:307–319

149. Hussel, L., Eichler, W., Liebisch, A., and Schneider, J., 1966, Die protozoären Blutparasiten der Haustiere in warmen Ländern, *S. Hirzel Verlag,* Leipzig.

150. Johnston. M.R.L., 1975., Distribution of *Pirhemocyton* Chatton & Blanc and other, possibly related, infections of poikilotherms, *J. Protozool.* **22**:529–535.

151. Jowett. W., 1910, Fowl diseases. Note on the occurrence of fowl spirochaetosis at the Cape, *Agric. J. Cape Good Hope* **37**:662–670.

151a. Kaiser, M.N., 1966, The subgenus *Persicargas* (Ixodoidea, Argasidae, *Argas*). 3. The life cycle of A. (*P.*) *arboreus*, and a standardized rearing method for argasid ticks, *Ann. Entomol. Soc. Am.* **59**:496–502.

152. Knowles, R., Das Gupta, B.M., and Basu, B.C., 1927, Preliminary observations on the morphology and life history of *Spirochaeta anserina. Trans. 7th Congr. Far East. Ass. Trop. Med.* **2**:573–581.

153. Knowles, R., Das Gupta, B.M., and Basu, B.C., 1932, Studies in avian spirochaetosis, parts I and II, *Indian Med. Res. Mem.* **22**:1–113.

154. Knuth. P., and Du Toit, P.J., 1921, Tropenkrankheiten der Haustiere, in Mense, C. (ed): Handbuch der Tropenkrankheiten, 2nd edition **6**:555–581.

155. Komarov, A., 1934, On the recovery of *Egyptianella pullorum* Carpano from wild *Argas persicus* Oken, *Trans. R. Soc. Trop. Med. Hyg.* **27**:525–526.

155a. Kraiss, A. and Gothe, R., 1982, The life cycle of *Argas (Argas) africolumbae* under constant abiotic and biotic conditions, *Vet. Parasitol.* **11**:365–373.

156. Kreier, J.P., Domingue, N., Krampitz. H.E., Gothe. R., and Ristic, M., 1981, The hemotrophic bacteria: The families Bartonellaceae and Anaplasmataceae, in Starr, M.P., Stolp, H., Trüper, H.G., Balows, A., and Schlegel, H.G. (eds): The Prokaryotes. A handbook on habitats, isolation, and identification of bacteria, Springer-Verlag, Berlin, Heidelberg, New York, pp. 2189–2209.

157. Kreier, J.P., and Gothe, R., 1976, Aegyptianellosis, eperythrozoonosis. grahamellosis and haemobartonellosis, *Vet. Parasitol.* **2**:83–95.

158. Kreier, J.P., Gothe, R., Ihler, G.M., Krampitz, H.E., Mernaugh, G., and Palmer, G.H., 1992, The hemotrophic bacteria: The families Bartonellaceae and Anaplasmataceae. in Balows, A., Trüper, H.G., Dworkin, M., Harder, W., Schleifer, K.-H. (eds.): The Prokaryotes. A handbook on the biology of bacteria: Ecophysiology, isolation, identification, applications. 2nd. edition Springer-Verlag, Berlin, Heidelberg, New York. Volume IV, 3994–4022.

159. Kreier, J.P., and Ristic, M., 1972, Definition and taxonomy of *Anaplasma* species with emphasis on morphologic and immunologic features, *Tropenmed. Parasitol.* **23**:88–98.

160. Kreier, J.P., and Ristic, M., 1973, Organisms of the family Anaplasmataceae in the forthcoming 8th edition of Bergey's manual. *Proc. 6th Nat. Anaplasmosis Conf.* Las Vegas, Nevada 1973 pp. 24–28.

161. Lämmler, G., and Gothe, R., 1967, Zur Chemotherapie der *Aegyptianella pullorum*-Infektion des Huhnes, *Tropenmed. Parasitol.* **18**:479–488.

162. Lämmler, G., and Gothe, R., 1969, Untersuchungen über die therapeutische und prophylaktische Wirsksamkeit oral applizierter Tetracycline gegen *Aegyptianella pullorum* Carpano, 1928, im Huhn, *Zentralbl. Veterinaermed.* **B16**:663–670.

163. Laird, M., and Lari, F.A., 1957, The avian blood parasite *Babesia moshkovskii* (Schurenkova, 1938), with a record from *Corvus splendens* Vieillot in Pakistan, *Can. J. Zool.* **35**:783–795.

164. Laveran, A., 1907, From Balfour, A., A peculiar blood condition, probably parasitic, in Sudanese fowls. *Br. Med. J.* **2445**:1330–1333.

165. Leeflang, P., 1977, Tick-borne diseases of domestic animals in northern Nigeria. I. Historical review, 1923–1966, *Trop. Anim. Health Prod.* **9**:147–152.

166. Leeflang, P., and Ilemobade, A.A., 1977, Tick-borne diseases of domestic animals in northern Nigeria. II. Research summary, 1966 to 1976, *Trop. Anim. Health Prod.* **9**:211–218.

167. Leger, A., and Le Gallen, R., 1917, Spirochétose des poules au Sénégal. Son évolution clinique, *Bull. Soc. Path. Exot.* **10**:435–438.

168. Levine, N.D., 1961, Protozoan parasites of domestic animals and of man, 1st. edition Burgess Publishing Company, Minneapolis.

169. Levine, N.D., 1968, Ecology and host-parasite relationship, in Weinman, D., and Ristic, M. (eds): Infectious blood diseases of man and animals Volume I, Academic Press, New York, London, pp. 3–21.

170. Levine, N.D., 1971, Taxonomy of the piroplasms, *Trans. Am. Microsc. Soc.* **90**:2–33.

171. Levine, N.D., 1973, Protozoan parasites of domestic animals and of man, 2nd. edition Burgess Publishing Company, Minneapolis, Minnesota.
172. Levine, N.D., 1985, Veterinary Protozoology, 1st ed., Iowa State University Press, Ames, Iowa.
173. Lingard, A., and Jennings, E., 1904, A preliminary note on a pyroplasmosis found in man and some of the lower animals, Indian Med. Gaz. **39**:161–
174. Macfie, J.W.S., and Johnston, J.E.L., 1914, A note on the occurrence of spirochaetosis of fowls in southern Nigeria, *Ann. Trop. Med. Parasitol.* **8**:41–51.
175. Manuel, M.F., and Tongson, M.S., 1967/1968, Occurrence of *Aegyptianella*-like organism in Philippine domestic fowls, *Philipp. J. Anim. Sci.* **4/5**:99–106.
176. Marchoux, E., and Chorine, V., 1933, Les formes jusqu'alors invisibles des spirochètes ne filtrent pas. *Bull. Acad. Vét. Fr.* **6**:48–49.
177. McNeil, E., and Hinshaw, W.R., 1944, A blood parasite of the turkey, *J. Parasitol.* Suppl. **30**:9.
178. Mesnil, F., 1930, No title, *Bull. Inst. Pasteur* **28**:125–126.
179. Mlinac, F., 1937, Početna istraživanja piroplazmoza živine u Južnoj Srbiji, *Jugosl. Vet. Glasn.* **17**:92–94.
180. Mohteda, S.N., 1947, Aegyptianellosis in fowls, *Indian Vet. J.* **24**:167–169.
181. Morcos, Z., 1931, *Babesia* in fowls, *Vet Rec.* **11**:217.
182. Morcos, Z., 1932, A preliminary note on the results of some experiments on avian spirochaetosis in Egypt, *J. Egypt. Med. Ass.* pp. 292–298.
183. Morcos, Z., 1935, Preliminary studies in fowl spirochaetosis in Egypt, *Vet. J.* **91**:161–171.
184. Anonymous, 1950. Kenya. Department of Veterinary Services Annual Reports, 1948, Government Printer, Nairobi, p. 30.
185. Anonymous, 1986, Other blood Sporozoa of birds, in The Merck Veterinary Manual, 6th edition, Merck & Co., Inc., Rayway, New Jersey p. 1251–1252.
186. Neitz, W.O., 1967, The epidemiological pattern of viral, protophytal and protozoal zoonoses in relation to game preservation in South Africa, *J.S. Afr. Vet. Med. Ass.* **38**:129–141.
187. Nikolsky, S.N., and Vodyanov, A.A., 1969, Pathogenicity of *Aegyptianella pullorum, Proc. 3rd Int. Congr. Protozool.*, Publishing house NAUKA, Leningrad Branch, Leningrad, 1969, pp. 276–277.
188. Nikolsky, S.N., and Vodyanov, A.A., 1970, Significance of *Aegyptianella pullorum* in poultry diseases (in the USSR), *Tr. Vses. Inst. Eksp. Vet.* **38**:277–284.
189. Ostertag, von R., and Kulenkampff, G., 1941, Tierseuchen und Herdenkrankheiten in Afrika, Walter de Gruyter & Co, Berlin.
190. Paikne, D.L., Dhake, P.R., and Sardey, M.R., 1974, *Aegyptianella pullorum* infection in domestic fowl, *Indian Vet. J.* **51**:575.
191. Pavlov, P., 1964, Die Rolle der Wildvögel bei der Verbreitung von *Argas persicus* in Bulgarien, *Angew. Parasitol.* **5**:167–168.
192. Peirce, M.A., 1972, *Rickettsia*-like organisms in the blood of *Turdus abyssinicus* in Kenya, *J. Wildl. Dis.* **8**:273–274.
193. Peirce, M.A., 1973, *Nuttallia balearicae* sp. n., an avian piroplasm from crowned cranes (*Balearica* spp.), *J. Protozool.* **20**:543–546.
194. Peirce, M.A., 1975, *Nuttallia* Franca, 1909 (Babesiidae) preoccupied by *Nuttallia* Dall, 1898 (Psammobiidae): A reappraisal of the taxonomic position of the avian piroplasms, *Int. J. Parasitol.* **5**:285–287.

195. Peirce, M.A., and Bevan, B.J., 1977, Blood parasites of imported psittacine birds, *Vet. Rec.* **100**:282–283.

196. Pfeifer, R., 1990, Zur Ökologie von *Argas* (*Persicargas*) *walkerae* Kaiser und Hoogstraal, 1969, *Inaug. Diss.* Munich.

197. Poisson, R., 1953, Sporozoaires incertains. Super-famille des Babesioidea nov. in Grassé, P.-P., (ed): *Traité de Zoologie*, Volume I, Masson et Cie, Paris, pp. 935–975.

198. Porter, A., 1915, On *Anaplasma*-like bodies in the blood of vertebrates, *Ann. Trop. Med. Parasitol.* **9**:561–568.

199. Pusat, M.M., 1955, Memleketimizde ilk defa tesbit edilen *Aegyptianella pullorum* vak'asi, *Türk. Vet. Hekim. Dern. Derg.* **25**:2487–2490.

200. Qureshi, M.I., and Sheikh, A.H., 1978, Studies on blood protozoan parasites of poultry in Lahore district, *Pak. J. Sci.* **30**:165–167.

201. Raether, W., and Seidenath, H., 1972, Verhalten der Infektiosität verschiedener Protozoen-Spezies nach längerer Aufbewahrung in flüssigem Stickstoff, *Tropenmed. Parasitol.* **23**:428–431.

202. Raether, W., and Seidenath, H., 1977, Survival of *Aegyptianella pullorum, Anaplasma marginale* and various parasitic Protozoa following prolonged storage in liquid nitrogen, *Z. Parasitenkd.* **53**:41–46.

203. Rageau, J., and Vervent, G., 1959, Les tiques (acariens Ixodoidea) des iles francaises du pacifique, *Bull. Soc. Path. Exot.* **52**:819–835.

204. Receveur, P., 1947, Note sur la répartition géographique d'*Aegyptianella pullorum, Rev. Élev.* **1**:54.

205. Receveur, P., and Thomé, M., 1948, Nouvel hôte d'*Aegyptianella pullorum* et mensurations de quelques globules rouges d'oiseaux, *Rev. Élev.* **2**:239–243.

206. Reichenow, E., 1953, in Doflein, F., and Reichenow, E. (eds): *Lehrbuch der Protozoenkunde* Volume II, VEB Gustav Fischer Verlag, Jena, pp. 949–966.

207. Reshetnyak, V.Z., Bartenev, V.S., Voronyanskii, V.P., Rubanov, A.A., and Firsov, N.F., 1971, Aegyptianellosis in poultry, *Veterinariva, Moscow* **1**:66–68.

208. Richardson, U.F., 1948, Veterinary Protozoology, Oliver & Boyd, Edinburgh, London.

209. Richardson, U.F., and Kendall, S.B., 1957, Veterinary Protozoology, Oliver & Boyd, Edinburgh, London.

210. Ristic, M., and Kreier, J., 1974, Genus III. *Aegyptianella* Carpano 1929, in Buchanan, R.E., and Gibbons, N.E. (eds): *Bergey's Manual of Determinative Bacteriology*, Williams & Wilkins Company, Baltimore, pp. 909–910.

211. Robinson, E.M., and Coles, J.D.W.A., 1932, A note on *Aegyptianella pullorum* in the fowl in South Africa, *18th Rep. Dir. Vet. Serv. Anim. Ind.*, Government Printer Onderstepoort, Pretoria, pp. 31–34.

212. Rousselot, R., 1947., Parasites du sang de divers animaux de la region de Tehran., *Arch. Inst. Hessarek* **5**:62–72.

213. Rousselot, R., 1953, Note de Parasitologie tropicale, Vigot Frères, Paris, Tome I et II.

214. Sambon and Terzi, 1907, Reports. The Society of Tropical Medicine and Hygiene, *J. Trop. Med. Hyg.* **10**:380–388.

215. Schellack, C., 1908, Uebertragungsversuche der *Spirochaete gallinarum* durch *Argas reflexus* Fabr., *Zentralbl. Bakteriol. I. Abt. Orig.* **46**:486–488.

216. Schrecke, W., and Gothe, R., 1972, Zur Pathogenität südafrikanischer Stämme von *Aegyptianella pullorum* Carpano, 1928, *Tropenmed. Parasitol.* **23**:406–410.

217. Schurenkova, A., 1938, *Sogdianella moshkovskii* gen. nov., sp. nov.- a parasite belonging to the Piroplasmidea in raptororial bird—*Gypaëtus barbatus* L., *Med. Parazitol. (Moscow)* 7:932–937.

218. Senadhira, M.A.P., 1966, The parasites of Ceylon. I. Protozoa. A host check list, *Ceylon Vet. J.* 14:65–78.

219. Sergent. E., 1935, Rapport sur le fonctionnement de l'Institut Pasteur d'Algérie en 1934, *Arch. Inst. Pasteur Algér.* 13:418–450.

220. Sergent, E., 1941, *Aegyptianella pullorum* Carpano, 1929 *Aegyptianella granulosa penetrans* (Balfour, 1911) chez la caille, *Coturnix coturnix coturnix* (L.) en Algérie, *Arch. Inst. Pasteur Algér.* 19:26–28.

221. Sharat el Din. H., 1937, Piroplasmosis in geese in Egypt, *Rep. Vet. Serv. Egypt.* 1932–1933:83–96.

222. Soulsby, E.J.L., 1968, Helminths, arthropods and protozoa of domesticated animals, Baillière, Tindall and Cassell, London.

223. Sparapani, 1934, *Riv. Agric.* Roma, 17:1.

224. Toumanoff, C., 1940, Le parasite sanguin endoglobulaire du héron cendré de l'Indochine (*Ardea cinerea* var. *rectirostris* Gould), *Rev. Med. Fr. Extreme-Orient* 19:491–496.

225. Tsanov, T.S., 1983, Epidemiology of *Aegyptianella pullorum* infection in fowls, *Veterinarnomed. Nauki* 20:41–46.

226. Tyzzer, E.E., 1938, *Cytoecetes microti*, n. g., n. sp., a parasite developing in granulocytes and infective for small rodents, *Parasitology* 30:242–257.

227. Tyzzer, E.E., and Weinman, D., 1939, *Haemobartonella*, n. g. (*Bartonella* olim pro parte), *H. microti*, n. sp., of the field vole, *Microtus pennsylvanicus*, *Am. J. Hyg.* 30:141–157.

228. Umarji, G.Y., and Vaishnav, T.N., 1969, A record of avian piroplasmosis in Gujarat state, *Gujarat Coll. Vet. Sci. Anim. Husb. Mag.* 2:11–13.

229. Vodyanov, A.A., 1969, The morphology and development of *Aegyptianella pullorum* in hens, *Proc 3rd Int. Congr. Protozool.*, Publishing house NAUKA, Leningrad Branch, Leningrad, 1969, pg. 285.

230. Vodyanov, A.A., 1972, Epizootological importance of *Argas persicus* during *Aegyptianella pullorum* infection of birds, *Tr. 13, Mehzdunar. Entomol. Kongr. Moskva*, 1968, p. 275.

231. Wasielewski, T. von, 1913, Zur Kcnntnis der Halteridien-Krankheit der Vögel, *17th Int. Congr. Med. London*, pp. 245–249.

232. Weiss, E., and Moulder, J.W., 1984, Order I. Rickettsiales Gieszczkiewicz 1939, in Krieg, N.R., and Holt, J.G. (eds): *Bergey's Manual of Systematic Bacteriology* Volume 1:687, Williams & Wilkins, Baltimore, London, p. 687.

233. Wenyon, C.M., 1926, Protozoology, Baillière, Tindall, Cox, London.

234. Yakimoff, W., 1933, Les corps de Balfour au Caucase, *Bull. Soc, Path. Exot.* 26:606.

235. Yakunin, M.P., and Krivkova, A.M., 1971, New bloodparasite species of the family Babesiidae (Piroplasmidae) in birds, *Parazitologiya*, Leningrad 5:462–465.

236. Yasarol. S., 1974, Parasitic diseases in Turkey, University Press, Ege University lzmir.

237. Huchzermeyer, F.W., Horak, Z.G., and Braack, L.E.O., 1991, Isolation of *Aegyptianella* sp. (Rickettsiales: Anaplasmataccac) from helmeted guineafowls in the Kruger National Park, *S. Afr. J. Wildl. Res.* 21:15–18.

238. Earlé, R.A., Horak, Z.G., Huchzermeyer, F.W., Bennett, G.F., Braack, L.E.O., and Penzhorn, B.L., 1991, The prevalence of blood parasites in helmeted guinea-fowls, *Numida meleagris*, in the Kruger National Park, *Onderstepoort J. Vet. Res.* **58**:145–147.

239. Huchzermeyer, F.W., Horak, Z.G., Putterill, J.F., and Earlé, R.A., 1992, Description of *Aegyptianella botuliformis* n. sp. (Rickettsiales: Anaplasmataceae) from the helmeted guineafowl *Numida meleagris* with notes on geographical distribution and possible vectors, *Onderstepoort J. Vet. Res.* (in press)

240. Huchzermeyer, F.W., Earlé, R.A., Horak, Z.G., and Braack, L.E.O., 1991, Blood parasites of helmeted guineafowls in the Kruger National Park, *Proc. Congr. Parasitol. Wildl.* p. 83.

241. Huchzermeyer, F.W., and Putterill, J.F., 1991, Light and transmission electron microscopy of an *Aegyptianella* sp. of the helmeted guineafowl. *Proc. Congr. Parasitol. Wildl.* p. 84.

4
Immunity in Haematophagous Insect Vectors of Parasitic Infection

Peter J. Ham

Introduction

Since the early specialist studies of Salt on invertebrate immunity (eg. 175, 176) when it was widely accepted that invertebrates are able to recognise infections as foreign, there has been an upsurge in the study of insect immunity. A number of comprehensive reviews have recently been compiled on the subject, to the extent that volumes exclusively devoted to the topic of invertebrate immunity have been edited (eg. 22, 72, 114). This article is specifically concerned with the progress that has been made in our understanding of the immunity of potential insect vectors of parasitic infection, to the pathogens that they may transmit. The word 'potential' is deliberately used in view of the fact that successful immunity may render an insect a non-vector. It is intended, that by covering the whole subject of immunity in haematophagous vectors the reader will be pointed to sections of the literature which will help further in-depth study, rather than cover the entire field in depth as well as breadth. In addition the intention is to introduce some findings of this laboratory, which may stimulate research along similar lines.

Clearly, much of the community of insect immunologists, is concerned with model systems that lend themselves to the large scale preparation of material, where insects are either relatively large, such as several members of the Lepidoptera (including the Silk moth, *Bombyx mori* and *Hyalophora cecropia*) and the Orthoptera, or those which are smaller but easy to rear, and as a result are genetically well defined, such as *Drosophila melanogaster*, the fruit fly. This makes for easier isolation of molecules of potential interest. For example, approximately 0.5 of a microlitre of haemolymph can be perfused out of a single *Aedes aegypti*, the laboratory vector of many species of filariae,

Peter J. Ham, Vector Immunity Group, Department of Medical Entomology, Liverpool School of Tropical Medicine, Pembroke Place, Liverpool L3 5QA, United Kingdom. Present address: Professor of Vector Biology, Centre for Applied Entomology and Parasitology, Department of Biological Sciences, University of Keele, Keele, Staffordshire, United Kingdom.

or from an individual British *Simulium ornatum* a vector of bovine onchocerciasis. Moreover when working with West African *S. damnosum* approximately half this amount can be obtained, and yet a further half again when
dealing with Central American species such as *S. ochraceum*, purely due to the
small size of the flies themselves. When compared with the volumes (mls)
obtainable from the larger non vector species of insects mentioned above,
the problems encountered by the vector biologist trying to understand the
mechanisms of immunity, can be better understood.

These larger insect systems are nonetheless proving to be essential models
for work on the more fastidious and often minute vectors of parasitic infection. Much of the research direction concerning mosquitoes and blackflies,
for example is following along lines that have been laid down by groups
applying relatively new technical procedures to insect immunology, with
considerable achievement. Therefore the vector biologist is now acquiring
molecular biological, biochemical and immunological skills to crack the un-
answered questions surrounding the mechanisms of immunity in 'their be-
loved insects'.

In this field I believe considerable care should be taken not to clutch at
parallels with the vertebrate immune systems, and particularly with those of
mammals, too readily. It is clear that many similarities of function do occur.
The clear division of immunity into two interlinked humoral and the cellular
arms in insects is one parallel. However, true memory and molecular specificity such as occurs in the classic antibody response, is not yet known in any
arthropods. A caveat to this is that recently an immunoglobulin like molecule
has been isolated which may play some role in recognition with opsonin like
properties (192). Parallels in function between vertebrates and invertebrates
may be exploited to see if analogies in mechanism occur. An example of this is
the prophenoloxidase pathway leading to melanin production, as an immune
mechanism. This extremely rapid cascade reaction has many functional similarities to the cascade involving complement in vertebrates. However, similarities of mechanism should by no means be assumed. Current evidence is
that insect immunity, including that of haematophagous vectors of disease,
maybe fairly non specific in its afferent as well as its efferent arms.

Innate Immunity and Infection Acquired Immunity

Immunity in insects can be thought of on two levels. Innate immunity refers
to the inherent refractory nature of a potential vector to infection. This takes
no regard of the previous history of disease in the organism and assumes a
naivety in terms of infection. Clearly vectors differ markedly in their innate
immunity even within a species complex such as *S. damnosum* sl of West
Africa (51, 78). The second level of immunity is that which is acquired upon
infection. This is very much determined by the previous history of infection
within the insect. It indicates the presence of immune molecules of either

humoral or cellular origin, that have been upgraded as a result of a prior infection. Both heterologous and homologous immunity occur and as of yet, the clearest evidence of its activity is limited to the haemocoel. The nature and level of specificity are still poorly understood.

Vector/Parasite Systems

There are a number of categories of insects, mostly haematophagous Diptera, that are vectors of parasitic pathogens of both medical and veterinary significance. Many of these pathogens differ widely in their development within the differing vectors.

Of the protozoa and the helminth parasites that infect insects as an intermediate host, the most widely studied with respect to vector immunity have been the filarial nematodes (83, 147). As far back as 1921 Manson-Bahr (135) noted that melanotic capsules form around the worm within the thorax of mosquitoes (see 26). Since then there have been a plethora of studies on the melanisation response of insects against infection, particularly in relation to filariasis. There has been good reason; this being the most obvious sign of an immune response against a medically important organism in the vector. The possibilities of utilizing such information in future mosquito control strategies has not been lost on the research worker. The bulk of the work has been performed on *Aedes* and *Anopheles* spp. with comparatively little work being done on *Culex* species, despite the fact that *C. quinquifasciatus* is a major vector species transmitting bancroftian filariasis. There has been expediency here too. The elegant pioneering work of MacDonald and co-workers from the 1960's onwards (128–132), defining the genetic basis of susceptibility to filariae, was carried out in the relatively easily reared insect, *Ae. aegypti*, the yellow fever mosquito. Selection of refractory and susceptible strains of this laboratory vector have meant that it has been used as an *at hand* manipulable surrogate, but not necessarily always appropriately. Particular strains such as the 'Black Eye', Rockerfeller (RKF) or the Ref[m] are used around the world and it is not always clear what the lineage of these selected strains is, and how true they are to their original form. More difficulty was evidently encountered with *Culex quinquefasciatus*, the main intermediate host of *W. bancrofti* (131), (see review 196).

Fortunately, as rearing techniques have developed and as knowledge has been gleaned from the *Brugia/Ae. aegypti* model, studies have diversified into other systems. Understanding of vector immunity is now increasing for many species of important mosquitoes, as well as other vectors of filariae, including the vectors of onchocerciasis, the Simuliidae or blackflies.

It has been clear for years that mosquitoes vary in their innate susceptibility to filariae and early systematic research concentrated on three main areas. First was the efficiency of microfilariae to migrate through and out of the blood meal and through the mid-gut epithelium into the haemocoel (eg.

Ewert 1965a). Second was the rate of microfilariae exsheathment before, during and after migration. Third was the rate of melanisation and encapsulation, either in the haemocoel, the flight muscles or in the malpighian tubules, in the case of *Dirofilaria* spp.

These three elements of migration, exsheathment and melanisation were and to some extent still are thought to be the main components determining vector susceptibility. The first two are intimately associated with the gut. The third is known to be linked to the humoral and/or cellular defences of the haemocoel. The fact that melanization of *Dirofilaria* species occurs in the malpighian tubules is probably due to the interchange of haemocoel humoral immune molecules across a weakened syncytial tubule barrier. There is increasing evidence that the rate of migration is determined by gut factors such as proteases, agglutinins or other carbohydrate binding molecules. Linked to this is the possibility that the peritrophic membrane may play an role in determining parasite migration.

Vector susceptibility to onchocerciasis has been examined using a variety of model systems. These include the use of mosquitoes (13, 14, 43, 126, 212, 213), biting midges (14, 43, 140) and blackflies (55, 125, 164, 169). A model system using *O. lienalis* infections in British species of blackflies has been developed as a parallel for the *S. damnosum/O. volvulus* system of West Africa. The main vector of bovine *O. lienalis* in the UK is *S. ornatum.*, sl a species complex of varying susceptibility to the parasite (195). Different species of simuliid, when inoculated with doses of microfilariae will exhibit innate variation in susceptibility to *O. lienalis* (77) and *O. volvulus* (76, 78) and *O. ochengi* (127). Melanisation is not a phenomenon normally associated with *Onchocerca* infections in blackflies. In fact although the pathway for melanisation, utilizing phenoloxidase appears to be present in the haemolymph, it is not expressed as a melanotic capsule around dead, dying or non living material of non-self origin (74, 77, 125). Conversely, it has long been known that *S. damnosum* sl. from West Africa, is susceptible to more than one species of filariae (148).

Interest has also developed in the immunity of mosquitoes to malaria and investigations have taken place into the mechanisms of susceptibility of vector species. It has been known for several years that susceptibility is, as with filariae, a heritable trait and that strains and species differ in innate immunity to infection (48, 64, 87, 88, 106). Frizzi et al. (64) supported Huffs original assertion that immunity was expressed at a single genetic locus. Collins et al. (42) selected strains of *A. gambiae* for susceptibility and refractoriness to the simian malaria *P. cynomolgi*. These lines differed in their susceptibility to other species of malaria, but it was not possible to show a clear monofactorial inheritance of immunity/refractoriness. Development of *P. falciparum* in mosquitoes has been described by Sinden and co-workers as well as others (see 139, 183 for brief reviews).

Of course several other animal models have been used in addition to the human *P. falciparum*, including the rodent species *Plasmodium berghei* and *P.*

yoelii. Also used frequently is the avian malaria, *P. gallinaceum* (eg. 61, 67, 171, 203). The advantages of the different model systems include the abllity to look at the responses of both *Aedes* and *Anopheles* species infected with either filarial parasites or *Plasmodium.* The stimulation of different aspects of the mosquito physiology, with different parasites, provides valuable information as to the nature and specificity of the induced immune response. Furthermore the specificity of induction can be examined by examining the effects of the one upon the other. Much of the experimental work thus far performed has compared naive susceptible and refractory insects with regard to known areas of immune function, for example the presence or specificity of midgut agglutinins, or the level of prophenoloxidase and phenoloxidase in the haemolymph. On the other hand studies have also been applied to the responses of insects to a single infection with a parasite and how immunity is switched on, enhanced, or even downgraded. Furthermore comparisons have been made as to how insects of different innate immunity to a particular infection, respond or acquire immunity. Few however are the studies which investigate how acquired immunity affects a subsequent or secondary infection with either the same organism or one of a different species. It is of considerable importance to try and link the physiological processes going on in an insect in response to infection, with a functional immunity that expresses itself as a deterioration of the parasites ability to develop, or in the complete failure of the parasite.

Mosquitoes constitute the major group of haematophagous vectors of parasitic infection in their role as vectors of filariasis and malaria. Neverthe-

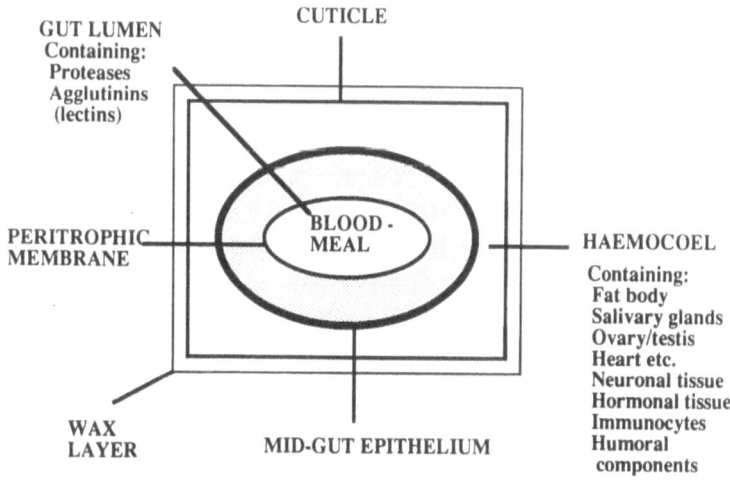

Schematic of Insect Compartmentation

FIGURE 4.1. Stylized diagram of the barriers confronted by parasitic pathogens

less, the last few years have seen dramatic improvements in the techniques applied to colonisation of tsetse flies (*Glossina*) and sandflies (Phlebotomines). This together with the availability of triatomid bugs such as *Rhodnius* and *Triatoma* have meant that the study of immunity in vectors of the Trypanosomatidae has flourished. This subject has recently been reviewed comprehensively (143).

This article is broadly organised according to the life-cycle of the parasite in the insects, in terms of the distinct microhabitats occupied (see Figure 4.1). In this way it is hoped that most situations in which the host poses a barrier to infection, will be encountered, and discussed. The role of the gut and its environs will be broken down into the different groups of parasites, as all are required to pass through, or remain in this structure, when infection results from taking an infecting blood meal. The haemocoel mediated mechanisms are classed according to the molecules elicited therein.

Initial Barriers to Infection

Pharyngeal and Cibarial armatures

Such armatures comprise a series of cuticular spines or barbs in the pharynx and cibarium, close to the cibarial pump which is responsible for ingesting the blood (7 (see 133), 25, 27). These teeth, together with spines in the anterior foregut, have the effect of breaking up the blood cells in extreme cases, and their main effect against parasites has been noted for filariae (6, 44).

The difference in transmission competence of *Onchocerca volvulus*, by *Simulium metallicum* and *S. ochraceum* in Central America, is thought to be partially due to the presence of these teeth in *S. ochraceum*, a vector of poor efficiency. They result in lethal damage to the microfilariae as they are ingested and so prevent ongoing development to infective third-stage larvae (152). Cibarial armatures have been detected in other *Simulium* species (164). McGreevy et al. (133) have shown that the armatures are responsible for the mortality of up to 96% of microfilariae in *A. gambiae*. Considerable variation was observed between species and genera as might be expected, with *Culex pipiens fatigans* killing only 6% of microfilariae. These cuticular protrusions of the pharyngeal and cibarial pumps may assist in exsheathment of microfilariae prior to passage through the gut epithelial cell wall. Furthermore the sheath may provide protection to the worm (133).

The Gut Lumen

Most parasitic infections enter the insect via an infecting blood meal, with a life cycle stage that is precisely suitable for gaining entry and establishing itself within the potential vector. As such they have to pass into, and as is often the case with malaria and filaria, beyond the mid-gut. *Plasmodium* species pass through the mid-gut epithelium as ookinetes and establish themselves within but on the haemocoel side of the mid gut wall. Filariae either pass through the mid-gut completely, usually prior to peritrophic membrane formation (156), pass through the haemocoel and into the cells of the thoracic flight muscle, as in most species of filariae, or continue through the mid-gut into the malpighian tubules lying at the junction of the mid and hind-gut. Some species of *Dirofilaria* undertake such a migration route and never enter the haemocoel, but induce a syncytium in the malpighian tubules in which it bathes while taking up nutrients as it develops. Yet others develop in the fat body of mosquitoes (eg. 84, 160) or tabanids (eg. 45).

Surrounding the blood meal lies the midgut and the peritrophic membrane, structures reviewed in the literature recently (15, 170). Their role in regulating infection and determining susceptibility, has therefore been discussed and will not be covered in detail here. Affects on vector susceptibility to infection seems to vary depending on the insect and the parasite involved. As far back as 1964, Duke and Lewis (52) suggested that the peritrophic membrane played a prominent role in determining *S. damnosum* susceptibility to *O. volvulus*. The knowledge we have concerning the carbohydrate moieties of the peritrophic membrane, and the fact that midgut agglutinins and lectins are important in vector susceptibility, imply that it may indeed play a part in determining innate susceptibility to infection.

Filariae

Iyengar (96) was one of the earlier workers to study the migration of microfilariae from the blood meal into the body cavity of mosquitoes. Furthermore, working with *Anopheles*, *Aedes* and *Culex*, the rate of exsheathment was linked to susceptibility (59, 60). Owen (153, 154) suggested that the rate of migration over the initial 3 hours following a blood meal was a good indicator of innate susceptibility. It appeared that in refractory species there was a tail off in migration after about 1 hour so that a differential was clearly observed between groups of mosquitoes. Owen suggested that exsheathment is only one component in limiting microfilariae migration through the midgut. Sutherland et al. (193) and LaFond et al. (117) went on to show that the passage of microfilariae through the mid-gut was a factor in the avoidance of haemocoel derived induced immunity. Rates of melanization in *Aedes* species, of *Brugia pahangi* microfilariae, were reduced following migration through the mid-gut epithelium, immediate dissection, washing and intrathoracic inoculation. Microfilariae inoculated directly without prior passage through the mid-gut were melanized at a much higher rate. They propose that the microfilariae are protected from immune recognition as non-self by a coating of host material. However, no change was found in the level of susceptibility to infection if the gut was bypassed by intrathoracic inoculation of artificially exsheathed microfilariae (177).

More recently, it was demonstrated that exsheathment rates in vitro of *B.*

pahangi, during incubation with abdomen homogenates correlated with susceptibility in vivo (95). The dogma has been that exsheathment in vivo occurs in the mid-gut lumen, prior to passage through the mid-gut. Clearly however, it is not essential that it occurs at this stage in order to allow parasite development. Sheathed *Brugia* microfilariae inoculated into the thorax of *Aedes aegypti* will undergo complete development to the infective stage, with the necessary migrations into and out of the thoracic flight musculature, that this requires (eg. 41).

Yamamoto et al. (209) working with *Armigeres subalbatus* and Christensen and Sutherland (41) observed exsheathment in the haemocoel of the mosquitoes. *B. pahangi* microfilariae allowed to migrate out of *Aedes aegypti* (Black Eye strain) mid-guts in vitro were largely sheathed, and it was postulated that exsheathment occurs predominantly within the haemocoel (41). Indeed, it was shown that exsheathment of a proportion of *B. malayi* microfilariae in the same 'Black Eye' strain of *Aedes aegypti* occurs during penetration of the mid-gut, and not only within the mid-gut lumen (1). The worms leave the sheath behind as they emerge into the haemocoel. Furthermore 70% of microfilariae observed at EM level retained their sheath. They suggest that sheaths cast in the haemocoel may act to decoy the immune system of the mosquito, so conserving the escaping microfilariae. This is in contrast with earlier workers (eg. Feng 1936 (see 41) 58, 60) who postulated that exsheathment is a gut lumen phenomenon. Chen and Shih (33) demonstrated that in both susceptible and refractory strains of *Ae. aegypti*, exsheathment occurred both in the haemocoel and in the midgut. The site of exsheathment appears to be defined by the period taken to escape from the blood meal and its surroundings. Indeed cast sheaths were found to be encapsulated in the haemocoel of both *Ae. aegypti* and *A. quadrimaculatus* (31).

One class of molecule in the gut of insects are the agglutinins or if defined by their carbohydrate specificity, the lectins. These are found throughout the living community. They are proteins with agglutinating properties, that have specific affinities for particular carbohydrate molecules (see 68). Because of the possible link between exsheathment, midgut penetration and the presence of gut lectins, much attention has been paid to the role of carbohydrate receptors on the surface of developing filarial worms (34, 49, 65, 81, 105, 166, 182, 211). Recent studies have shown that gut lectins may prevent migration into the haemocoel by *Ae. aegypti* of *Brugia* microfilariae (157). However, their role in gut protozoan infections is better defined.

MALARIA

The gut lumen forms a much longer lasting home for many other parasitic infections, particularly some of the protozoa. For malaria parasites it is the site of zygote formation between the macro and microgametocytes. The ookinete develops from the fertilised macrogametocyte within the gut lumen prior to attachment to the midgut wall and penetration through to the haemocoel side of the gut. It appears that migration is either intracellular as for *P. berghei* and *P. yoelii*, or intercellular in the cases of *P. falciparum* and *P. gallinaceum*. The different carbohydrate receptors on the midgut and the peritrophic membrane in different species of mosquitoes, may play a role in determining vector susceptibility to specific infections (173). There are interesting homologies between the surface of *P. falciparum* and *P. gallinaceum* such as a 25kDa protein (104, 139). Transition to the oocyst occurs prior to its final break down and release of sporozoites. Hence *Plasmodium*, like the filarial nematode, occupies and has to develop in close proximity to both the gut lumen and the haemocoel. Immunity can therefore be expressed against the parasite at both of these sites. However *Plasmodium* spp. spend the majority of time in the barrier between the two sites, enjoying some protection from both.

Despite this, melanisation of oocysts occurs in refractory mosquitoes. In the work of Collins et al. (42), Paskewitz, and others, the immunity is expressed against the late ookinete/early oocyst as it lies within the midgut wall, structurally isolated from both the gut lumen and the haemocoel by the basal lamina and a newly formed lamina protecting it from the haemocoel. Is the failure of the ookinete itself, to establish in the gut, a mechanism of immunity? The topic of stage specific blockage in malaria development has recently been reviewed (200). In 1973 Yoeli (210) postulated that a number of factors halted or reduced sporogonic development of *P. berghei*. By examining development in a number of mosquitoes, and depending on the species used, parasite development was either arrested at the exflagellation or ookinete formation stage, at penetration of the midgut wall, or in the production of abnormal and degenerate oocysts and a lower production of sporozoites. Interestingly strain selectivity of *An quadrimaculatus* according to visual genetic markers, resulted in lines with differing susceptibilities to *Plasmodium*, (207). It was suggested that selection of one gene for a homozygous form, affected others which were responsible for determining susceptibility.

The predominant molecules in the insect midgut are proteases, (12, 24, 70, 119—trypsin like predominantly) and agglutinins/lectins. Of the latter it is true that they are involved in both eliminating parasites from the gut, and in being a means of establishment. The peritrophic membrane and the midgut, are known to contain carbohydrate, and it is also known that the moieties differ with mosquito (173). So the PM of *Ae. aegypti* contains N-acetyl-D-glucosamine type sugars (see also 86), whereas that of *A. stephensi* contains N-acetyl-D-galactosamine (173). These carbohydrates form potential binding sites for gut lectins, which in turn may act as opsonins between the parasite surface and the insect tissue. For some species of malaria that form ookinetes after the normal time required for peritrophic membrane formation, it is clearly necessary for the parasite to penetrate the membrane prior to invasion of the midgut epithelium and successful establishment of the oocyst stage. Recently, working with *P. gallinaceum* in the Liverpool Black Eye strain of *Aedes aegypti*, an ookinete specific chitinase has been detected (86). Interestingly this enzyme is not secreted by the early zygote but only by the stage that will be required to penetrate the peritrophic membrane. For protozoa, these gut lectins may also result in their demise by self agglutination (as for *Leishmania* in phlebotomine sandflies, and for trypanosomes in tsetse).

TRYPANOSOMIASIS AND OTHER HAEMOFLAGELLATES

The Trypanosomatidae as one of the group of organisms that spend their time in the gut of the host, interact with potential immune molecules at this site more than many others. There is sometimes involvement of the haemocoel and the salivary glands, but this is infrequent.

Since the review by Molyneux et al. (143) one of the major contributions to the knowledge of how immunity is expressed has been by Maudlin and co-workers. They have postulated that susceptibility to infection in *Glossina* is linked to a number of midgut factors. Their recent investigations with tsetse flies have yielded an interesting story concerning an interaction between a host symbiont and successful trypanosome development. The presence of rickettsial like organisms (RLO's) in the midgut cells appears to be linked with enhanced susceptibility (136). It is suggested that these organisms produce chitinases which result in increased carbohydrate concentrations in the midgut lumen. It is further proposed that these glucosamine-like sugars bind and effectively mop up midgut lectin and so prevent the lectin binding to the trypanosome surface. This in turn blocks parasite agglutination, allowing establishment and subsequent success of the trypanosome infection. Despite this inhibition of lectin activity, it is also proposed, conversely, that different carbohydrate specific lectins are positively involved in the successful establishment of infection in the midgut (137, 138). Unfortunately at present it has not proved possible to eliminate RLO's from infected tsetses to investigate such a change on trypanosome development. However, it is another instance in which midgut molecules render an insect innately immune to infection. Other workers have put such factors forward as possible mechanisms of determining susceptibility (141).

Lectins have been found in *Rhodnius*, a vector of *T. cruzi*, another trypanosome. These have been demonstrated in the haemolymph, the mid-gut and the crop, and have the same carbohydrate specificity as the surface of the epimastigotes forms of the parasite that infect this intermediate host (155). Clearly, then there is a potential mechanism for susceptibility/immunity in this vector/parasite relationship, based on the carbohydrate-lectin interactions.

Other haemoflagellates include the *Leishmania* (see reviews 107, 109, 142). It is thought that on ingestion of an infected blood meal containing the infectious amastigates, the *Leishmania* stay in the gut lumen, without traversing the peritrophic membrane. They fail to establish themselves in the midgut epithelial wall until the peritrophic membrane has broken down (199). There appears to be no real evidence that attachment of the promastigotes, the gut dwelling form of *Leishmania*, is lectin mediated (201). Indeed the mechanism of attachment is at present unknown (109). It appears however, that the metacyclic promastigotes, or the infective forms, do not attach to the gut wall. There are surface carbohydrate differences between the non-infective and infective promastigotes and these have been determined by the use of labelled lectins (eg. 174). There is no evidence as yet that the change in sugar is related to attachment to the gut wall, although this would be an obvious conclusion, if it were not for the fact that carbohydrate inhibition tests do not affect gut attachment. Because sandflies feed off plant saps, which contain sugars and lectins, it has been postulated that the sugar/lectin interaction is probably important in the establishment of infections in the various segments of the gut (108, 142). This is particularly thought to be so in view of other evidence involving lectin-associated attachment of parasites both in vitro (181) and in vivo (155). Nevertheless the flagellar membrane-insect cuticle attachment in the fore and hind-gut are thought to be via hemidesmosomal junctions with the involvement of specific receptor molecules (142). The flagella attachment of the nectomonad promastigote to the midgut epithelium is not thought to be via hemidesmosomes, as with the later haptomonads. There is however pronounced interdigitation of the flagellum into the distal and proximal portions of the epithelial microvilli.

Clearly therefore, the innate immunity of sandflies to *Leishmania* spp. is most likely to be effected in the gut. This may be in the lumen of the mid-gut as the parasites await the breakdown of the peritrophic membrane. Proteases may play a role in reducing the early parasitic gut dwelling forms. Working with *Phlebotomus papatasi* a sandfly which is highly susceptible to *L. major* but not to *L. donovani*, a correlation has been found with the proteolytic enzymes in the gut (19, 179). *L. major* infections result in reduced enzyme activity, whereas *L. donovani* appears to enhance it. Conversely, in order to survive the prolonged period in the mid-gut, any gut parasites must be resistant to some proteolytic activity as it is a necessary prerequisite of blood digestion. It has also been suggested that DNAases in the nucleated erythrocytes of birds result in parasite degradation (180). Immunity may also be expressed in the ability to prevent attachment of the promastigote forms to either the mid-gut microvilli, and their attachment to the fore- and/or the hind-gut.

Haemocoel Mediated Immunity

The parasite enters the haemocoel from the blood meal either directly through the mid-gut within a few minutes of haematophagy, as with the filariae, or indirectly as with *Plasmodium* which occupy the boundary between gut and haemocoel for some days. Whatever the route, the entrance of the invader into the haemocoel, usually initiates a series of major changes in the metabolism of the insect. There are two major series of events which are most clearly noticeable. The first is the initiation of the prophenoloxidase (PPO) enzyme cascade reaction which may result in the deposition of the pigment melanin onto the surface of the parasite. The second is the rapid increase in the number of circulating cells or haemocytes within the haemolymph. However, in addition to these more obvious effects of parasitic infection, there are a number of other molecules secreted and released into the haemocoel.

The basis of this haemocoel mediated immunity is the ability of the insect host to recognise the parasitic organism, that is seeking to establish itself, as 'host' (37, 62). Therefore immunity must have important recognition and regulatory mechanisms, that allow initiation, while also being switched off, when necessary. Furthermore recognition, although important is only of use if the response is in some way discriminatory, which requires the presence/ absence of specific receptors on host tissue, hence preventing it from being rejected. The use of tissue transplants, either xenographs or allographs, have provided useful information as to the level of non-self recognition in insects (114, 115).

Humoral Immunity

It is the understanding of these areas of vector immunity which rely heavily on studies initiated and worked through in insects other than those of medical or veterinary importance as vectors. There are several groups of immune proteins, the functions and origins of which are described in the literature, which are now beginning to be identified in the haematophagous mosquitoes, simuliids and hemipteran bugs, as well as others.

A variety of techniques are now being used to disentangle the complex mass of descriptive and sometimes rather unfocussed reports on the various susceptibilities of this or that mosquito, to this or that filarial nematode or malaria parasite etc. These include a more systematic approach to examining susceptibility, by using genetically selected strains of insect (MacDonald and co-workers, 128–132, and 134, 191) as potential laboratory vectors, and comparing their responses to differing infection. This may be by direct infection, now often by intrathoracic inoculation for more specific quantification. Parabiotic twinning of susceptible and refractory species has been used (eg. 202, 204). Passive transfer of haemolymph components also provides extremely useful information regarding the relevance of a response to immunity against a parasite (74, 196). They also include the examination of the acquired response and its effectivity against subsequent infection in the same individual insect (195). It appears clear that the parasite is initially recognise as foreign when it passes through the basement membrane of the cuticle or the mid-gut epithelium, into the haemocoel (190).

Prophenoloxidase/Phenoloxidase

As mentioned above this enzyme and its precursor (prophenoloxidase-PPO), are initiated during infection, as well as at other times, and give rise to the production of melanin via the oxidation of phenols to toxic quinones. In fact it is reported that the 'melanin-like pigment' is a complex comprised of proteins and polyquinones (69). The inactive and nontoxic proenzyme (PPO) prevents excessive formation of these toxic products (see review 186). It has been shown that serine proteases are partly responsible for the initiation of this pathway.

These enzymes form an early part of the cascade reaction leading to the activation of prophenoloxidase to phenoloxidase (eg. 3, and Soderhall and co-workers—as in the excellent review, 186). Many other factors such as heat, detergents, lipids and microbial substances initial the switch from the proenzyme (PPO) to active PO. However, in many insects serine proteases appear to act by cleavage of the PPO. The normal end product of the cascade reaction, melanin, is the pigment that can be seen coating sporozoites of *Plasmodium* and filarial microfilariae in refractory strains of mosquitoes such as *Ae. aegypti*. Encapsulation of the microfilariae may also involve cells but not essentially so (32).

The role of melanization, however, although well studied and biochemically fairly well determined, is still rather ambiguous. It is clear that foreign material is melanized in conjunction with a preceding encapsulation response, which begins to isolate the organism. This encapsulation step may or may not involve the circulating immune cells of the haemolymph, the haemocytes (see reviews 159, 190 for responses against nematodes in particular). The fact that melanization requires the presence of haemocytes is denied by the occurrence of melanization within the malpighian tubules, of *Dirofilaria* species filaria (21, 103, 159). This worm occupies these tubules throughout its life cycle, never entering the haemocoel. Whether humoral haemolymph proteins involved in the encapsulation and melanisation process pass across the gut wall is another matter. It is possible that they can, as large molecules such as immunoglobulins are known to pass from the mid-gut into the haemocoel intact (197). However it seems unlikely that the haemocytes themselves are able to penetrate.

Furthermore it is quite possible that melanisation is merely the end point in the immune response rather than the effector response itself. It can be postulated that encapsulation and/or melanization are physiological mopping up operations of what may be non viable parasites. In other words some other factor(s) may be responsible for the killing or weakening of the foreign organisms and liberating molecules that are recognised as non-self, whether protozoa or helminths. The deposition of the melanin often occurs prior to cellular encapsulation in insects (69). It seems that melanization occurs, whether or not cells are involved.

Recent studies with simuliids have shown that the pathway for melanin production exists within the haemolymph. The use of iodotyrosine as a substrate for phenoloxidase in a simple in vitro blot assay, shows melanin formation using haemolymph perfused from adult female individuals of *S. ornatum* sl. (see Figure 4.2). *Onchocerca* microfilariae, although killed in vitro (see Figure 4.3) or in vivo are not melanized as part of this immune reaction.

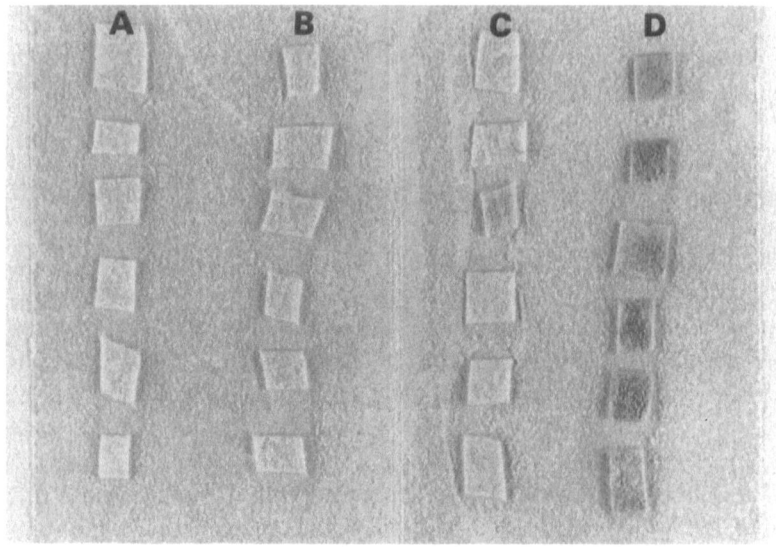

```
ANALYSIS OF VARIANCE
SOURCE     DF          SS         MS        F         p
FACTOR      3    17583310    5861104    22.03     0.000
ERROR      20     5321150     266058
TOTAL      23    22904460
                                          INDIVIDUAL 95 PCT CI'S FOR MEAN
                                          BASED ON POOLED STDEV
  LEVEL     N        MEAN      STDEV     ----+---------+---------+---------+--
PBS/PBS     6       115.2      116.8     (---*----)
PBS/iodo    6       243.5      324.0     (---*----)
HAEM/PBS    6      1066.7      770.5             (----*---)
HAEM/iod    6      2263.0      593.2                        (----*---)
                                          ----+---------+---------+---------+--
POOLED STDEV =     515.8                     0      1000      2000      3000
```

FIGURE 4.2. Photograph of melanin assay, using haemolymph from individual *S. ornatum* females on each filter paper. Underneath is the densitometric data quantifying the level of melanin formation (statistical analysis—one way ANOVA). As seen in C some automelanization occurs, but addition of iodotyrosine significantly increases this, demonstrating that humoral components are present in the haemolymph of *Simulium*. (A) PBS control, no substrate or haemolymph (B) iodotyrosine substrate, no haemolymph (C) PBS control with haemolymph (D) iodotyrosine with haemolymph

The PPO pathway, as described above, is reported to involve serine proteases (172). The use of protease substrate SDS poly-acrylamide gels (0.2% casein or gelatin) have shown that trauma induces a strong release of active proteases into the haemolymph of *Simulium ornatum*. Within an hour or two proteolytic activity can be clearly seen in the coomassie stained gels after the SDS has been washed out with triton X and then buffer (see Figure 4.4). A number of bands are visible, although some of these are clearly more diffuse than others,

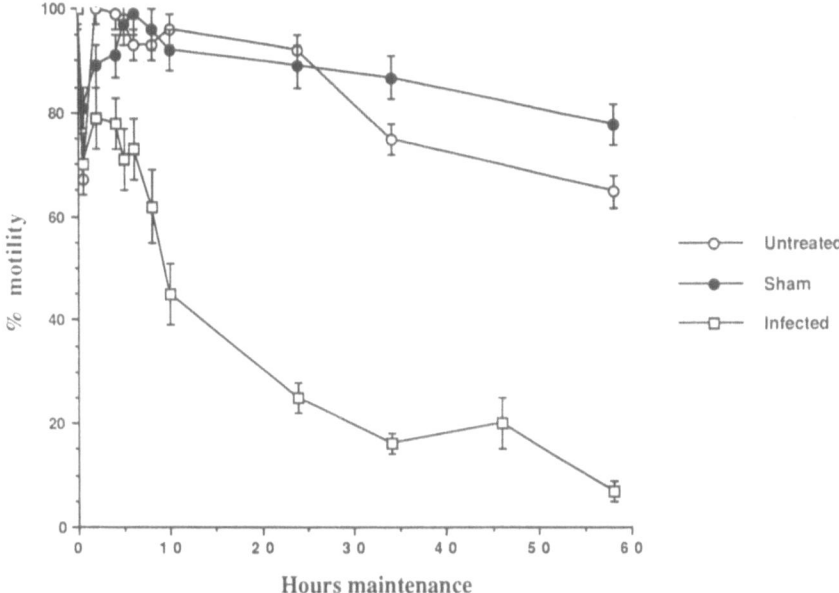

FIGURE 4.3. Effect of haemolymph from filariae infected, sham inoculated and untreated *Simulium ornatum* on *Onchocerca lienalis* microfilariae in Terasaki microtitration plates.

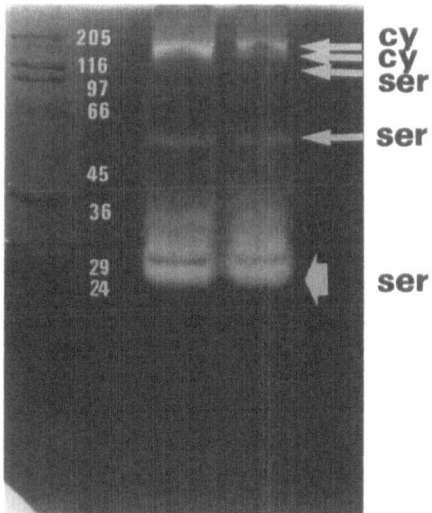

FIGURE 4.4. Protease substrate SDS polyacrylamide gel, following electrophoretic separation of immune haemolymph from *Simulium ornatum*. SDS was washed out with triton X, and PBS at Ph 7.2, prior to overnight incubation at 37°C and staining for protein. Clear zones indicate proteolytic activity. Arrows denote presence of protease bands. Where the use of specific inhibitors allows the class of protease to be determined, this has been shown (cy—cysteine protease, ser—serine protease).

making exact molecular weight determination difficult. Nevertheless these bands do not appear in haemolymph from untreated flies. Furthermore the use of specific protease inhibitors shows that some of these bands have serine protease activity, while two very large proteolytic molecules, forming thin discrete clear zones on the substrate, have cysteine protease activity. Both of these are significant in that the serine proteases may be involved in the activation of prophenoloxidase in vivo.

Molecular weight fractionation of immune haemolymph from *S. ornatum* was performed to investigate if the killing activity (82) could be isolated. Interestingly there was considerable variation in the survival of microfilariae in vitro, and this appeared to correlate with the presence of protease activity (see Figure 4.5). The fractions containing cystine and serine protease activity reduced the microfilariae motility, as did a fraction containing protein at 20-23kDa. It was not possible to assay activity of haemolymph below around

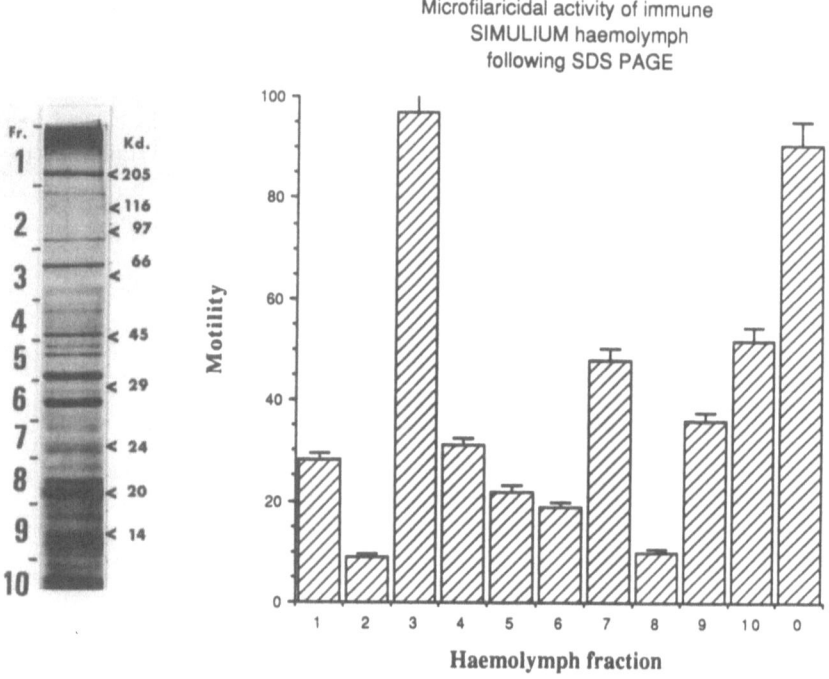

Microfilaricidal activity of immune
SIMULIUM haemolymph
following SDS PAGE

FIGURE 4.5. Microfilaricidal properties of immune *Simulium ornatum* haemolymph in vitro, following excision, electroelution and dialysis from a preparative SDS-polyacrylamide gel (see left hand panel). Fraction numbers to the left of the gel correlate with numbers along the horizontal axis of the histogram in the right hand panel. Fractions below 10kDa were not assayed using this system. The zones of activity in fractions 1/2 and 4/5/6 correlate with the presence of cysteine and serine proteases respectively. The zone of activity in fraction 8 covers molecules between ~18–23kDa, which includes the range of the attacins.

10kDa, although at this level there was no toxicity whatsoever. The fact that serine proteases and PO, are present in the haemocoel of blackflies indicates that the initial constituents for the PPO/PO pathway are present. Why the normal end product of melanin is not observable is rather unclear. Perhaps the necessary substrates for its formation as a result of PO activity, are not present in the constituents of the haemocoel, but only in the cuticular area, allowing deposition there.

It is proposed, therefore, in the light of this preliminary information, that proteolytic activity in simuliid haemolymph is perhaps not restricted to cleavage of PPO but may also play a direct role in attacking the parasite. Clearly further work must be carried out to clarify the mechanism further, but haemolymph proteolytic activity is very strong following trauma and infection in blackflies, indeed much stronger than that induced in *Ae. aegypti* and *Anopheles stephensi*, other vectors of both filariae and malaria (Ham, unpublished observations).

There is some dispute as to the origin of PPO in invertebrates with some workers having reported activity in haemocytes and others in the non cellular haemolymph (see review 186). However, preparation of cell free haemolymph without incidental activation of the haemocytes is rather difficult, despite the use of special solutions to perfuse the insects with. Therefore care has to be taken in interpreting results demonstrating PPO activity derived in the haemolymph itself. Notwithstanding this, there is a considerable body of evidence that PPO activity is not limited to the cells in the haemocoel in insects (see 186 for details).

Investigations into the nature of the stimuli necessary for initiation of the PPO cascade have been carried out by the inoculation in vitro and in vivo of a variety of agents, into insects or onto cell lines (121, 162). For example, haemocytes of *Galleria* exhibit enhanced phagocytosis of bacteria in vitro following treatments including laminarin (a fungal wall component, and lipopolysaccharide (LPS), but this effect could be inhibited if a serine protease inhibitor was incorporated.

This work was repeated (196) in vivo using *Ae. aegypti*, the laboratory vector of Brugian filariasis. In these experiments the use of the passive transfer technique was made, first described for vectors of onchocerciasis, the blackflies (74). After infecting susceptible *Aedes* with a variety of treatments including either *B. pahangi* microfilariae, TC199 medium, *E. coli*, laminarin, dextran or LPS, they withdrew the haemolymph and transferred it by inoculation to recipients which had just before been infected with *Brugia*.

They found that haemolymph from donors given all treatments except the dextran group, conferred some degree of immunity to infection in the recipients. It is known that dextran does not stimulate the PPO activation, but that laminarin and LPS do. Hence it was shown that these inducers give rise to an effective immunity against the eukaryotic filariae. However, the most effective stimulant for immunity was an injection with living microfilariae. This difference may be purely due to a quantitative difference in stimulation. However, it would be interesting to know whether the response against eukaryotes had some qualitative differences from that expressed against prokaryotes.

Having said this, there is precious little work on the mechanistics of the PPO/PO cascade in haematophagous vectors of parasitic infections. There is a wealth of descriptive and some analytical studies on the melanisation of filariae (see review 35). This has been published over a considerable period of time by a variety of groups. There has also been considerable emphasis on the comparative innate immune responses (eg. 10, 38) and PO activity between different strains or species of filariae vectors. For example, it is known that the melanisation response to *Dirofilaria immitis* and *Brugia pahangi* in *Ae. trivittatus* and *Ae. aegypti* (Liverpool strain) differ markedly and that older mosquitoes mount a reduced response (40). Furthermore, using a radiometric assay, haemocyte monophenoloxidase activity has been demonstrated in these two species of mosquitoes (124). Interestingly, the authors state that this is greater for *Ae. trivittatus* than for *Ae. aegypti* despite the fact that baseline haemocyte counts are similar, and that it is in *Ae. aegypti* that there is an induced cellular inflammation. Their implication from this is that there is enhanced monophenoloxidase synthesis or activation within the cells of *Ae. trivittatus* following immune stimulation. There are problems in discerning enzyme activity within and outside haemocytes which may or may not be covered in adherent humoral haemolymph components. Analysis using transmission electron microscopy and other techniques giving more resolution of the phenomena, may help to elucidate this. Metabolic labelling with ^{35}S methionine in insects of synthesized haemolymph proteins in *Ae. aegypti* inoculated with *D. immitis* demonstrate an 84kDa protein which appears to be synthesised at an increased rate in the sham and in the infection injected flies. There seemed to be no alteration in the amount of this protein in the untreated control insects, although there must have been active synthesis, for label to have been incorporated. Haemocytes are suggested as a possible source of this molecule, but at this stage its function remains unclear. Clearly though, a role in wound healing and general immune reactivity is likely.

By pulse labelling simuliids in a similar way, but at intervals following infection it has been possible to identify several molecules that are induced in response to injection and infection. Three particular proteins appear on SDS PAGE, and it is possible that these are subunits of the same molecule. The main difference between the *Onchocerca* injected flies and the sham inoculated insects is that, like the work with *Aedes* the synthesis is upgraded for a prolonged period in response to an active infection. Another difference in the studies, is that *O. lienalis*, used by this laboratory, is a normal infection which undergoes development. However, it has the property of inducing an acquired response which results in resistance to reinfection within the same fly (195). Furthermore haemolymph from these flies confers immunity to naive susceptible recipients following passive transfer (74). The use of tissue culture techniques and metabolic labelling has demonstrated that injection of blackflies induces synthesis and release of 4 proteins visible on non-reducing PAGE. These are at $570+$, $250+$, 66 and 44kDa. Of these, we have raised monoclonal antibodies to the $250+$ and the 66kDa molecules, in the hope of using them as tools to study the specificity and dynamics of synthesis. Although uncommon, it is not without precedent that reproductive tissues secrete inducible proteins (188). However, the fat body and haemocytes are clearly important targets for analysis. Their role in the synthesis of the humoral antibiotic molecules, attacins and cecropins, are also being investigated (see next section).

Intriguingly, investigations with the mosquito *Armigeres subalbatus* have shown a differential ability to mount a successful immune response against two very similar parasites, *B. pahangi* (which survives) and *B. malayi* (which is melanized rapidly) (208). This implies, not so much that the response can discern one species of parasite from another, but that the *B. pahangi* can evade the immune response. Clearly it is in the mosquitoes best interests to eradicate both species. The presence of the microfilariae in the midgut or its passage through the mid-gut epithelium may be involved in engineering some change whereby the mosquito cannot recognise the parasite as foreign, before it has begun developing and entered the cells of the flight muscle.

However, recent work is suggestive of a PPO/PO activation in larval *Ae. aegypti* (4). It was deduced that as with other insects, the proenzyme is present in the mosquitoes which when activated in the presence of a bivalent (Ca^{2+}) cation (eg. 121), and limited proteolytic action, gave rise to PO. Indeed, evidence was that serine proteases are involved in this activation. Using an antibody to silk worm PPO and Western blot analysis, they were able to show bands of PPO (at 74 and 66kDa) and PO (at 63kDa) homology.

Extracellular melanization then, is well reported as part of the response which may lead on to encapsulation, either with or without cells. However, intracellular melanization has also been reported, (111, 120) as part of the response by mosquitoes against filariae which develop within the cells of tissues such as the thoracic flight muscle. Although melanization in the haemocoel normally involves microfilariae, in the case of filarial infections, it was reported that the developing first stage larvae were melanized intracellularly (146). It has been suggested that this melanization is acellular, as haemocytes cannot penetrate the muscle cells (120). As mentioned earlier, melanization of *D. immitis* also occurs, within the malpighian tubules of the excretory system (eg. 21).

Antibacterial Molecules

Insects of non-medical importance have been extensively studied with regard to their immunity, in particular to bacteria. This response is usually acquired on infection and involves the rapid synthesis of a battery of potent antibacterial molecules. The expression of these is acute and titres of protein rise at a phenomenal rate following infection, often leaving the insect concerned with some degree of transient acquired immunity to subsequent bacterial infection (see review 53).

There are now known to be an array of small proteins that are synthesised and/or released into the haemolymph, usually very rapidly, following infection (53), or trauma such as injury (17). Initial studies showed that such molecules with antibiotic function such as lysozyme, were produced in response to bacterial stimulation (eg. 187). Over the past few years, thanks to some superb work performed predominantly, but not exclusively by Boman and his colleagues with diapausing pupae of *Hyalophora cecropia*, and Natori and colleagues studying the flesh fly, *Sarcophaga peregrina*, several other potent antibacterial molecules have been isolated and purified (see reviews 115, 196), eg. The cecropins (89, 91), the attacins (90), the sarcotoxins (I, II and III) which have many similarities to the cecropins and the attacins (150 and others), and the diptericins and the defensin like peptides isolated from *Phormia terranovae* (eg. 50, 122). In addition a tiny peptide (2.1kDa) with antibacterial function has been described from bees, and appropriately named Apidaecin (29).

Furthermore, the so called P4 protein (48kDa) has also been described from *H. cecropia*, the function of which is still largely unknown. However, recent studies (192) have shown that this molecule, now named as hemolin, has extremely interesting structural homologies. They suggest that hemolin, which has similarities to members of the immunoglobulins, is a recognition molecule and it one of the first to be involved in the immune response. Their studies were with *H. cecropia*, but the implications for this, in terms of recognition, and other possible homologies with more sophisticated immune systems, are considerable. The significance in relation to specificity remains to be seen, but the authors suggest that there is a basis for believing that hemolin may be involved in pattern recognition.

Amino acid and genomic sequence determinations have been carried out for representatives of most of these molecules, and many of them appear to be controlled pre-translationally at the level of transcription (53). As of yet, these molecules have not been widely identified in haematophagous vector species, but this is probably because they have not generally been looked for. This oversight has been for two good reasons. First, the molecules are thought to be exclusively antibiotic, and not functional against eukaryotes. Second, there has been a preoccupation with the most visible sign of immune activity in the haemolymph of vector species, the enzymic pathway leading to encapsulation and melanisation.

Although these molecules have been identified in non vector species of insects, emphasis has been placed on them here because evidence is beginning to appear that they may have some significance in immune responses against eukaryotes. Although the evidence is sparse, some of these molecules may have a role in the immune elimination of parasitic infection (47, 54, 94, 189). Further description will be confined to the work on the cecropin-sarcotoxin I groups and the attacin-sarcotoxin II groups of proteins. Distinct homologies between these pairs of families exist, despite being isolated from different species, as mentioned above.

CECROPINS

This family of molecules were first described by Hultmark et al. (89), as being small antibiotic peptides of approximately 3.8kDa. They have strong activity against gram positive and negative bacteria and the molecule is characterized by being bipolar, with a basic N-terminal region and a predominantly hydrophobic region in the C-terminal part. It is suggested (18) that the members of this family isolated by his group and co-workers from *Hyalophora cecropia* (89, 90) are derived through gene duplication. The nature of the amino acid sequence suggests that the molecule is partially helical and that this is amphipathic in nature. It is possible that such a structure would enable alignment onto/into the cell membrane under attack, resulting in rupture and lysis. Furthermore, it has been postulated that a group of cecropin molecules will have the property of channel or pore formation again resulting in cell lysis and death. Investigations using cholesterol have given rise to the hypothesis that this reduces the effectivity of cecropins. This in turn is put forward as an explanation as to why cecropins are ineffectual against eukaryotic organisms such as the protozoan and metazoan parasites under discussion here (36).

Notwithstanding this hypothesis, there is recent evidence that cecropins may indeed be active against some eukaryotic organisms. In vitro activity against trypomastigotes of *T. cruzi* and *P. falciparum* in infected erythrocytes in culture, was demonstrated using synthetic cecropin derivatives (98). Investigations with *Plasmodium* infected *Anopheles* also demonstrated some toxicity of cecropin to the parasites, which was expressed by reduced development of oocysts (73). Administration was by inoculation, and the toxicity of the compound to the mosquitoes at the concentrations used, was very high. Nevertheless a slight effect on the parasites was observed.

The presence of an inducible antibacterial molecule from the triatomid vector of *T. cruzi*, *Rhodnius prolixus* has been described (5). It was suggested that this was lytic and of a low molecular weight. Studies indicated that it was not lysozyme, and suggestions were put forward that it may be a cecropin like molecule. Activity against bacteria was differential.

A proposed mode of action of cecropins against prokariotic cells has been described above. However, the mechanism whereby it functions against eukaryotes is entirely unclear. We do know that the molecule does result in parasite death. Evidence that these antibacterial molecules may be involved in immune mediated killing of filaria in insects that are innately and/or have acquired immunity can be seen in that synthetic cecropin A and B are both microfilaricidal to *Brugia pahangi* (Chalk, R., Townson, H., Ham, P.J.—

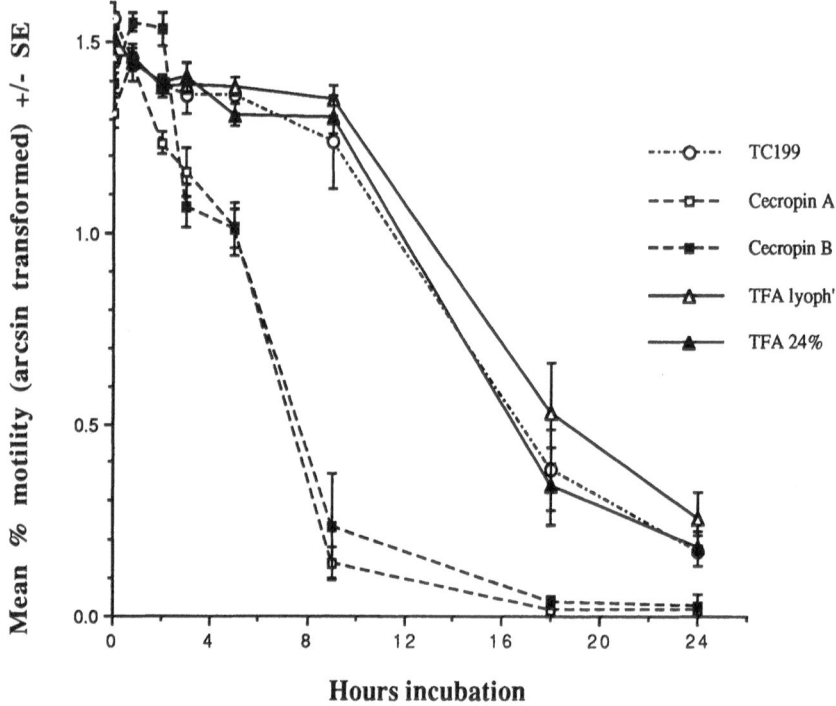

FIGURE 4.6. Motility rates over a 24 hour period in Terasaki microtitration plates TC199 denotes medium in wells, TFA Iyoph' and TFA 24% are controls for the maximum possible quantity of Trifluoroacetic acid (salt) present in the cecropin preparations; They had no affect on motility, whereas cecropin was toxic.

unpublished data) and *Onchocerca lienalis* microfilariae in vitro (Ham, P.J. unpub. see Figure 4.6). Perhaps just as important is the fact that these synthetics are also bactericidal in vitro. Activity against the microfilaria motility has been observed in blind trials, and although much slower to take effect than in bacteria, can be seen after a few hours incubation in vitro. In addition, in vivo activity has been demonstrated, with significant reductions in the proportion of microfilariae of *B. pahangi* and *O. lienalis* developing in the flight muscle of the mosquito *Ae. aegypti* and the simuliid *S. ornatum*, respectively.

Southern blot analysis of *Aedes aegypti* and *Simulium* genomic DNA, probed with oligonucleotides constructed from consensus sequences of cecropins (75, 110) and a *Drosophila* CDNA clone for cecropin A (Baxter, A.J., Ham, P.J., and Maingon, R.-unpublished), have demonstrated that cecropins are likely to be present in these vector species. Indeed clones have been isolated from genomic libraries of different blackfly species and are currently being characterized. In addition Western blot analysis of *Aedes aegypti* and

Simulium ornatum haemolymph samples probed with antibody to sarcotoxin IA (cecropin like peptide) and IIA (attacin-like protein) revealed bands following SDS PAGE (polyacrylamide gel electrophoresis). Furthermore the use of these antibodies in in vitro killing assays in conjunction with immune haemolymph from these insects shows that anti-sarcotoxin IA and IIA antibody block the microfilaricidal effect of the haemolymph, when compared to the effects of normal rabbit serum (NRS).

THE ATTACINS

This group or family of molecules are larger than the cecropins, having a molecular weight of around 20–23kDa. Secretion has been reported to be from the haemocytes and the fat body. Again, they are potent antibacterial

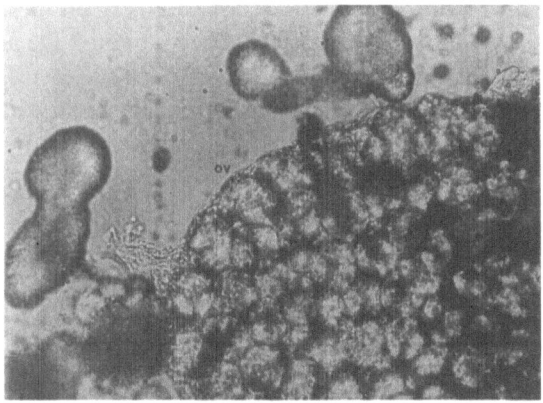

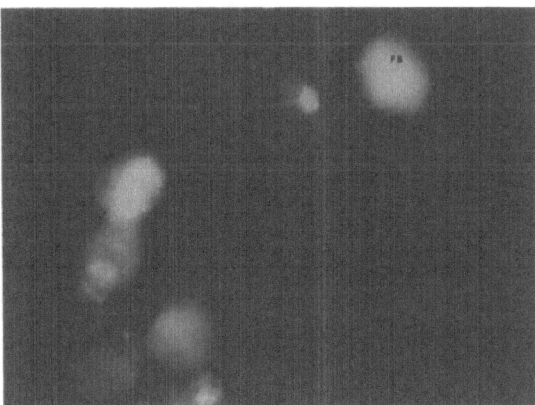

FIGURE 4.7. Paired light and UV fluorescence photomicrographs, showing binding of FITC-conjugated sarcotoxin IIA antibody to the fat body tissue (FB) of immune *Simulium ornatum*, following in vivo incubation. Also included but unstained is part of the ovary (OV).

molecules, and their presence has been described in a number of insects. As with cecropins they constitute a protein family, of six molecules (56).

At present there appears to be very little literature concerning these molecules in vector species of insects. However, Kaaya et al. (100) report the occurrence of attacin like molecules in tsetse flies. Furthermore, as reported above for cecropin like molecules (sarcotoxin 1A), antibody to attacin (sarcotoxin IIA) like proteins blocks microfilarial killing by immune haemolymph from *Aedes aegypti* and from *Simulium*. Furthermore, if FITC labelled antibody to sarcotoxin IIA is injected into living *Simulium*, and allowed to circulate for 10 minutes, followed by dissection and rigorous washing of the tissues, there is a preferential binding to the fat body of infected flies (see Figure 4.7). In addition fat body from immunised flies taken 24 hours post infection, show phenomenal R.E.R. activity and protein production (see Figure 4.8).

As with cecropins, the possible mode of action in eukaryotes, is unclear, and of course much more work needs to be carried out on these molecules in relation to this large group of organisms. The kinetics of the action must be examined, and the large size and diversity of cell type or size may also be important factors in determining whether or not the molecules can act. The synergistic effects of different immune molecules may also be important, bearing in mind that immune activation involves a panoply of molecules,

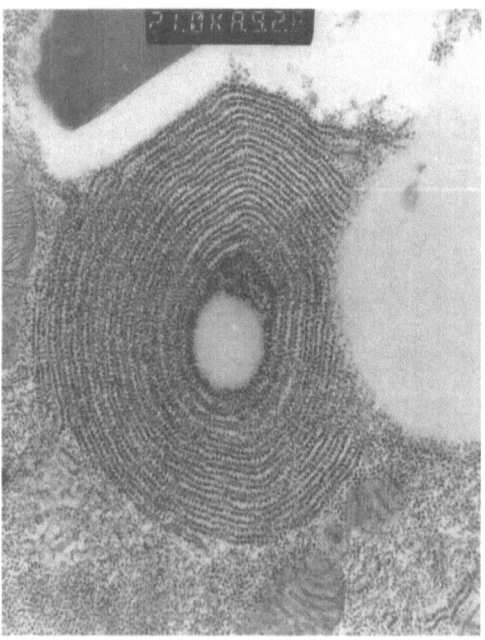

FIGURE 4.8. Electron micrograph of section of fat body from *Simulium ornatum* 24 hours post infection, showing protein (centre) surrounded by coils of rough endoplasmic reticulum (RER). Magnification at top.

ranging from the low molecular weight peptides, to the carbohydrate specific lectins and the substrate dependent PPO/PO system. For example, it is believed that attacins may act on the outer wall of *E. coli* and allow cecropins, and the more digestive lysozyme, to penetrate (56, 115). It has been reported that cecropins are susceptible to protease activity (eg. 16, 97), and that such enzymes may play a modulatory role in controlling cecropin activity. Clearly, the immunity stimulated by a foreign body, must be equally efficiently switched off, and surplice peptide neutralized. There may be a role being played, by some of the proteases that have been demonstrated in the haemolymph of immunized simuliids.

Agglutinins/Lectins

Humoral haemolymph agglutinins are now recognised to be a major component of invertebrate immunity and have been reported from a number of insects. As mentioned earlier, excellent reviews have been written on their distribution, physicochemical properties and possible roles in defense (eg. 172). The reports of agglutinin function in the gut of haematophagous vectors are relatively common but there are fewer pertaining to the haemocoel. The likely roles that they play are rather different. Midgut lectins as described above are predominantly involved in the agglutination of protozoal infections particularly those belonging to the Trypanosomatidae. There may be a function in terms of metazoan parasites, such as the regulation of microfilariae migration through the midgut wall into the haemocoel. However, studies indicate that haemocoel lectins are predominantly involved indirectly, and that agglutination is not so common. Lackie (115) in a brief review on insect immunity describes the role of this group of molecules as controversial because of the fact that assays have involved diluted haemolymph. However, it is thought that lectins may play a role in recognition and perhaps in differential recognition due to their carbohydrate specificities. Hence there exists a possible mode or mechanism of cell differentiation. The use of a lectin by an insect as part of its immune battery, would enable it to differentiate between cells with and without a particular carbohydrate moiety or moieties. There is therefore the potential here for recognition of self from non-self (172).

Particular studies on those haemocoel agglutinins or lectins of haematophacous vectors have shown them to occur in the tsetse flies. *Glossina morsitans morsitans, G. palpalis palpalis, G. fuscipes fuscipes* (92, 93, 205, 206), triatomine bugs (155), simuliid blackflies (82, 184) and anopheline mosquitoes (178).

However, it is with other generally more accessible and larger insects that most progress has been made in isolation and characterisation of immune lectins. The flesh fly, *Sarcophaga peregrina*, has been used by Komano et al. (112, 113) to isolate a 'sarcolectin'. This injury induced molecule is secreted by the fat body of the insect. Others have found humoral lectins, however their role in the insects internal defence is not clearly understood. Agglutinins are known, as mentioned above, to be synthesised in insects by the fat body and the haemocytes and the reproductive tissues such as ovary and testis (188). However it is not always clear, despite labelling experiments whether the agglutinin is fat body or haemocyte derived (172).

Concerning vectors of infection: The studies on tsetse flies, intermediate hosts of trypanosomes, appear to indicate a diversity of lectin function. Mid-gut lectins, as mentioned earlier have the ability to render the insect immune. They also appear to be involved in maturation. Similarly the haemocoel lectin has been suggested to be involved in maturation of the parasite (206). Alpha-D-melibiose fed to *G. m. morsitans* infected with either *T. congolense* or *T. brucei rhodesiense* resulted in reduced transmission indices. This is an index that takes into account either the salivary gland infection (*T. b. rhodesiense*) or the hypopharynx infection (*T. congolense*), prior to disease transmission. Because the mid-gut infection was not affected by the sugar, it was assumed that the mid-gut lectin was not inhibited. The reduction in the transmission index was therefore deduced to be due to the inhibition of the haemocoel lectin. As further studies are carried out, the role of lectins in both parasite maturation as well as vector immunity, appear to be significant. However, there is still much work to be done on isolation and characterization of these lectins. However, in view of the fact that trypanosomes penetrate the midgut of *Glossina* (Evans and Ellis 1975) their role must be of significance.

Little work on the agglutinins of anophelene mosquitoes has been reported. A significant rise in haemolymph titre in *P. yoelii nigeriensis* infected *A. stephensi* has been demonstrated (178). This was observed at around 9/10 days post infective blood meal, when sporozoites were released into the haemocoel. In addition, the proportion of insects demonstrating positive haemagglutination, rose following infection. It appears then, that this is in response to the sudden release of parasite material from the growing oocysts. It was interesting that no rise in titre was observed, due to infection, until the oocysts ruptured. This indicated that as far as agglutinin production was concerned, the oocysts were not recognised as foreign, and were perhaps in a privileged site.

Work with *Simulium ornatum*, the natural vector of bovine onchocerciasis (*Onchocerca lienalis*) has shown that agglutinins are secreted into the haemocoel following intrathoracic inoculation of microfilariae. These parasites develop intracellularly in the thoracic flight muscle over a period of 7–10 days depending on temperature. Using a micro-haemagglutination assay with cat erythrocytes, that enables the titration of haemolymph from individual blackflies, it was possible to identify elevated agglutinin levels in the infected flies at 5 days post infection (82). This was accompanied by a greater proportion of flies exhibiting agglutinating activity at days 1, 5 and 7. There was however, a transient increase in agglutinin titre in injury control groups as found with Sarcophaga (112, 113).

Later studies (184) demonstrated that this haemagglutinating activity
could be inhibited by particular carbohydrates, according to the stage of
development of the infection within the insect. During the initial phase of
development up to day 5/6, the agglutination could be inhibited by glu-
cose/mannose type sugars including N-acetyl-D-glucosamine. However, later
on during infection from days 6 through to 7/8, these sugars were increasingly
ineffective at blocking agglutination, whereas sugars such as N-acetyl-D-
galactosamine, were inhibitory. It appeared then, that a switch in the carbohydrate specificity of the induced haemocoel agglutinins was occurring
during infection. This confirmation of a lectin(s) in the haemocoel of blackflies
is significant, in that the surface of the developing *Onchocerca* larvae acquire
lectin binding sites in the form of surface carbohydrates. Indeed the nature of
these sugars was determined by the use of commercial FITC labelled lectins,
and found to be developmentally compatible with the blackfly derived molecules in the haemolymph (81, see Figure 4.9). Importantly, then, there are
receptors on the parasite surface that the induced lectins could bind to.
The specificity of these lectins is unknown. Clearly there is developmental
specificity within the blackfly but studies with locusts have shown that agglutinins stimulated in locusts by *Leishmania*, have a broad specificity and
appear to be active against several species of protozoa. It is possible that

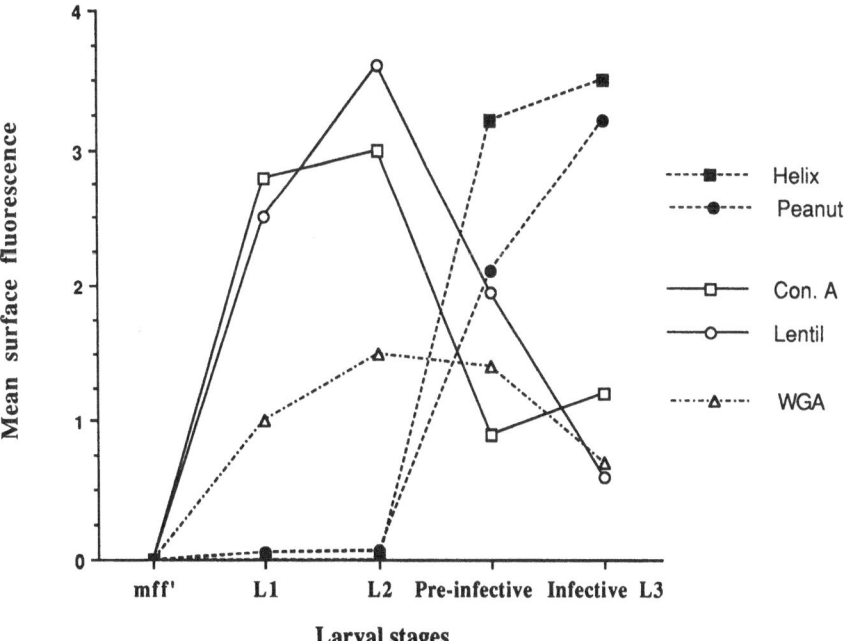

FIGURE 4.9. Binding intensities of five FITC labelled lectins to the general body surface
of developing larvae.

blackfly lectins are responsible for the differential immunity expressed against different species or geographical strains of *Onchocerca* (51, 76, 77). At this stage, however, this cannot be assumed. There is preliminary evidence that the surface carbohydrates of the developmental stages of different *Onchocerca* species vary, with infective larvae of *O. ochengi*, an african cattle parasite, being negative for *Helix pomatia* (Hagen, H. pers. comm.), whereas *O. lienalis* from the same flies is positive (81). The interaction between filariae and inducible lectins is therefore an important area for study in terms of specific recognition.

Presently the site of synthesis of these *Simulium* lectins is unclear. However, antibody to the molecule, sarcolectin, supplied by Professor Natori, Japan, results in the partial amelioration of mortality in the presence of immune *Ae. aegypti* haemolymph. This is indicative that the lectins secreted by blackflies in response to infection do have some epitope homology with the *Sarcophaga* lectin, and that this may be functionally involved in immunity. However, these studies are still preliminary, and need to be repeated with *Simulium*.

Nonetheless, the blackfly response is unusual in that no encapsulation or subsequent melanisation of parasites has been reported. In addition in in vitro assays of immune killing (80, 82) there has been no indication that haemocytes bind the parasite surface or that the parasites self agglutinate in immune haemolymph. This is perhaps to be expected in view of the fact that carbohydrates do not appear on the surface until development has continued for 24–48 hours within the fly (81, 194). Currently the in vitro assays have utilised living microfilariae with no surface sugars.

In general it is thought that lectins may act as opsonins between the foreign body, such as the microfilariae, sporozoite or oocyst, and the immune competent cells or haemocytes that circulate in the haemocoel (169). However, opsonins may not necessarily be agglutinins (see review 172). It is clear that haemocytes do have carbohydrate receptors on their surface (eg. 145). The use of lectin tools enables different subsets of such haemocytes to be identified (123), indicating that an affinity for humoral lectins is feasible. They may also act by agglutinating parasites themselves, although there do not appear to be any reports of this in the haemocoel of haematophagous vectors infected with eukaryotic parasites. A third possible mode of action in immune defence may be via enhanced cell orientated recognition of the parasites caused by haemocyte bound lectins. Olafsen (151) reported that there is very little evidence of this in any insect system studied.

Haemocytes of Vectors

A number of excellent reviews exist, covering various aspects of insect haemocytes. Indeed, the volume edited by Gupta (71), is a major source of information that has been drawn together from a number of eminent workers in the field. Further reviews have been written in the later volumes edited by

Gupta (72), Lackie (114) and Brehelin (22). One of the major problems surrounding this very important aspect of insect immunology, has been the
general diversity in methods of classification of the cells themselves (163). In
the main, vectors of parasitic infection belong to the holometabolous groups,
such as the mosquitoes, tsetse flies, phlebotomine sandflies, simuliid blackflies, etc. However, the triatomid bugs which transmit *Trypanosoma cruzi*
belong to the hemimetabolous group.

Various reports with differing groups of insects describe different haemocyte morphs, and there is considerable confusion when trying to draw up
analogies between them (2). The general consensus appears to be that at least
three discernable types exist in most insects, termed the prohaemocytes,
(which are thought to be the stem cells or progenitors of the other types), the
granulocytes or granular haemocytes (which as their name suggests contain
densely staining granules which are lost or degranulated when they contact
foreign material, and the plasmatocytes. The oenocytes (which are often rare
and are believed to contain PPO, spherulocytes/adipohaemocytes (of unknown function), and the cystocytes/coagulocytes (which are part of the
coagulation response involved in wound healing) are others that have been
described (see review 163). A rationalization of these haemocyte types has
been put forward. This classification includes thrombocytes and 4 classes of
granular haemocytes in addition to those mentioned above (23). Indeed the
identification of haemocytes is a veritable minefield. Those of particular
medical importance have been described by Kaaya and Ratcliffe (101). With
the advent of pertinent reagents such as monoclonal antibodies raised against
specific humoral proteins, together with the use of oligonucleotide DNA and
cDNA probes specific for particular molecules, the morphological classification of haemocytes will give way to molecular markers for their function in
immunity.

Already it is known that haemocytes are a source of PPO prior to the
formation of the active enzyme PO, in the haemolymph (see above). They are
also involved in the secretion of other immune proteins such as some of the
antibiotic molecules described above. The activation of haemocytes is of
particular interest in the control or otherwise of parasites. Degranulation of
the granulocytes, in which PPO occurs, is thought to occur in response to
sugars on the surface of foreign cells, resulting in the coating of the invader
with host material, in turn leading to the trophic migration of plasmatocytes
onto the parasite. The plasmatocytes, being phagocytic may ingest the minute
organisms, or together as a population, may result in its encapsulation (121,
161).

There appear to be four main cellular responses; phagocytosis, nodule
formation, encapsulation and segregation. Phagocytosis and nodule formation are thought to be limited to small particles such as viruses, bacteria,
protozoa and cellular debris and were reviewed as far back as 1962 by Jones
(99), and as well as in many other reports, also more recently in the medically
important tsetse (102). Again this procedure involves several steps including

the attraction of the cell, the binding or attachment of the cell, the phago-cytosis by the cell and the death of the invader (163). As mentioned earlier the encapsulation response occurs on larger organisms and where cells are utilised, these are predominantly the plasmatocytes (see section on PPO/PO). These flatten to cover organisms such as microfilariae, in several layers of cells.

However, acellular encapsulation is common in many medically important vectors (eg. 158) and it is only relatively recently that haemacytes were incrim-inated as described previously (eg. 8). It is thought that cells often come in after humoral encapsulation has occurred as described in the analytical work of Chen and Laurence (32). Further study (30) showed that plasmatocyte like cells were involved, as early as 2 hours after infection of *A. quadrimaculatus* with *B. pahangi*, and were similar to those described in other insects such as the orthoptera (116). The capsule around the microfilariae consisted of an inner acellular and an outer cellular layer, which appeared complete at be-tween 24 to 48 hours post infection. A week later the outer layer was degraded but still distinguishable. These findings differ to those in another system, using *D. immitis* infections in the thorax of *Ae. aegypti*. This parasite, which normally dwells in the malpighian tubules in this intermediate host, was observed as eliciting an initial cell mediated response. Perhaps this was due to the trauma of the inoculation procedure used. The homeostatic mechanism, that halts the second phase of encapsulation by the plasmatocytes, is thought to be related to a different E.M. morphology in the cells on the outside of the 'cuff' (63), but this is yet to be proved.

The numbers of circulating haemocytes under normal conditions varies tremendously, both with species and with age. Whether large or small num-bers exist, is thought to influence the ability to mount a successful immune response. Others believe that the haemocyte numbers are not fundamental to levels of immunity (eg. 116). Working with *Rhodnius prolixus* and cockroaches and locusts, looking at their ability to clear trypanosomes, it was concluded that the number of phagocytic haemocytes is more important in determining haemocyte mediated immunity (143). On this theme, microfilarial infection of *Ae. aegypti*, by intrathoracic inoculation, have demonstrated enhanced haemocyte densities, as well as a rise in the proportion that appear to be stimulated (144). Furthermore irradiation of mosquitoes to prevent cellular division resulted in reduced melanotic encapsulation of the microfilariae (39). This was despite the fact that no reductions in haemocyte populations oc-curred, as could be observed. There was reduced monophenoloxidase activity within the cells themselves. This is perhaps further evidence that cell numbers are not as important as the activity of the cells present. Indeed, the continua-tion of this work has shown that the proportion of immune activated cells, as defined by wheat germ agglutinin binding, correlates with immune activity (123). However, comparisons between two mosquitoes, as performed here, must be cautious, in view of the enormous differences in the genetic make up that occur between what are after all, insects that are genetically defined from each other.

Another group of filaria vectors, the Simuliidae, exhibit a quantitative but not as far as can be understood at present, a qualitative haemocytic reaction to infection with *Onchocerca*. Investigations on the early phase of infection of *S. ornatum* and *S. equinum* show that cell numbers increase. As the mode of infection was by inoculation, sham inoculated control (17) insects were given clean tissue culture medium with no parasites. The sham initiated a rise in cell numbers per fly, but the infection gave rise to a larger increase (see Figure 4.10). Interestingly the use of a mitotic inhibitor, colchicine, administered via the sugar meal for 24 hours prior to, and during infection, resulted in inhibition of cellular increases due to parasitic infection (Philip and Ham unpublished observations). This technique may also have uses in determining the lineage of the haemocytes during the processes of the immune response. Despite this haemocytic inflammation during *Onchocerca* infection, melanization and cellular encapsulation, have not been observed, as explained earlier. This would indicate that the cellular involvement gives rise to other components in the immune response, such as the lectins, proteases, and possibly the antibacterial proteins, described later. Clearly further studies are

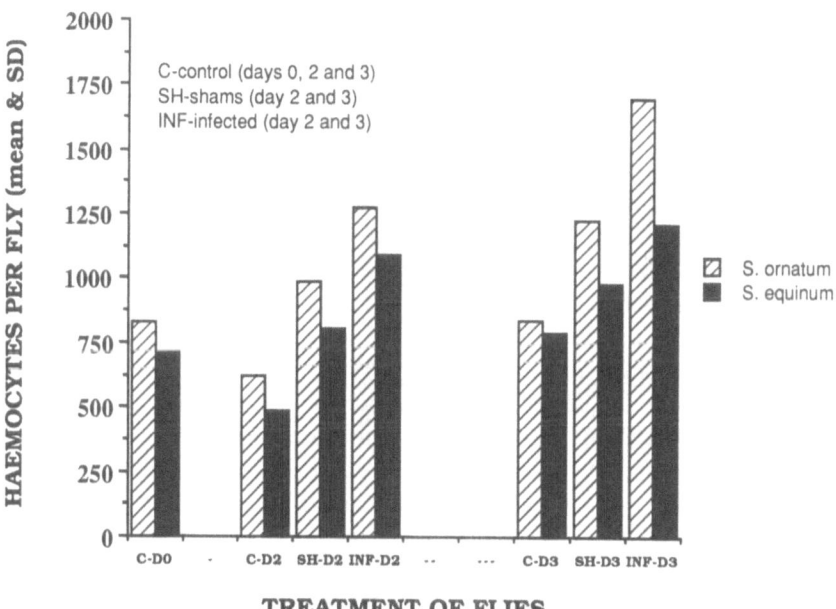

FIGURE 4.10. Histogram showing mean total haemocyte counts from two species of *Simulium* at day 0 (DO), day 2 (D2), and day 3 (D3) post injection with either 30 *O. lienalis* microfilariae (INF), or clean tissue culture medium (SH) or completely untreated flies.

necessary to determine the route of immunity and the exact role of the haemocytes following immune stimulation. As with many other insects, the haemocyte counts drop naturally with age in the flies, probably due to the requirement for strong immunity during pupation and just after eclosion. The break up of the internal structures in these holometabolous insects during the transformation over a three day period, from larvae to pupae, means that challenge by bacteria from the gut must be a real problem requiring potent agents for control.

Acquired Immunity—Homologous and Heterologous Resistance to Reinfection

The earlier sections of this article have outlined the general mechanisms of immunity in vector species. The innate and the acquired response can be separated into both humoral and cellular components, both of which are clearly stimulated by a variety of parasitic infections. As with almost all studies on insect immunity, the innate responses to a single infection have been discussed here. However, in the field, where real infections of real vectors are taking place, there is the continual reality that reinfection of the same insect is possible. High insect mortality combined with low parous rates and low infection rates in the definitive host, mean that the probability of reinfection within the same vector is low. Nevertheless, multiple infections do occur.

This has been known for some time in the transmission of onchocerciasis by *S. damnosum* cytospecies of West Africa. Female insects caught coming to bite man, have been dissected and found to contain existing filariae infections, some of which are clearly of non human origin. Furthermore, wild anophelenes (*A. punctulatus*) carrying dual infections of both malaria and filariae have been recovered in Papua New Guinea (28). In this instance the mosquitoes carrying *Plasmodium* were carrying fewer *Wuchereria bancrofti* larvae than those uninfected with malaria. The authors suggest that this may be due to different sub-populations in the community with differing opportunities to become infected. Another explanation could be that microfilarial damage of the mid-gut epithelium has resulted in easier penetration by the malaria ookinetes, and enhanced infection rates. Whatever the explanation, it is clear that in general, when analysing field data, it would help to have information resulting from some experimental backup. The controlled conditions, would allow the investigator to make reasoned predictions as to how the vectors will react in the natural situation.

There are a few studies on the interactive effects of repeated homologous infections or concomitant heterologous infections in haematophagous vectors. Observations have been made with *Ae. aegypti* infected with the gregarine parasite *Ascogregarina culicis* and *D. immitis*. In one study there appeared to be a reduction in the number of filariae when the heterologous infection was present (185), bot this was not confirmed elsewhere (11) using a different vector *Ae. triceriatus*. Superinfections of *Ae. togoi* with *B. malayi* indicated that there was no effect on the development of the filariae, although in this trial the blood meal itself seemed to be a factor not taken into account (20). Indeed more recent studies have shown that the blood meal itself significantly increases the rate of filaria development in vector mosquitoes (8).

Investigations in this laboratory have demonstrated that *Simulium* infected with *Onchocerca* demonstrate strong immunity to reinfection within the same individual (195). Following injection of counted doses of microfilariae at an interval of 5 days, a clear reduction in the development of the second dose was shown (between 34 and 55% reduction in 4 trials with *S. ornatum*, a natural vector of *O. lienalis*). Resistance to reinfection was much lower if a more refractory species was used (14–22% reduction with *S. erythrocephalum*). Equally interesting and perhaps epidemiologically more significant was the fact that on administration of the second infection in *S. erythrocephalum*, the primary infection was almost completely prevented from developing. There was also a significant reduction in the primary infection in *S. ornatum* in two out of four trials (see Figure 4.11 for results).

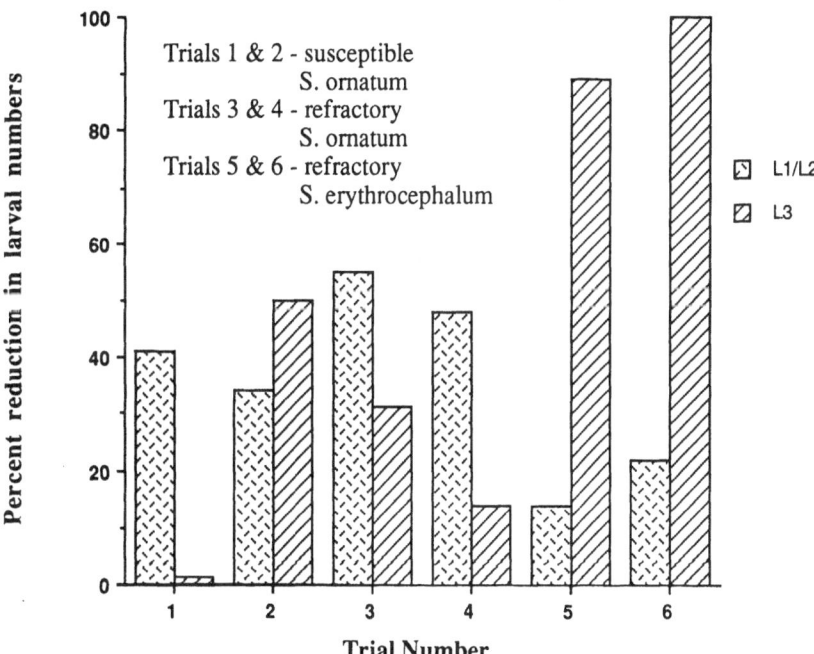

FIGURE 4.11. Histogram blocks represent the percent drop in the number of *O. lienalis* larvae developing in *Simulium* given two spaced infections as opposed to a single infection.

Density dependent nutritional factors may have a role in regulating development of *Onchocerca*. This is unlikely to be the case here because it has been demonstrated that increasing doses of microfilariae do not lead to reduced development rates (77, 126). When *Brugia* microfilariae from a secondary infection enter a muscle fibre already occupied by a primary infection, they only continue developing while the cell is intact. When the first larvae completes development and leaves the cell, destroying it in the process, the second larvae is deprived of nutrients and dies (9). However, in the series of experiments mentioned above, the infection dose was only 30 microfilariae, and it seems unlikely that this would result in the same fibre being infected twice.

Therefore, one can imagine a scenario in the field where perhaps an existing *O. volvulus* infection is prevented from continuing to infectivity, following a further infecting blood meal, or even following ingestion of an African bovine species of *Onchocerca*, which may not develop (eg. 79) but may initiate a secondary immune response. It is quite possible that in some geographical areas, *S. damnosum* may indeed pick up both human and bovine species of the filaria. Some cytospecies such as *S. soubrense* are facultatively zoophilic, feeding on ungulates as well as man (66). Identification of pathogenically unimportant bovine species (198) from the disease causing *O. volvulus* has proved to be a big problem in epidemiological surveys. Clearly then, the extent of this problem, and the importance of immunity in modifying transmission of onchocerciasis in vectors is unknown at present. The experimental work shows that there is a physiological basis to believing that such interactions do occur in the field.

The studies described above refer to homologous infections. Working with strains of *Ae. aegypti* we have shown that interactions between the vector and malaria and filaria superinfections also occur (Massaninga, F., Ham, P.J. unpublished data). In a filaria susceptible (REf^m) strain of the mosquito a preexisting infection of *P. gallinaceum* at the early oocyst stage, results in the reduced development of a secondary infection with *B. pahangi* (see Figure 4.12). However, there appears to be no concomitant effect of the *Brugia* infection on subsequent development of the oocysts to maturity. However if the same experiment is carried out using filaria refractory (Porto Novo—PN) mosquitoes, the challenge of a secondary dose of microfilariae, does result in the reduced development rate of the pre-existing oocysts despite the fact that the worms will not develop and are melanized. It is likely that the induction of melanization by the microfilariae in the refractory strain has something to do with the reduced oocyst numbers. There appears to be a cross specificity between the filaria and the malaria that does not occur in the susceptible strain of *Aedes*. It is also interesting that a parasite dwelling in the mid-gut wall, such as *P. gallinaceum* should result in the reduced development of a parasite that has as its microhabitat, the cells of the thoracic flight musculature (ie. *B. pahangi*). It is possible that an enhanced immune state caused by the oocysts, is effective against the microfilariae before they can penetrate the muscle cells, while still in the haemolymph. If so, then the oocysts themselves

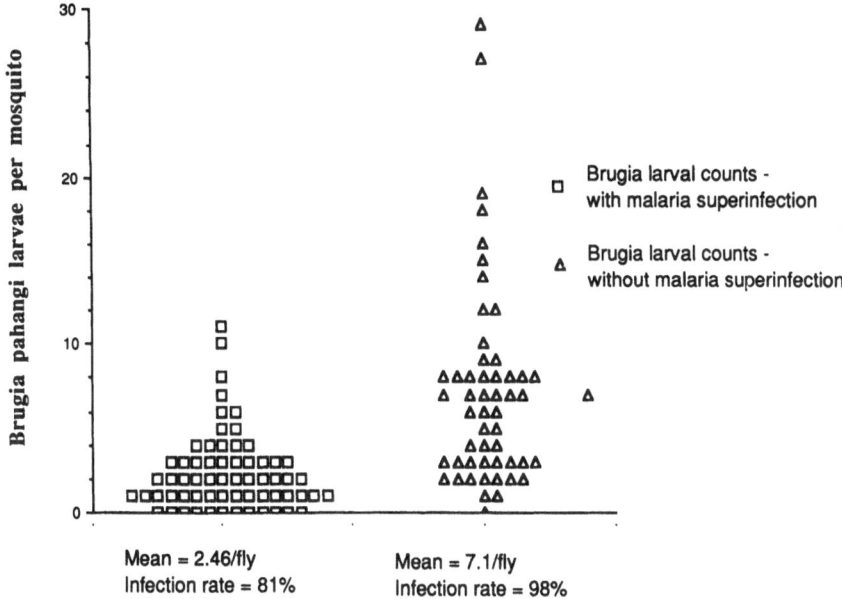

FIGURE 4.12. Scatterdiagram of parasite counts from individual *Aedes aegypti* following infection with avian malaria and subsequently *B. pahangi* microfilariae.

are somehow protected from this. Further investigation into this interaction is underway, and hopes to elucidate the questions raised above.

Of course in any situation where superinfections may be a factor, the longevity of the insect concerned is also of utmost importance. This appears to be related to climatic variation in the case of *S. damnosum* cytospecies (118). In savanna regions, estimates are that up to 35% of flies may survive to their third blood meal, which is period necessary to be potentially infective. Clearly then these other factors contribute to transmissibility in vectors of parasites.

Conclusions and Applications

At the beginning of this article it was mentioned that insect immunity has been a discipline of legitimate research for many years. Concerning the transmission of parasitic infections, the work of MacDonald and colleagues was heralded as a breakthrough in understanding why one species or strain of a vector was susceptible to infection, and another not. However, as stated much more recently by Townson and Chaithong in their paper on the influences of the mosquito on development of filariae, 'we are little nearer an understanding of the underlying processes by which such genes [the *f*m gene system] influence the development of the parasite'.

The relatively new findings concerning the identification and isolation of immune molecules from insects and other invertebrates mean that some physiological mechanisms may soon be attached to the genetical analysis of previous years. Potent molecules, possibly having a wide range of interactions deleterious to the parasite, are now known, some of which have been mentioned in this review. These together with what I believe is a much more stringent experimental approach to vector biology, should enable the fundamentals of immunity/susceptibility to be understood.

There are many recent developments in the field of molecular biology, immunology (eg. the use of monoclonal antibody technology), and biochemistry that entomologists and parasitologists are now applying to their trade. These, together with a good grasp of the basic zoology involved, mean that studies can be carried out on the often minute vectors of parasitic infection. It is important that as well as molecular geneticists using insects as tools to study molecular genetics, there are also vector biologists using molecular genetics to study the key questions of vector biology. These techniques offer us the potential of understanding the biology of an infected mosquito or sandfly, and how it resists, or copes with infection by parasites. In addition though, molecular genetics also puts forward further methods of control, which in the future could be added to the battery of methods already available.

It does not take much reasoning to realise that if a gene for immunity in an insect is isolated and characterised, there is a possible tool for part of an integrated control strategy. The subject of genetic transformation of insects, for pest control has been reviewed already (46, 149). Sufficient to say that methods envisaged for the agricultural market, may also be applied to the control of medically and veterinary important insects. Clearly many hurdles need to be overcome before this type of intervention can be considered, not least the practical one of inserting a gene for immunity into a wild germ line successfully, in mosquitoes of medical importance such as *Anopheles gambiae*, rather than the surrogate *Aedes aegypti*. Prior to this, suitable genes need to be identified and characterized.

Should all this be working well, there still remains the problem of rearing mosquitoes in enough numbers to have an impact on the wild population. This is likely to require some type of population 'drive'. Should all this be overcome, there is still the ethical problem, probably the biggest hurdle, of releasing manipulated insects into the field, even if they are stable. Notwithstanding these difficulties, this is an area that is currently receiving financial support and is seen as having genuine possibilities in the long term future.

A number of molecules seem interesting, but before they are seized upon by molecular biologists as potential control agents, the importance of characterizing them cannot be overstressed. For example, antibiotic inducible proteins may be identified in our vector species, but demonstration or otherwise, of function against the worm or protozoa must be carried out. In addition, the complex mechanisms involved, such as their specificity, site of synthesis, action on other tissues etc. must be investigated. This is where the vector biologist is playing a fundamentally important role. It is to be hoped that more entomologists will take hold of some of these powerful techniques and pull the field of experimental vector biology into the 21st century.

Acknowledgments. Some of the work described here was funded by research grants from the Wellcome Trust, the British Medical Research Council, the World Health Organization and the British Council. P.J. Ham is a Wellcome Trust Senior Research Fellow in Basic Biomedical Science. Thanks also go to Garry Nolan, Elise Winbolt and Peter Young for their excellent technical assistance, and to Professors W.W. Macdonald and G. Webbe, in whose departments my studies were carried out.

References

1. Agudelo-Silva, F., and Spielman, A., 1985, Penetration of mosquito midgut wall by sheathed microfilariae, *J. Invert. Path.* **45**:117–119.
2. Arnold, J.W., 1979, Controversies about hemocyte types in insects, in Gupta, A.P. (ed): Insect Hemocytes. Development, forms, functions, and techniques. Cambridge University Press, England, pp. 231–258.
3. Ashida, M., and Dohke, K., 1980, Activation of prophenoloxidase by the activating enzyme of the silkworm, *Bombyx mori. Ins. Biochem.* **10**:37–47.
4. Ashida, M., Kinoshita, K., and Brey, P.T., 1990, Studies on prophenoloxidase activation in the mosquito *Aedes aegypti* L. *Eur. J. Biochem.* **188**:507–515.
5. de Azambuja, P., Freitas, C.C., and Garcia, E.S., 1986, Evidence and partial characterization of an inducible antibacterial factor in the haemolymph of *Rhodnius prolixus*, *J. Insect. Physiol.* **32**:807–812.
6. Bain, O., Durette-Desset, M., and De Leon, R., 1974, Onchocercose au Guatemala: l'ingestion des microfilaries par *Simulium ochraceum* et leur passage dans l'hemocele de ce vecteur, *Ann. Parasit. Hum. Comp.* **49**:467–487.
7. Barraud, P.J., and Covell, G., 1928, The morphology of the buccal cavity in anopheline and culicine mosquitoes, *Ind. J. Med. Res.* **15**:671–679.
8. Bartlett, C.M., 1984, Development of *Dirofilaria scapiceps* (Leidy, 1886) (Nematoda: Filarioidea) in *Aedes* spp. and *Mansonia perturbans* (Walker) and responses of mosquitoes to infection, *Can. J. Zool.* **62**:112–129.
9. Beckett, E.B., 1990, Species variation in mosquito flight-muscle damage resulting from a single filarial infection and its repercussions on a second infection, *Parasitol. Res.* **76**:606–609.
10. Beerntsen, B.T., Luckhart, S., and Christensen, B.M., 1989, *Brugia malayi* and *Brugia pahangi*: inherant difference in the immune activation of the mosquitoes *Armigeres subalbatus* and *Aedes aegypti.*, *J. Parasit.* **75**:76–81.
11. Beier, J.C., 1983, Effects of gregarine parasites on the development of *Dirofilaria immitis* in *Aedes triseriatus* (Diptera: Culicidae), *J. Med. Entomol.* **20**:70–75.
12. Berner, R., Rudin, W., and Hecker, H., 1983, Peritrophic membranes and protease activity in the midgut of the malaria mosquito, *Anopheles stephensi* (Liston) (Insecta: Diptera) under normal and experimental conditions, *J. Ultrastruct. Res.* **83**:195–204.

13. Bianco, A.E., and El Sinnary, K., 1980, Infection of *Aedes aegypti* with *Onchocerca gutturosa. J. Helm.* **54.** 105–107.

14. Bianco, A.E., Townson, S., El Sinnary, K., and Nelson, G.S., 1979, Successful development of *Onchocerca* from cattle in *Aedes, Anopheles, Culex* and *Culicoides* sp., *Parasitol.* **79**:35.

15. Billingsley, P.F., 1990, The midgut ultrastructure of hematophagous insects., *Ann. Rev. Entomol.* **35**:219–248.

16. Boman, H.G., 1982, Humoral immunity in insects and the counter defence of some pathogens, *Fortschritte der Zoologie* **27**: 211–222.

17. Boman, H., Boman, A., and Pigon, A., 1981, Immune and injury responses in cecropia pupae—RNA isolation and comparison of protein synthesis in vivo and in vitro., *Ins. Biochem.* **11**: 33–42.

18. Boman, H.G., Faye, I., Hofsten, P.v., Kockum, K., Lee, J.-L., Xanthopoulos, K.G., Bennich, H., Engstrom, A., Merrifield, B.R., and Andreu, D., 1986, Antibacterial immune proteins in insects—A review of some current perspectives, Brehelin, M. (ed): Immunity in Invertebrates. Cells, molecules, and defense reactions. Springer-Verlag, pp. 63–73.

19. Borovsky, D., and Schlein, Y., 1987, Trypsin and chymotrypsin-like enzymes of the sandfly *Phlebotomus papatasi* infected with *Leishmania* and their possible role in vector competence. *Med. Vet. Ent.* **1**:235–242.

20. Bosworth, W., and Ewert, A., 1973, Superinfection of *Aedes togoi* with *Brugia malayi, J. Med. Entomol.* **10**:217–219.

21. Bradley, T.J., and Nayar, J.K., 1985, Intracellular melanization of the larvae of *Dirofilaria immitis* in the malpighian tubules of the mosquito, *Aedes sollicitans, J. Invert. Path.* **45**:339–345.

22. Brehelin, M., 1986, Immunity in invertebrates. Cells, molecules and defense reactions, *Springer-Verlag.* pp. 233.

23. Brehelin, M., and Zachary, D., 1986, Insect haemocytes: A new classification to rule out the controversy, in Brehelin, M. (ed): Immunity in invertebrates. Cells, molecules and defense reactions, Springer Verlag, pp. 36–48.

24. Briegel, H., 1975, Excretion of proteolytic enzymes by *Aedes aegypti* after a blood meal, *J. Ins. Physiol.* **21**:1681–1684.

25. Bruce-Chwatt, L.J., 1945, The morphology of the pharyngeal armature in *Anopheles gambiae* var. *melas* from Southern Nigeria, *Ann. Trop. Med. Parasit.* **39**:124–128.

26. Brug, S.L., 1932, Chitinization of parasites in mosquitoes, *Bull. Ent. Res.* **23**: 229–231.

27. Bryan, J.H., Oothuman, P., Andrews, B.J., and McGreevy, P.B., 1974, Effects of pharyngeal armatures of mosquitoes on microfilariae of *Brugia pahangi., Trans. Roy. Soc. Trop. Med. Hyg.* **68**:14–15.

28. Burkot, T.R., Molineaux, L., Graves, P.M., Paru, R., Battistutta, D., Dagoro, H., Barnes, A., Wirtz, R.A., and Garner, P., 1990, The prevalence of naturally acquired multiple infections of *Wuchereria bancrofti* and human malarias in anophelines, *Parasit.* **100**:369–375.

29. Casteels, P., Ampe, C., Jacobs, F., Vaeck, M., and Tempst, P., 1989, Apidaecins: Antibacterial peptides from honeybees, *EMBO J.* **8**:2387–2391.

30. Chen, C.C., 1988, Further evidence of both humoral and cellular encapsulation of sheathed microfilariae of *Brugia pahangi* in *Anopheles quadrimaculatus, Int. J. Parasit.* **18**:819–826.

31. Chen, C.C., and Laurence, B.R., 1985a, The encapsulation of the sheaths of microfilariae of *Brugia pahangi* in the haemocoel of mosquitoes, *J. Parasit.* **71**:834–836.

32. Chen, C.C., and Laurence, B.R., 1985b, An ultrastructural study on the encapsulation of microfilariae of *Brugia pahangi* in the haemocoel of *Anopheles quadrimaculatus, Int. J. Parasit.* **15**:421–428.

33. Chen. C.C., and Shih, C.M., 1988, Exsheathment of microfilariae of *Brugia pahangi* in the susceptible and refractory strains of *Aedes aegypti., Ann. Trop. Med. Parasit.* **82**:201–206.

34. Cherian, P.V., Stromberg, B.E., Weiner, O.J., and Soulsby, E.J.L., 1980, Fine structure and cytochemical evidence for the presence of polysaccharides surface coat of *Dirofilaria immitis* microfilariae, *Int. J. Parasit.* **10**:227–233.

35. Christensen, B.M., 1986, Immune mechanisms and mosquito-filarial worm relationships, *Symp. Zoo. Soc.* London, **56**:145–160.

36. Christensen, B., Fink, J., Merrifield, R.B., and Mauzerall, D., 1988, Channel-forming properties of cecropins and related model compounds incorporated into planar lipid membranes, *Proc. Natl. Acad. Sci.* USA, **85**:5072–5076.

37. Christensen, B.M., Forton, K.F., LaFond, M.M., and Grieve, R.B., 1987, Surface changes on *Brugia pahangi* microfilariae and their association with immune evasion in *Aedes aegypti, J. Invert. Pathol.* **53**:216–219.

38. Christensen, B.M., and Gleason, L.N., 1984, Defense reactions of mosquitoes to filarial worms: Comparative studies on the response of three different mosquitoes to inoculated *Brugia pahangi* and *Dirofilaria immitis* microfilariae, *J. Invert. Pathol.* **44**:267–274.

39. Christensen, B.M., Huff, B.M., and Jianyong, L., 1990, Effect of irradiation on the hemocyte-mediated immune response of *Aedes aegypti* against microfilariae, *J. Invert. Path.* **56**:123–127.

40. Christensen, B.M., LaFond, M.M., and Christensen, L.A., 1986, Defense reactions of mosquitoes to filarial worms: Effect of host age on the immune response to Dirofilaria immitis, J. Parasit. **72**: 212–215.

41. Christensen, B.M., and Sutherland, D.R., 1984, *Brugia pahangi*: Exsheathment and midgut penetration in *Aedes aegypti. Trans. Am. Micr. Soc.*, **103**:423–433.

42. Collins, F.H., Sakai, R.K., Vernick, K.D., Paskewitz, S., Seeley, D.C., Miller, L.H., Collins, W.E., Campbell, C.C., and Gwadz, R.W., 1986, Genetic selection of a *Plasmodium* refractory strain of the malaria vector *Anopheles gambiae*, Science **234**:607–610.

43. Collins, R.C., and Jones, R.H., 1978, Laboratory transmission of *Onchocerca cervicalis* with *Culicoides variipennis, Am. J. Trop. Med. Hyg.* **27**:46–50.

44. Coluzzi, M., and Trabucchi, R., 1968, Importanza dell'armatura buccofaringea in *Anopheles* e *Culex* in relazione alle infezioni con *Dirofilaria, Parassitologia,* **10**:47–59.

45. Connal, A., and Connal, S.L.M., 1922, The development of *Loa loa* (Guyot) in *Chrysops silacea* (Austen) and in *Chrysops dimidiata* (van der Wulp), *Trans. Roy. Soc. Trop. Med. Hyg.* **16**:64–89.

46. Crampton, J., Morris, A., Lycett, G., Warren, A., and Eggleston, P., 1990, Transgenic mosquitoes: A future vector control strategy, *Parasit. Today* **6**:31–36.

47. Croft, S.L., East, J.S., and Molyneux, D.H., 1982, Anti-trypanosomal factor in haemolymph of *Glossina, Acta Trop.* **39**:293–302.

48. Curtis, C.F., and Graves, P.M., 1983, Genetic variation in the ability of insects to

transmit filariae, trypanosomes and malarial parasites, *Current Topics in Vector Research,* **1**:31–62.

49. Devaney, E., 1985, Lecitin-binding characteristics of *Brugia pahangi* micro-filariae, *Trop. Med. Parasit.* **36**:25–28.

50. Dimarcq, J.L., Keppi, E., Dunbar, B., Lambert, J., Reichhart, J.M., Hoffman, D., Rankine, S.M., Fothergil, J.E., and Hoffmann, J.A., 1988, Insect immunity. Purification and characterisation of a family of novel inducible antibacterial proteins from immunised larvae of the dipteran *Phormia terranovae* and complete amino acid sequence of predominant member, dipericin A, *Eur. J. Biochem.* **171**:17–22.

51. Duke, B.O.L., 1968, Studies of factors influencing the transmission of onchocerciasis. VI. The infective biting potential of *Simulium damnosum* in different bioclamatic zones and its influence on the transmission potential, *Ann. Trop. Med. Parasit.* **62**:164–170.

52. Duke, B.O.L., and Lewis, D.J., 1964, Studies on factors influencing the transmission of onchocerciasis. II. Observations on the effect of the peritrophic membrane in limiting the development of *Onchocerca volvulus* microfilariae in *Simulium damnosum, Ann. Trop. Med. Parasit.* **58**:83–88.

53. Dunn, P.E., 1990, Humoral immunity in insects, *BioSci.* **40**:738–744.

54. East, J., Molyneux, D.H., Maudlin, I., and Dukes, P., 1983, Effects of *Glossina* haemolymph on salivarian trypanosomes in vitro, *Ann. Trop. Med. Parasit.* **77**:97–99.

55. Eichler, D.A., 1973, Studies on *Onchocerca gutturosa* (Neumann 1910) and its development in *Simulium ornatum* (Meigen 1818). 3. Factors affecting the development of the parasite in its vector, *J. Helm.* **47**:73–88.

56. Engstrom, P., Carlsson, A., Engstrom, A., Tao, Z.J., and Bennich, H, 1984, The antibacterial effect of attacins from the silk moth *Hyalophora cecropia* is directed against the outer membrane of *Escherichia coli, EMBO J.* **3**:3347–3351.

57. Evans, D.A., and Ellis, D.S., 1975, Penetration of mid-gut cells of *Glossina morsitans morsitans* by *Trypanosoma brucei rhodesiense, Nature,* London, **258**: 231–233.

58. Esslinger, J.H., 1962, Behaviour of microfilariae of *Brugia pahangi* in *Anopheles quadrimaculatus, Am. J. Trop. Med. Hyg.* **11**:749–758.

59. Ewert, A., 1965a, Comparative migration of microfilariae and development of *Brugia pahangi* in various mosquitoes, *Am. J. Trop. Med. Hyg.* **14**:254–259.

60. Ewert, A., 1965b, Exsheathment of the microfilariae of *Brugia pahangi* in susceptible and refractory mosquitoes, *Am. J. Trop. Med. Hyg.* **14**:260–262.

61. Eyles, D.E., 1951, Studies on *Plasmodium gallinaceum* 1. Characteristics of the infection in the mosquito *Aedes aegytpi, Am. J. Trop. Med. Hyg.* **55**:386–391.

62. Ferro, E.T., and Theis, J.H., 1984, Antigenic sharing between filarial worms and their vectors, *Mosq. Cont. Res. Ann. Rep.* University of California, Davis, pp. 124–125.

63. Forton, K.F., Christensen, B.M., and Sutherland, D.R., 1985, Ultrastructure of the melanization response of *Aedes trivittatus* against inoculated *Dirofilaria immitis* microfilariae, *J. Parasit.* **71**:331–341.

64. Frizzi, G., Rinaldi, A., and Bianchi, U., 1975, Genetic studies on mechanisms influencing the susceptibility of anopheline mosquitoes to plasmodial infection, *Mosq. News* **35**:505–513.

65. Furman, A., and Ash, L.R., 1983, Analysis of *Brugia pahangi* microfilariae surface carbohydrates: comparison of the binding of a panel of fluoresceinated lectins to

mature in vivo—derived and immature in utero—derived microfilariae, *Acta Trop.* **40**:45–51.

66. Garms, R., 1983, Studies on the transmission of *Onchocerca volvulus* by species of the *Simulium damnosum* complex occurring in Liberia, *Z. angew. Zool.* **70**:101–117.

67. Gass, R.F., and Yeates, R.A., 1979, In vitro damage of cultured ookinetes of *Plasmodium gallinaceum* by digestive proteinases from susceptible *Aedes aegypti*, *Acta Trop.* **36**:243–252.

68. Goldstein. I.J., Hughes, R.C., Monsigny, M., Osawa, T., and Sharon, N., 1980. What should be called a lectin? *Nature* **285**:66.

69. Gotz, P., and Vey, A., 1974, Humoral encapsulation in Diptera (Insecta): defense reactions of *Chironomus* larvae against fungi, *Parasit.* **68**:193–205.

70. Graf, R., Raikhel, A.S., Brown, M.R., Lea, A.O., and Briegel, H., 1986, mosquito trypsin: immunocytochemical localization in the midgut of blood-fed *Aedes aegypti*, *Cell Tiss. Res.* **245**:19–27.

71. Gupta, A.P. (ed), 1979, Insect Hemocytes. Development, forms, functions and techniques, Cambridge University Press, England.

72. Gupta, A.P., 1986, Hemocytic and Humoral Immunity in Arthropods, John Wiley and Sons, pp. 533.

73. Gwadz, R.W., Kaslow, D., Lee, J.-Y., Maloy, L., Zasloff, M., and Miller, L.H., 1989, Effects of magainins and cecropins on the sporogonic development of malaria parasites in mosquitoes, *Inf. Imm.* **57**:2628–2633.

74. Ham, P.J., 1986, Acquired resistance to *Onchocerca lienalis* infections in *Simulium ornatum* Meigen and *Simulium lineatum* Meigen following passive transfer of haemolymph from previously infected simuliids (Diptera, Simuliidae), *Parasit.* **92**:269–277.

75. Ham, P.J., Baxter, A.J., Bockarie, M., Thomas, P., and Chalk, R., 1989, Vector immunity to onchocerciasis, *Trop. Med. Parasit.* **40**:84.

76. Ham, P.J., and Bianco, A.E., 1983a, Development of *Onchocerca volvulus* from cryopreserved microfilariae in three temperate species of laboratory reared blackflies, *Tropenmed. Parasit.* **34**:137–139.

77. Ham, P.J., and Bianco, A.E., 1983b, Screening of some British simuliids for susceptibility to experimental *Onchocerca lienalis* infection, *Z. Parasitenkd.* **69**: 765–772.

78. Ham, P.J., and Garms, R., 1985, Development of forest *Onchocerca volvulus* in *Simulium yahense* and *Simulium sanctipauli* following intrathoracic injection and ingestion of microfilariae, *Trop. Med. Parasit.* **36**:25.

79. Ham, P.J., and Garms, R., 1987, Failure of *Onchocerca gutturosa* to develop in *Simulium soubrense* and *Simulium yahense* from Liberia, *Trop. Med. Hyg.* **38**:135–136.

80. Ham, P.J., and Garms, R., 1988, The relationship between innate susceptibility to *Onchocerca*, and haemolymph attenuation of microfilarial motility in vitro using British and West African blackflies, *Trop. Med. Parasit.* **39**:230–234.

81. Ham, P.J., Smail, A.J., and Groeger, B.K., 1988, Surface carbohydrate changes on *Onchocerca lienalis* larvae as they develop from microfilariae to the infective third-stage in *Simulium ornatum*, *J. Helm.* **62**:195–205.

82. Ham, P.J., Zulu, M.B., and Zahedi, M.B., 1988, In vitro haemagglutination and attenuation of microfilarial motility by haemolymph from individual blackflies (*Simulium ornatum*) infected with *Onchocerca lienalis*, *Med. Vet. Entomol.* **2**:7–18.

83. Hawking, F., and Worms, M., 1961, Transmission of filarioid nematodes, *Ann. Rev. Entomol.* **6**:413–432.

84. Ho, B.C., and Kan, S.P., 1971, Evidence of the intracellular development of *Breinlia sergenti* (Dipetalonematidae) in the fat cells of mosquitoes, *J. Parasit.* **57**:1145–1146.

85. Ho, B.C., Yap, E.H., and Singh, M., 1982, Melanisation and encapsulation in *Aedes aegypti* and *Aedes togoi* in response to parasitization by a filarioid nematode (*Breinlia booliata*), *Parasitol.* **85**:567–575.

86. Huber, M., Cabib, E., and Miller, L.H., 1991, Malaria parasite chitinase and penetration of the mosquito peritrophic membrane, *Proc. Natl. Acad. Sci. USA,* **88**:2807–2810.

87. Huff, C.G., 1927, Studies on the infectivity of plasmodia of birds for mosquitoes, with special reference to the problem of immunity in the mosquito, *Am. J. Hyg.* **7**:706–734.

88. Huff, C.G., 1934, Comparative studies on susceptible and insusceptible *Culex pipiens* in relation to infection with *Plasmodium cathemerium* and *P. relictum, Am. J. Hyg.* **19**:123–147.

89. Hultmark, D., Steiner, H., Rasmuson, T., and Boman, H.G., 1980, Insect immunity. Purification and properties of three inducible bactericidal proteins from haemolymph of immunized pupae of *Hyalophora cecropia, Eur. J. Biochem.* **106**:7–16.

90. Hultmark, D., Engstrom, A., Andersson, K., Steiner, H., Bennich, H., and Boman, H.G., 1983, Insect immunity. Attacins, a family of antibacterial proteins from *Hyalophora cecropia, EMBO J.* **2**:571–576.

91. Hultmark, D., Engstrom, A., Bennich, H., Kapur, R., and Boman, H.G., 1982, Insect immunity. Isolation and structure of cecropin D and four minor antibacterial components from cecropia pupae, *Eur. J. Biochem.* **127**:207–217.

92. Ibrahim, E.A.R., Ingram, G.A. and Molyneux, D.H., 1984, Haemagglutinins and parasite agglutinins in haemolymph and gut of *Glossina, Tropenmed. Parasit.* **35**:151–156.

93. Ingram, G.A., and Molyneux, D.H., 1988a, Sugar specificities of anti-human ABO(H) blood group erythrocyte agglutinins (lectins) and haemolytic activity in the haemolymph and gut extracts of three *Glossina* species, *Ins. Biochem.* **18**:269–279.

94. Ingram, G.A., Molyneux, D.H., 1988b, Lectins (agglutinins) and lysins in the haemolymph and gut extracts of the tsetse fly, *Glossina fuscipes fuscipes*, Bog-Hansen, T.C., and Freed, D.L.J. (eds): Lectins. Biology, Biochemistry, Clinical Biochemistry Sigma, pp. 63–68.

95. Irungu, L.W., 1987, Studies on the in vitro exsheathment of *Brugia pahangi*-2. The in vitro exsheathment of *B. pahangi* microfilariae incubated with mosquito tissues and cells, *Ins. Sci. Appl.* **8**:49–51.

96. Iyengar, M.O.T., 1936. Entry of filaria larvae into the body cavity of the mosquito, *Parasit.* **28**:190–194.

97. Jarosz, J., and Glinski, Z., 1990, Selective inhibition of cecropin-like activity of insect immune blood by protease from american foulbrood scales, *J. Invert. Pathol.* **56**:143–149.

98. Jaynes, J.M., Burton, C.A., Barr, S.B., Jeffers, G.W., Julian, G.R., White, K.L., Enright, F.M., Klei, T.R., and Laine, R.A., 1988, In vitro cytocidal effect of novel lytic peptides on *Plasmodium falciparum* and *Trypanosoma cruzi. FASEB J.* **2**:2878–2883.

99. Jones, J.C., 1962, Current concepts concerning insect haemocytes, *Am. Zool.* **2**:209–246.

100. Kaaya, G.P., Flyg, G., and Boman, H.G., 1987, Induction of cecropin and attacin-like antibacterial factors in the haemolymph of *Glossina morsitans*, *Ins. Biochem.* **17**:309–315.

101. Kaaya, G.P. and Ratcliffe, N.A., 1982, Comparative study of hemocytes and associated cells of some medically important dipterans, *J. Morph.* **173**:351–365.

102. Kaaya, G.P., Ratcliffe, N.A., and Alemu, P., 1986, Cellular and humoral defences of *Glossina* (Diptera: Glossinidae): Reactions against bacteria, trypanosomes and experimental implants, *J. Med. Ent.* **23**:31–43.

103. Kartman, L., 1956, Notes on the encapsulation of *Dirofilaria immitis* in the mosquito *Aedes aegypti. Am. J. Vet. Med.* **17**:810–812.

104. Kaslow, D.C., Syin, C., McCutchan, T.F., and Miller, L.H., 1989, Comparison of the primary structure of the 25KDa ookinete surface antigens of *Plasmodium falciparum* and *Plasmodium gallinaceum* reveal six conserved regions, *Mol. Biochem. Parasit.* **33**:283–288.

105. Kaushal, N.A., Simpson, A.J.G., Hussain, R., and Ottesen, E.A., 1984, *Brugia malayi*: Stage specific expression of carbohydrates containing N-acetyl-D-glucosamine on the sheath surfaces of microfilariae, *Exp. Parasit.* **58**:182–187.

106. Kilama, W.L., Craig, G.B., 1969, Monofactorial inheritance of susceptibility to *Plasmodium gallinaceum* in *Aedes aegypti*, *Ann. Trop. Med. Parasit.* **63**:419–432.

107. Killick-Kendrick, R., 1979, The biology of *Leishmania* in phlebotomine sandflies, in Lumsden, W.H.R., and Evans, D.A. (eds): Biology of Kinetoplastida Volume 11, Academic Press, pp. 395–460.

108. Killick-Kendrick, R., 1987, The microecology of *Leishmania* in the gut and proboscis of the sandfly, *NATO ASI ser.* **11**:397–406.

109. Killick-Kendrick, R., 1990, The life-cycle of *Leishmania* in the Sandfly with special reference to the form infective to the vertebrate host, *Ann. Parasitol. Hum. Comp.* 65 Suppl. **2**:37–42.

110. Knapp, T., and Crampton, J.M., 1990, Sequences related to immune proteins in the mosquito *Aedes aegypti. Trans. Roy. Soc. Trop. Med. Hyg.* **84**:459.

111. Kobayashi, M., Ogura, N., and Yamamoto, H, 1986, Studies on filariasis. X: AS trial to analyse refractory mechanisms of the mosquitoes *Aedes aegypti* to the filarial larvae *Brugia malayi* by means of parabiotic twinning, *Dokkyo J. Med. Sci.* **13**:61–67.

112. Komano, H., Mizuno, D., and Natori, S., 1980, Purification of a lectin induced in the haemolymph of *Sarcophaga peregrina* larvae on injury, *J. Biol. Chem.* **255**:2929–2924.

113. Komano, H., Mizuno, D., and Natori, S., 1981, A possible mechanism of induction of insect lectin, *J. Biol. Chem.* **256**:7087–7089.

114. Lackie, A.M. (ed), 1986, Immune Mechanisms in Invertebrate Vectors, *Zool. Soc. Lond. Symp.* **56**:285.

115. Lackie, A.M., 1988, Immune mechanisms in insects, *Parasitology Today* **4**::98–105.

116. Lackie, A.M., Tackle, G.B., and Tetley, L., 1985, Haemocytic encapsulation in the locust *Schistocerca gregaria* (Orthoptera) and in the cockroach *Periplaneta americana* (Dictyoptera), *Cell Tiss. Res.* **240**:343–351.

117. LaFond, M.M., Christensen, B.M., and Lasee, B.A., 1985, Defense reactions of mosquitoes to filarial worms: Potential mechanisms for avoidance of the response by *Brugia pahangi* microfilariae, *J. Invert. Pathol.* **46**:26–30.

118. Le Berre, R., 1966, Contribution a l'etude biologique et ecologique de *Simulium damnosum* Theobald, 1903 (Diptera, Simuliidae), *Memoires O.R.S.T.O.M.* **17**:204.

119. Lehane, M.J., 1976, Digestive enzyme secretion in *Stomoxys calcitrans* (Diptera: Muscidae), *Cell Tiss. Res.* **170**:275–287.

120. Lehane, M.J., and Laurence, B.R., 1977, Flight muscle ultrastructure of susceptible and refractory mosquitoes parasitized by larval *Brugia pahangi, Parasitol.* **74**:87–92.

121. Leonard, C., Ratcliffe, N.A., Rowley, A.F., 1985, The role of prophenoloxidase activation in non-self recognision and phagocytosos by insect blood cells, *J. Ins. Physiol.* **31**:789–799.

122. LePage, P., Bitsch, F., Roecklin, D., Keppi, E., Dimarcq, J.-L., Reichhart, J.-M., Hoffmann, J.A., Roitsch, C., and Dorsselaer, A., 1991, Determination of disulfide bridges in natural and recombinant insect defensin A, *Eur. J. Biochem.* **196**:735–742.

123. Li, J.L., and Christensen, B.M., 1990, Immune competence of *Aedes trivittatus* hemocytes as assessed by lectin binding, *J. Parasit.* **76**:276–278.

124. Li, J.L., Tracy, J.W., and Christensen, B.M., 1989, Hemocyte monophenol oxidase activity in mosquitoes exposed to microfilariae of *Dirofilaria immitis, J. Parasit.* **75**:1–5.

125. Lok, J.B., Cupp, E.W., Bernardo, R.J., and Pollack, R.J., 1983, Further studies on the development of *Onchocerca* spp. (Nematoda: Filarioidea) in Nearctic blackflies (Diptera: Simuliidae), *Am. J. Trop. Med. Hyg.* **32**:1298–1305.

126. Lok, J.B., Cupp, E.W., Braide, E.I., and Bernardo, M.J., 1980, The development of *Onchocerca* spp. in *Simulium decorum* Walker and *Simulium pictipes* Hagen, *Tropenmed. Parasit.* **31**:498–506.

127. McCall, P.J., Trees, A.J., 1989, The development of *Onchocerca ochengi* in surrogate temperate Simuliidae, with a note on the infective larva, *Trop. Med. Parasit.* **40**:295–298.

128. MacDonald, W.W., 1962a, The selection of a strain of *Aedes aegypti* susceptible to infection with semiperiodic *Brugia malayi, Ann. Trop. Med. Parasit.* **56**:368–372.

129. MacDonald, W.W., 1962b, The genetic basis of susceptibility to infection with semi-periodic *Brugia malayi* in *Aedes aegypti, Ann. Trop. Med. Parasit.* **56**:373–382.

130. MacDonald, W.W., 1963, A preliminary cross-over value between the gene f^m (filaria susceptibility, *Brugia malayi*) and the sex locus in *Aedes aegypti, Ann. Trop. Med. Parasit.* **57**:461–465.

131. MacDonald, W.W., 1976, Mosquito genetics in relation to filarial infections, (Taylor, A.E.R., and Muller, R.L. (eds): in *Genetic Aspects of Host-Parasite Relationships, Symp. Brit. Soc. Parasit.* **14**:74–87.

132. MacDonald, W.W., and Sheppard, P.M., 1965, Cross-over values in the sex chromosome of the mosquito *Aedes aegypti* and evidence of the presence of inversions, *Ann. Trop. Med. Parasit.* **59**:74–87.

133. McGreevy, P.B., Bryan, J.H., Oothuman, P., and Kolstrup, N., 1978, The lethal effects of the cibarial and pharyngeal armatures of mosquitoes on microfilariae, *Trans. Roy. Soc. Trop. Med. Hyg.* **72**:361–368.

134. McGreevy, P.B., McClelland, G.A.H., and Lavoipierre, M.M.J., 1974, Inheritance of susceptibility to *Dirofilaria immitis* infection in *Aedes aegypti, Ann. Trop. Med. Parasit.* **68**:97–109.

135. Manson-Bahr, P.H., 1921, *Mansons Tropical Diseases* 7th Ed. London.

136. Maudlin, I., and Ellis, D.S., 1985, Association between intracellular rickettsia-like infections of midgut cells and susceptibility to trypanosome infections in *Glossina spp., Z. Parasit.* **71**: 683–687.

137. Maudlin, I., and Welburn, S.C., 1987, Lectin mediated establishment of midgut infections of *Trypanosoma congolense* and *Trypanosoma brucei* in *Glossina morsitans, Trop. Med. Parasit.* **38**:167–170.

138. Maudlin, I., and Welburn, S.C., 1988, The role of lectins and trypanosome genotype in the maturation of mid-gut infections in *Glossina morsitans, Trop. Med. Parasit.* **39**:56–58.

139. Meis, J.F.G.M., Pool, G., Gemert, G.J. van., Lensen, A.H.W., Ponnudurai, T., and Meuissen, J.H.E.T., 1989, *Plasmodium falciparum* ookinetes migrate intercellularly through *Anopheles stephensi* midgut epithelium, *Parasit. Res.* **76**:13–19.

140. Mellor, P.S., 1971, Studies on *Onchocerca cervicalis* (Railliet and Henry 1910) and its development in *Culicoides*, Latrielle, Ph.D. Thesis, University of London, England.

141. Moloo, S.K., and Kutuza, S.B., 1988, Comparative study on the infection rates of different laboratory strains of *Glossina* species by *Trypanosoma congolense, Med. Vet. Entomol.* **2**:253–257.

142. Molyneux, D.H., and Killick-Kendrick, R., 1987, Morphology, ultrastructure and life-cycles, in Peters, W; and Killick-Kendrick, R. (eds): The Leishmaniases in Biology and Medicine Academic, Press pp. 121–176.

143. Molyneux, D.H., Takle, G., Ibrahim, E.A., and Ingram, G.A., 1986, Insect immunity to Trypanosomatidae, *Symp. zool. Soc.* London, **56**:117–144.

144. Nappi, A.J., and Christensen, B.M., 1986, Haemocyte cell surface changes in *Aedes aegypti* in response to microfilariae of *Dirofilaria immitis, J. Parasit.* **72**:875–879.

145. Nappi, A.J., and Silvers, M., 1984, Cell surface changes associated with cellular immune reactions in *Drosophila, Science* **225**:1166–1168.

146. Nayar, J.C., Knight, J.W., and Vickery, A.C., 1989, Intracellular melanization in the mosquito *Anopheles quadrimaculatus* (Diptera: Culicidae) against the filarial nematode *Brugia* spp. (Nematoda: Filarioidea), *J. Med. Ent.* **26**:159–166.

147. Nelson, G.S., 1964, Factors influencing the development and behaviour of filarial nematodes in their arthropodan hosts, in Taylor, A.E.R. (ed): *Host-parasite Relationships in Invertebrate Hosts, 2nd Symp. Brit. Soc. Parasit.* p. 75.

148. Nelson, G.S., and Pester, F.R.N., 1962, The identification of infective filarial larvae in Simuliidae, *Bull. Wld. Hlth. Org.* **27**:473–481.

149. O'Brochta, D.A., 1990, Genetic transformation and its potential in insect pest control, *Bull. Ent. Res.* **80**:241–244.

150. Okada, M., and Natori, S., 1983, Purification and characterisation of an antibacterial protein from haemolymph of *Sarcophaga peregrina* (flesh fly) larvae, *Biochem. J.* **211**:727–734.

151. Olafsen, J.A., 1986, Invertebrate lectins: Biochemical heterogeneity as a possible key to their biological function, in Brehelin, M. (ed): Immunity in invertebrates. Springer, Berlin, pp. 94–111.

152. Omar, M.S., and Garms, R., 1975, The fate and migration of a guatemalan strain of *Onchocerca volvulus* in *Simulium ochraceum* and *S. metallicum* and the role of the buccopharyngeal armature in the destruction of microfilariae, *Tropenmed. Parasit.* **26**:183–190.

153. Owen, R.R., 1977, Differences in the migration patterns of *Brugia pahangi* micro-filariae in susceptible and refractory members of the *Aedes scutellaris* complex, *Trans. Roy. Soc. Trop. Med. Hyg.* **71**:110–111.

154. Owen, R.R., 1978, The exsheathment and migration of *Brugia pahangi* micro-filariae in mosquitoes of the *Aedes scutellaris* species complex, *Ann. Trop. Med. Parasit.* **72**:567–571.

155. Pereira, M.E.A., Andrade, A.F.B., and Ribeiro, J.M.C., 1981, Lectins of distinct specificity in *Rhodnius prolixus* interact selectively with *Trypanosoma cruzi.*, *Science* **211**:597–600.

156. Perone, J.B., and Spielman, A., 1986, Microfilarial perforation of the midgut of a mosquito, *J. Parasit.* **72**:723–727.

157. Phiri, J., and Ham, P.J., 1990, Enhanced migration of *Brugia pahangi* micro-filariae through the mosquito mid-gut following N-acetyl-D-glucosamine inges-tion, *Trans. Roy. Soc. Trop. Med. Hyg.* **84**:462.

158. Poinar, G.O., and Leutenegger, R., 1971, Ultrastructural investigation of the melanization process in *Culex pipiens* (Culicidae) in response to a nematode, *J. Ultrastr. Res.* **36**:149–158.

159. Poinar, G.O., 1974, Insect immunity to parasitic nematodes, in Cooper, E.L. (ed): Contemporary Topics in Immunobiology: Invertebrate Immunity pp. 167–178.

160. Prod'hon, J., and Bain, O., 1972, Developpement larvaire chez *Anopheles stephensi* d'*Oswaldofilaria bacillaris*, filaire de Caiman sud-americain, et re-description des adultes, *Ann. Parasit. Paris*, **47**:745–758.

161. Ratcliffe, N.A., and Gagen, S.J., 1977, Studies on the in vivo cellular reactions of insects: an ultrastructural analysis of nodule formation in *Galleria mellonella*, *Tissue and Cell* **9**:73–85.

162. Ratcliffe, N.A., Leonard, C., and Rowley, A.F., 1984, Prophenoloxidase activa-tion: Non-self recognision and cell cooperation in insect imrnunity, *Sci.* **226**: 557–559.

163. Ratcliffe, N.A., and Rowley, A.F., 1979, Role of hemocytes in defense against biological agents, in Gupta, A.P. (ed): Insect Hemocytes: Development, forms, functions and techniques, Cambridge University Press, England, pp. 331–414.

164. Reid, G.D.F., 1978, Cibarial armature of *Simulium* vectors of onchocerciasis, *Trans. Roy. Soc. Trop. Med. Hyg.* **72**:438.

165. Reid, G.D.F., 1979, The development of *Onchocerca volvulus* in two temperate blackfly species, *Simulium ornatum* Meigen and *S. Iineatum* Meigen, *Ann. Trop. Med. Parasit.* **73**:577–581.

166. Rao, U.R., Chandrashekar, R., Parab, P.B., Rajasekariah, G.R., and Subramanyam, D., 1987, Lectin binding characteristics of *Wuchereria bancrofti* microfilariae, *Acta Trop.* **44**:35–42.

167. Rao, U.R., Chandrashekar, R., Rajasekariah, G.R., and Subramanyam, D., 1987b, Wheat Germ agglutinin specifically binds to the surface of infective larvae of *Wuchereria bancrofti, J. Parasit.* **73**:1256–1257.

168. Reid, G.D.F., 1978, Cibarial armature of *Simulium* vectors of onchocerciasis, *Trans. Roy. Soc. Trop. Med. Hyg.* **72**:438.

169. Renwrantz, L.R., and Mohr, W., 1978, Opsonizing effects of serum and albumin gland extracts on the elimination of human erythrocytes from the circulation of *Helix pomatia, J. Invert. Pathol.* **31**:164–170.

170. Richards, A.G., and Richards, P.A., 1977, The peritrophic membrane of insects, *Ann. Rev. Entomol.* **22**:219–240.

171. Rosenberg, R., Koontz, L.C., Alston, K., and Friedman, F.K., 1984, *Plasmodium gallinaceum*: erythrocyte factor essential for zygote infection of *Aedes aegypti*, *Exp. Parasit.* **57**:158–164.

172. Rowley, A.F., Ratcliffe, N.A., Leonard, C.M., Richard, E.H., and Renwrantz, L., 1986, Humoral recognition factors in insects, with particular reference to agglutinins and the prophenoloxidase system, in Gupta, A.P. (ed): Hemocytic and Humoral Immunity in Arthropods, John Wiley and Sons, New York, pp. 381–406.

173. Rudin, W., and Hecker, H., 1989, Lectin binding sites in the midgut of the mosquitoes *Anopheles stephensi* and *Aedes aegypti* L. (Diptera: Culicidae), *Parasit. Res.* **75**:268–279.

174. Sacks, D.L., and Perkins, P.V., 1985, Development of infective stage *Leishmania* promastigotes within phlebotomine sandflies, *Am. J. Trop. Med. Hyg.* **34**:456–459.

175. Salt, G., 1956, Experimental studies in insect parasitism, IX. The reactions of a stick insect to an alien parasite, *Proc. Roy. Soc.* London, **B146**:93–108.

176. Salt, G., 1963, The defence reactions of insects to metazoan parasites, *Parasit.* **53**:527–642.

177. Sauerman, D. M., and Nayar, J.K., 1985, Characterization of refractoriness in *Aedes aegypti* (Diptera: Culicidae) to infection by *Dirofilaria immitis*, *J. Med. Ent.* **22**:94–101.

178. Scalzo-Lichtfouse, B., Townson, H., and Ham, P.J., 1990, Humoral and cellular responses of *Anopheles stephensi* to infection with *Plasmodium yoelii*, *Trans. Roy. Soc. Trop. Med. Hyg.* **84**:463.

179. Schlein, Y., and Romano, H., 1986, *Leishmania major* and *L. donovani*: effects on proteolytic enzymes of *Phlebotomous papatasi* (Diptera, Psychodidae), *Exp. Parasit.* **62**:376–380.

180. Schlein, Y., Warburg, A., Schnur, L.F., and Shlomai, J., 1983, Vector compatability of *Phlebotomus papatasi* dependent on differentially induced digestion, *Acta Trop.* **40**:65–70.

181. Schottelius, J., and da Costa, S.C.G., 1982, Studies on the relationship between lectin binding carbohydrates and different strains of *Leishmania* from the New World, *Mem. Inst. Oswaldo Cruz.* **77**:19–27.

182. Schraermeyer, U., Peters, W., and Zehner, H., 1987, Lectin binding studies on adult filaria, intrauterine developing stages and microfilariae of *Brugia malayi* and *Litomosoides carinii*, *Parasit. Res.* **73**:550–556.

183. Sinden, R.E., 1984, The biology of *Plasmosium* in the mosquito. *Experientia* **40**:1330–1343.

184. Smail, A.J., and Ham, P.J., 1989, *Onchocerca* induced haemolymph lectins in blackflies: Confirmation by sugar inhibition of erythrocyte agglutination, *Trop. Med. Parasit.* **39**:82–83.

185. Sneller, V., 1979, Inhibition of *Dirofilaria immitis* in gregarine-infected *Aedes aegypti*: preliminary observations, *J. Invert. Pathol.* **34**:62–70.

186. Soderhall, K., and Smith, V.J., 1986, Prophenoloxidase-activating cascade as a recognision and defense system in arthropods, in Gupta, A.P. (ed): Hemocytic and Humoral Immunity in Arthropods, John Wiley and Sons, New York, pp. 251–285.

187. Stephens, J.M., 1962, Bacteriocidal activity of the blood of actively immunized wax moth larvae, *Can. J. Microbiol.* **8**:491–499.

188. Stiles, B., Bradley, R.S., Stuart, G.S., and Hapner, K.D., 1988, Site of synthesis of the haemolymph agglutinin of *Melanoplus differentialis* (Acrididae: Orthoptera), *J. Ins. Physiol.* **34**:1077–1085 .

189. Stiles, J.K., Ingram, G.A., Wallbanks, K.R., Molyneux, D.H., Maudlin, I., and Welburn, S., 1990, Identification of midgut trypanolysin and trypanoagglutinin in *Glossina palpalis* sspp. (Diptera: Glossinidae), *Parasit.* **101**:369–376.

190. Stoffolano, J.G., 1986, Nematode induced host responses, in Gupta, A.P. (ed): Hemocytic and Humoral Immunity in Arthropods. John Wiley and Sons, New York, pp. 117–155.

191. Sulaiman, I., and Townson, H., 1980, The genetic basis of susceptibility to infection with *Dirofilaria immitis* in *Aedes aegypti*, *Ann. Trop. Med. Parasit.* **74**:635–646.

192. Sun, S.-C., Lindstrom, I., Boman, H.G., Faye, I., and Schmidt, O., 1990, Hemolin: An insect immune protein belonging to the immunoglobulin superfamily, *Science* **250**:1729–1732.

193. Sutherland, D.R., Christensen, B.M., and Forton, K.F., 1984, Defense reactions of mosquitoes to filarial worms: role of the microfilarial sheath in the response of mosquitoes to inoculated *Brugia pahangi* microfilariae, *J. Invert. Path.* **44**:275–281.

194. Taylor, D.W., Goddard, J.M., and McMahon, J.E., 1986, Surface components of *Onchocerca lienalis*, *Mol. Biochem. Parasit.* **18**:283–300.

195. Thomas, P.M., and Ham, P.J., 1993, Acquired immunity in *Simulium* following superinfection with *Onchocerca*, *Trop. Med. Parasit.* (in press)

196. Townson, H., and Chaithong, U., 1991, Mosquito host influences on development of filariae, *Ann. Trop. Med. Parasit.* **85**:149–163.

197. Vaughan, J.A., and Azad, A.F., 1988, Passage of host immunoglobulin G from blood meal into hemolymph of selected mosquito species (Diptera: Culicidae), *J. Med. Ent.* **25**:472–474.

198. Voelker, J., and Garms, R., 1972, Zur morphologie unbekannter filarienlarven aus dem Onchocercose-Ubertrager *Simulium damnosum* und aus *S. kenyae* in Liberia und zur frage der moglichen enwirte, *Z. Tropenmed. Parasit.* **23**:285–301.

199. Walters, L.L., Modi, G.B., Tesh, R.B., and Burrage, T., 1987, Host-parasite relationship of *Leishmania mexicana mexicana* and *Lutzomyia abonnenci* (Diptera: Psychodidae), *Am. J. Trop. Med. Hyg.* **36**:294–314.

200. Warburg, A., and Miller, L.H., 1991, Critical stages in the development of *Plasmodium* in mosquitoes, *Parasit. Today* **7**:179–181.

201. Warburg, A., Tesh, R.B., and McMahon-Pratt, D., 1989, Studies on the attachment of *Leishmania* flagella to sandfly midgut epithelium, *J. Protozool.* **36**:613–617.

202. Weathersby, A.B., 1962, Parabiotic twinning of mosquitoes, *Mosq. News*, **25**:44–45.

203. Weathersby, A.B., and McCall, J.W., 1968, The development of Plasmodium *gallinaceum* Brumpt in haemocoels of refractory *Culex pipiens pipiens* Linn. and susceptible *Aedes aegypti* Linn, *J. Parasit.* **54**:1017–1022.

204. Weathersby, A.B., and McCrodden, D.M., 1982, The effects of parabiotic twinning of susceptible and refractory mosquitoes on the development of *Plasmodium gallinaceum*, *J. Parasit.* **68**:1081–1084.

205. Welburn, S.C., Ellis, D.S., and Maudlin, I., 1989, Rate of trypanosome killing in midguts of different species and strains of *Glossina*, *Med. Vet. Entomol.* **3**:77–82.

206. Welburn, S.C., and Maudlin, I., 1990, Haemolymph lectin and the maturation of trypanosome infections in tsetse, *Med. Vet. Entomol.* **4**:43–48.
207. Wing, S.R., Young, M.D., Mitchell, S.E., and Seawright, J.A., 1985, Comparative susceptibilities of *Anopheles quadrimaculatus* mutants to *Plasmodium yoelii*, *J. Am. Mosq. Contr. Assoc.* **1**:511–513.
208. Yamamoto, H., Kobayashi, N., Ogura, N., Tsuruoka, H., and Chigusa, Y., 1985, Studies on filariasis Vl: The encapsulation of *Brugia malayi* and *B. pahangi* larvae in the mosquito, *Armigeries subalbatus*, *Jap. J. Sanit. Zool.* **36**:1–6.
209. Yamamoto, H., Ogura, N., Kobayashi, M., and Chigusa, Y., 1983, Studies on filariasis 11: Exsheathment of the microfilariae of *Brugia pahangi* in *Armigeres subalbatus*, *Jap. J. Parasit.* **32**:287–292.
210. Yoeli, M., 1973, *Plasmodium berghei*: Mechanisms and sites of resistance to sporogonic development in different mosquitoes, *Exp. Parasit.* **34**:448–458.
211. Zahedi, M., Denham, D.A., and Ham, P.J., 1990, Surface lectin binding characteristics of developing stages of *Brugia* in *Armigeres subalbatus*: I *Brugia pahangi*, *Japan. J. Trop. Med. Hyg.* **18**:271–283.
212. Zielke, E., 1977, Further studies on the development of *Onchocerca volvulus* in mosquitoes, *Trans. Roy. Soc. Trop. Med. Hyg.* **71**:546–547.
213. Zielke, E., Schulz-Key, H., and Albiez, E.J., 1977, On the development of *Onchocerca volvulus* in mosquitoes, *Tropenmed. Parasit.* **28**:254–257.

5
Role of *Circulifer/Neoaliturus* in the Transmission of Plant Pathogens

Meir Klein

Introduction

The beet leafhopper, *Circulifer tenellus* (Baker), has been mentioned in more scientific publications of the New World than any other Auchenorrhyncha species (221). This leafhopper was first described almost 100 years ago in Colorado (139) and New Mexico (6) in the USA. Its appearance in cultivated fields is invariably harmful to crops and large infestations may even prove disastrous to them. *C. tenellus* is by nature a desert and semi-desert leaf-hopper and it can migrate long distances when food becomes scarce. There-fore, it breeds in dry regions of southwestern states of the USA, causing, the most damage generally at sites adjacent to the breeding zones and can move also into northern and mideastern states and cross the Rocky Mountains. Most entomologists of the Auchenorrhychae agree that this species of leaf-hopper is not native to the New World but rather originated in the Old World (248, 349). Apparently, only a single species of the *Circulifer* complex of the Old World has penetrated the USA. However, the occurrence of this group of leafhoppers in Europe, in the Mediterranean countries and the Far East may not necessarily be followed by extensive damage to commercial crops, as in the case in western USA. This is probably the reason why this group of leafhoppers has never been studied thoroughly outside the USA. We focus our attention on the *Circulifer/Neoaliturus* complex mainly because certain species of this group have been found to be vectors of plant pathogens that are the cause of important diseases.

In the USA. *C. tenellus* is known to develop in cultivated fields very large populations which can cause extensive damage merely by feeding. However, much more important is the indirect damage caused by virus, spiroplasma

Meir Klein, Department of Entomology, Institute of Plant Protection, ARO, The Volcani Center, Bet Dagan 50250, Israel.
© 1992 by Springer-Verlag New York, Inc. *Advances in Disease Vector Research*, Volume 9.

and mycoplasma-like organisms (MLOs) transmitted to a broad range of plant varieties. Massive invasions of *Circulifer/Neoaliturus* species have not yet been reported in the Old World, where pathogens, similar to those found in the USA, are transmitted by leafhoppers of this group. However, the extent of the spread of these agents in nature is still unknown.

The beet curly top virus (BCTV), transmitted by *C. tenellus* in the USA (7, 299) and by *Circulifer opacipennis* in Turkey and Iran (28, 126, 164), is one of two or one of five geminiviruses transmitted by leafhoppers to dicotyledonous plants (158, 328). This group of geminiviruses comprises one or more additional viruses, such as the Summer Death of French beans in Australia (12, 323) also probably the curly leaf of sugarbeet virus in Argentina and the curly top of tobacco and of tomato in Brazil (18, 70). Neither the beet leafhopper nor any other *Circulifer* species does occur in South America (293) or Australia (Oman, P.W. personal communication).

Spiroplasma citri, the causal agent of citrus stubborn disease, is transmitted in the USA mainly by *C. tenellus* (244). Two species of *Circulifer*, *C. tenellus* (168, 173, 174, 264) and *Neoaliturus* (*C.*) *haematoceps* (114), and probably others, are known to transmit *S. citri* in North Africa, the Middle East and the Far East (34, 35).

Some MLOs are also transmitted by this group of leafhoppers: the beet leafhopper transmitted virescence agent (BLTVA) by *C. tenellus* in the USA (240) the safflower phyllody agent (SPA) by *Neoaliturus fenestratus* in Israel (167), the phyllody agent of *Vinca* (34) and the *Spiroplasma phoeniceum* probably by the *Circulifer/Neoaliturus* group in the Old World (274).

Techniques have been developed recently for isolating and cloning the DNA of MLOs. These will allow direct analysis and comparison of MLO genomes and will provide diagnostic reagents for detecting MLOs in plants and in insects. In addition, they will enable the identification of genes that are important in the pathogenesis process (76, 77, 165, 166, 189).

Circulifer spp. introduce pathogens into suitable plant sites in a highly specific way. The BCTV has to be introduced into parenchymal cells of the phloem (102, 229). However, it is very difficult to infect plants with conventional techniques of mechanical inoculation (25, 225, 282) which require a high vector specificity (43). A leafhopper becomes a well known vector due to the considerable damage caused by the pathogen it transmits or because of its ability to transmit different pathogens. Few leafhoppers are known to transmit a complex of pathogens of different types. Among these, the best known are *C. tenellus* (13), *Macrosteles fascifrons* (13, 239), *Dolbulus maidis* (206, 235), *Cicadulina mbila* (206, 269) and *Orosius argentatus* (11, 40). The virus vector specificity within the group known as geminiviruses is consistent with the vector-mediated endocytosis in that it adapts itself to feeding at sites where viruses congregate and at sites where viruses have to be discharged from the stylet (43, 220). Specificity of MLOs/spiroplasmas and of the vectors is apparently governed by different principles. MLOs and spiroplasmas have the ability to penetrate most leafhopper gut-walls. It seems that the subsequent process and feeding behavior plays a key role in the success of the insect as a vector (206, 224).

Circulifer/Neoaliturus Species

The definition of the beet leafhopper in the USA evolved significantly during the first fifty years after its identification. Baker described this species in 1896 as *Thamnotettix tenella* (6), and Forbes and Hart in 1900 identified it as *Eutettix tenella* (111). Oman decided in 1948 to include it in the genus *Circulifer* (Zakhvatkin) after a thorough study of Old World species (248), and he subsequently subdivided all known species belonging to this genus into two genera: *Neoaliturus*, for species with their aedeagal rami in a semi-circle, and *Circulifer*, for species with their aedeagal rami in a full circle (249). European taxonomists, however, advocate the classification of all these leafhoppers under *Neoaliturus* only (Wilson, M.R. and Oman, P.W., personal communications). Due to similarities in some species of *Circulifer* and *Neoaliturus*, male genitalia can no longer serve as the only element for segregation among the species. Therefore, the concept of a complex of species for *C. tenellus* (in the Old World), *C. dubiosus*, *C. haematoceps* and *C. opacipennis* (Nielson, M.W. and Oman, P.W., personal communications, 168, 171, 172) will be used until better clues are found.

The Beet Curly Top Virus and its Transmission

The Virus

The BCTV is a small, isometric particle (101, 182, 226, 328). BCTV is restricted to the nuclei of phloem parenchymal cells (102, 103, 229) of dicotyledon plants only. In purification the virus appears as monomers or dimers and only the latter are infectious (182). Only recently has a good level of knowledge of the geminiviruses been attained (312). All these viruses are characterized by a genome of circular single stranded DNAs ((ss) DNA) encapsidated in geminated quasi-isometric particles. The geminiviruses are found among viruses transmitted by whiteflies and leafhoppers. The genome of the geminiviruses transmitted by whiteflies is made up of two (ss) DNAs, while the genome of those transmitted by the leafhoppers consists of only one (312). Infectious clones of the BCTV genome were constructed in the laboratory and their nucleotide sequence was determined (313).

The genome of the BCTV resembles in part only one of the two (ss) DNAs in the whitefly geminiviruses. The presumed coding region of the coat protein, very similar to the codings of the other few geminiviruses transmitted by leafhoppers, is an exception. Therefore, the coat protein is important for vector specificity (313). The role of this virus component in the infection process in plants was studied by means of the *Agrobacterium*-mediated inoculation technique (43). This technique is a big improvement over the injection method of mechanical transmission used in the past in the study of the BCTV (225). The new technique evolved mutants of the coat protein gene and demonstrated the essential character of the coat protein in the spread of the virus in the plants (43).

It was also discovered recently that the nucleotides sequences of all geminiviruses contain inverted repeats that may form stem-loop structures. The sequence of nucleotides in the stem loop of the BCTV is quite similar to that of only one of the geminiviruses transmitted by whiteflies, the bean golden mosaic virus. However, no similarity was observed between this structure of the BCTV and all other viruses transmitted by leafhoppers to monocotyledons. Therefore a third group of geminiviruses was created. This group is transmitted by leafhoppers to dicotyledons (158). The genomic characterization of phenotype variants of the BCTV has been reported recently (315, 316).

In Australia, Summer Death of the French bean virus, also called the tobacco yellow dwarf (12, 323), was purified and shown to be of a gemini nature. This virus is serologically related to the BCTV in the USA (323), but only distantly according to immunosorbent electron microscopy (IEM) examinations (267). The very short incubation periods (IPs) of the Australian virus in the leafhopper vector *Orosius argentatus* (41) are reminiscent of the BCTV in *C. tenellus* (279). Therefore, it is believed to belong together with the BCTV to the same group of geminiviruses transmitted by leafhoppers to dicotyledons (158). Unfortunately, not enough details are available at present on the structures of the viruses and the similarity between the curly leaf of sugar beet virus in Argentina and the curly top of tomato and tobacco virus in Brazil to allow their classification. The only similarity between these viruses and the BCTV in the USA is provided by the symptoms of the disease in different plants and in the short IPs of the viruses in their vectors (18, 26, 33, 70).

Transmission Characteristics

ACQUISITION ON PLANTS

Transmission tests with the beet leafhopper and the beet curly top virus started very early. It has been known since 1910 that a short feeding on BCTV-infected sugar beets was sufficient to elicit in *C. tenellus* its vectorial nature (299, 309). Severin proved in 1921 that a 1- to 2-minute feeding on infected plants was enough to induce infectivity in a small number of tested leafhoppers (279). Transmission rates were directly related to the length of the acquisition access period (AAP). Very low rates of transmission (3%) resulted from leafhoppers which had had only a 1–20 minute AAP. Maximum transmission (33%) with individual leafhoppers was obtained from those which had fed on BCTV-infected plants for at least 4 h. It was found that a minimum latent period (LP) of 4 h of the virus within the vector at 38°C is required before the leafhopper transmits the virus to healthy plants. The IP of the virus in the plant is longer, and at least 5 days elapse before symptoms can be detected (279). However, IP in plants is even shorter than 5 days when small seedlings were inoculated (280). The LPs of the BCTV was found to be longer in *C. tenellus* males than in the females (282). Freitag (119) confirmed earlier results of Severin (279). According to Bennett and Wallace (29) the maximum ability to transmit BCTV is gained by a relatively long AAP (2 days) on infected plants, whereby a maximum virus charge is acquired. Contrary to non-persistent transmissions of viruses by aphids, fasting prior to the AAP in *C. tenellus* may interfere with the initial process of virus charge. Leafhoppers are more charged with BCTV on plant species showing severe symptoms and with a high concentration of the virus. They can transmit the virus longer than leafhoppers fed on plant species with mild symptoms and known to be weak sources (21). A very short inoculation access period (IAP) may take sometimes as little as 1 min to infect healthy plants (119, 277). Transmission rates become gradually lower over a period of 8-10 weeks, regardless of the amount of time leafhoppers have fed on virus source (29). Ability to transmit BCTV is not lost by the insect migrating over long distances. The entire transmission process of the virus can be carried out through nymphs and is not obliterated by moulting. The very short AAPs and the comparatively short IPs in the vector enable even the smallest nymphs to acquire and transmit the virus while at the same nymphal stage. In laboratory tests, no signficant differences were obtained in transmission rates among insects of the five nymphal instars of the beet leafhopper or between nymphal instars and adults (127, 277).

ACQUISITION FEEDING THROUGH MEMBRANES

Feeding through a membrane on either juice extract from infected plants or on extracts of viruliferous leafhoppers was probably tested first on *C. tenellus* (56, 58, 291). Extracts from viruliferous leafhoppers were found to constitute a very good source of infectivity for non-viruliferous ones which had fed on them (17). More and more leafhoppers became inoculative by increasing the AAPs on these extracts and about half of the insects transmitted the BCTV following an 8–16 h AAP (22). These results closely matched those obtained following AAPs on infected sugar beets (287). However, IPs in test plants differed significantly depending on the virus source: 4 days IP for virus from plants (287) and 10 days IP for virus from leafhoppers (22). Transmission rates of nymphs fed on leafhopper extracts were appreciably lower than those obtained for young adults, contrary to that observed when following AAPs on plants (127, 277). These results seem to indicate that virus concentrations in the tested plants (127, 277) are higher than in the leafhoppers' extracts (22).

The feeding process through a membrane provides a suitable technique for studying the relationship among BCTV isolates (95).

Injections of plant and leafhopper extracts, and of purified virus solutions were also used in the study of BCTV transmission (202). This method facilitates the control of titers of virus concentration. It bears out the relationships between the virus dose and the transmission rates and between the virus dose and the LPs in vector, and those between the virus dose and the IPs in plants, on the one hand, and the retention period of the virus in the leafhoppers on the other (197, 200, 202).

The results obtained from tests of the leafhopper's capability to transmit BCTV (as in sections 1, 2 and 3 above) suggest that, contrary to the situation in plants, the BCTV does not seem to multiply within the vector's body. The BCTV is known to multiply in a special type of plant cell, the nuclei of the parenchymal cells of the phloem (102, 103, 229, 330). No special tissue can be identified which harbors the virus within the leafhopper (290).

Epidemiology of the Beet Curly Top Virus—Connection with the Biology and Ecology of the Insect Vector

Biological and ecological properties of the beet leafhopper were investigated early in the 20th Century (8, 9), a period marked by the heavy damage inflicted to sugar beets, tomatoes, snap beans, etc., by this leafhopper. Some success became apparent when resistant varieties were introduced, but many difficulties still have to be overcome.

The introduction of resistant varieties of sugar beet in western USA saved the local sugar industry from utter ruin. Nevertheless, severe outbreaks of C. tenellus, originating in its winter breeding sites, were the cause of heavy damage to sugar beet crops and invasion by this leafhopper resulted always in heavy BCTV infection (295). Resistant sugar beet varieties did not, in these particular cases, provide a solution to the problem.

The beet leafhopper was thoroughly investigated in the laboratory and in several of its breeding areas in western USA. As expected, data from these investigations were sometimes contradictory due to the different environmental and climatic conditions at the study sites. Ball, a pioneer in this kind of research in the USA, claimed that the beet leafhopper is a single-brood species in nature (10). Having found in California that this species of leafhoppers produces two generations in nature, Stahl and Severin reported a few years later two broods (280, 310). Carter obtained results indicating three generations a year (60), and Cook reported three to five generations (69).

Severin, doing research in California, found that C. tenellus exists in the USA in two distinct morphological morphs (280). The summer morph is conspicuous for its bright colors, as opposed to the dark winter morph. Adults of the summer generation survive for 3–4 months and females outlive males by approximately one month. Adults of the winter generation live longer than those of the summer one-males 4 months and females 7–8 months-and produce a very dense population (280).

C. tenellus is a polyphagous insect and can adapt itself to new hosts very quickly. Among such new hosts are weeds from various botanical families, cultivated plants, winter as well as summer plants, annual and perennial (60, 69, 87, 89, 116, 151, 152, 185, 288). *C. tenellus* has the ability to regain lost vitality after finding favorable hosts (68). The winter generation grows usually on winter weeds thus remaining innocuous (295). The summer generation of *C. tenellus* grows mainly in irrigated fields. In the autumn, when temperature starts to drop, the adults migrate to the foothills of the deserts in the southwestern states. Factors which limit the size of this migrating population and interfere with its hibernation may reduce the migrating population in the following spring and, as a result, decrease virus incidence as well.

The spring migration sometimes covers fairly short distances of only a few miles (e.g. 2, 67, 69, 71, 83, 84, 88, 115, 152, 185, 288, 347), and sometimes very great ones, of several hundred miles (83). During long periods of drought, migration may extend over thousands of miles, carrying BCTV to the north and east of the USA (84).

Conditions leading to or hindering springtime movements of the beet leafhopper and virus transmission (2, 60, 152, 283, 288) are as follows:

(1) Mild winters—Massive flights of the summer generation adults were detected after mild winters during which many suitable breeding hosts were recorded. If winter started with no breeding hosts present, the winter generation leafhoppers would have to rely on "hold over" hosts and the next generation would be too sparse and weak to endanger cultivated crops in the early summer (116, 347).

(2) Severe winters—Freezing temperatures in winter effect the mortality of the younger nymphs but not that of the adults which can withstand relatively long periods under the snow. Thus, the insect does not multiply early in the spring. During mild winters in snowy areas, the population can rise to huge numbers and present a serious problem later on (91, 117).

(3) Precipitation in autumn—Sufficient precipitation in September is conducive to a good start for hibernation. A dry September means a bad start for hibernation, since insects need to change host at least twice before they hibernate.

(4) Dry spring—Dry weather in early spring (March)—This does not allow nymphs to mature on their winter hosts, and leads to premature death.

(5) A long dry spring (April)—This is best for nymphal maturation, as regards vegetation. Typical heavy outbreaks of *C. tenellus* occur when such conditions are present and especially when they are followed by optimal flying temperatures and winds. Then, migration is of short duration—only a few days—involving nevertheless huge masses of insects.

(6) A long wet spring—In some of the northwestern states of the USA, rain, fog, heavy dew and low temperatures interfere with the development of a

desert leafhopper such as *C. tenellus*. These conditions reduce the population directly and increase its sensitivity to various entomo-pathogens. Long wet springs and heavy summer precipitation in the middle and eastern states of the USA appear to be the reason why *C. tenellus* is unable to settle there on a permanent basis.

(7) Dry winds in the spring—In certain years these conditions kill many nymphs prior to their maturation and migration.

In a study conducted recently concerning the flight performance of field-collected *C. tenellus* (147), it was found that both males and females have a propensity to fly for extended periods, with females in greater numbers and flying longer. Pre-reproductive females constitute mainly the mass of spring and autumn migration. Induction to these flights appears to be associated more with environmental factors than to originate from an internal signal. Leafhoppers leave plants as the latter start to dry. Leafhoppers reared on different host plants differ significantly in their lipid content, that is in the amount of flight fuel reserves (147).

The direction of the main leafhopper movement in the spring is different from that in the autumn and is determined by winds prevailing in the breeding areas at the time. The leafhoppers drift passively each year, usually in the same direction (184). The most dangerous leafhopper movement is undoubtedly the early migration in summer, the time when small seedlings are in the field. This plant stage is the most sensitive to inoculation and injury by viruses, because of the virus's extremely rapid multiplication rate (88). In some areas, like Arizona, the time for planting sugar beets has been moved to the end of summer. The danger from most BCTV infected areas is expected to come in the autumn from leafhoppers migrating to their winter sites. In the spring sugar beets have developed sufficiently and have reached the stage of relative resistance (71).

The biology of *C. tenellus* was studied once again in 1970 (253), in order to find an alternative method of controlling this vector. The attraction of the insect to monochromatic electromagnetic radiation was greatest in the vicinity of 350 and 500–550 nanometers. This information might be used in future in a technique to control *C. tenellus*. The mating and reproductive behavior of *C. tenellus* investigated by Perkes (253) did not portend a bright future for the control of the insect.

Host Plants for the Virus

Cultivated plants—Since 1905, sugar beet has been known to be affected by the curly top virus and to have a connection with the beet leafhopper (7). The link existing between BCTV and beets and tomatoes has been recognised since 1906 (308). Not until 1927 did it become clear that tomato (219), severely infected with the summer blight disease, and sugar beet (285), with BCTV during the same period, had both been affected by the same virus transmitted by *C. tenellus*. The connection between BCTV and the bean blight was defined in a publication which appeared in 1925 (50). Severin published in 1927 a list of crops naturally infected by the BCTV in California (284) among which were solanaceous plants, cruciferous plants and parsley. Pepper plants were severely infected in New Mexico by a disease defined in 1927 as BCTV (72). In the same year, BCTV was found to be the cause of a highly destructive disease of squash (218). In 1928, Severin and Henderson published a list of field and garden plants of three plant families in California (Chenopodiaceae, Leguminosae and Cucubitaceae) all very susceptible to BCTV in nature (294). These include beets of all kinds, spinach, field and garden beans, cowpea, alfalfa, pumpkin, squash, watermelon, cucumber, muskmelon and cantaloupe. A year later, Severin added a few more economically important plants, including potato and tomato which are naturally infected by BCTV in California (286) and Utah (297). To that list he added many important plants which were infected in laboratory transmission tests. Fourteen species of ornamental flowering plants (13 genera belonging to ten families) have been found to be infected in the fields (292) and the BCTV was experimentally transmitted to 92 additional species (73 genera belonging to 33 plant families) (121). In New Mexico, tobacco was first reported in 1927 to harbor BCTV by natural infection (72). The occurrence of BCTV in flax was first observed in California in 1947 (131).

Non-cultivated plants—Many weeds become naturally infected by BCTV in the USA (289). Some of these weeds are known to grow side by side with cultivated crops as well as in cultivated areas. A considerable number of weeds were proved to be hosts for the virus in laboratory experiments. However, some among them showed no symptoms at all. Many of the tested plants did not become infected in laboratory tests with viruliferous leafhoppers (318). Of interest was *Oxalis stricta*, a plant known for its high acidic sap content, which nevertheless became infected in Virginia when grown near sugar beet fields (314). BCTV overwinters mainly in wild annual and perennial plants growing in the winter breeding areas of the vector and in regions adjacent to cultivated fields. It may overwinter even in some commercially grown plants (289).

BCTV Strains and *C. tenellus*

Strains of BCTV were recorded as early as 1938. The process started with four strains which were distinguished on sugar beet by differential reaction, i.e., severity of symptoms and transmission rate (128). In 1944, ten strains of BCTV had already been identified (130). They required hosts other than sugar beet for segregation (130). Ten years later, strains 11 and 12 were added to the list, the former extremely virulent against resistant lines of sugar beet (134, 231) and the latter harmful to field potatoes grown for certified seeds (135).

The 13th strain of BCTV, designated yellow vein virus, was isolated from tobacco plants and described as the mutant of a strain regularly infecting tobacco plants (19). In the early 1960s, strains isolated from sugar beets in the western USA were more virulent than those known previously (23). Four or five additional strains, identified in the late 1960s, were segregated on the resistant sugar beet varieties (198, 199, 228, 319). During the last 20 years, these strains were the apparent cause of the increased frequency of BCTV outbreaks in western USA. In 1982 another strain of the virus was found infected with the brittle root disease in horseradish plants in the eastern USA. This strain was unique owing to its infection of cruciferous species only. It reacted serologically to typical BCTV strains from the western, and its transmission by *C. tenellus* was similar to that of the most common BCTV isolates (96).

Cross-protection, was not detected among BCTV strains in plants or in the insect vector (24, 132). This constitutes additional evidence that BCTV does not multiply in *C. tenellus*. The occurrence of so many virus strains in BCTV suggests virus instability, but all attempts to prove this have failed (133). Pure cultures of most strains, kept on plants in the laboratory for fairly long periods, and frequent transmission tests during the early stages of rapid virus multiplication conducted on young seedlings, did not reveal any instability trends (138).

Resistance and Sensitivity to BCTV in Plants and the Association of Resistancy with the Vector Behavior

The development of sugar beet varieties resistant to BCTV saved the sugar industry in the dry western States of USA from utter ruin. The success with sugar beet led to research in order to develop resistant varieties of tomatoes and beans. Thorough selections, using a number of BCTV strains, produced in the early 1930s the first resistant sugar beet variety, U.S. No. 1 (54). Very soon afterwards, a similar technique was adopted for squash (*Cucurbita maxima*), which was very susceptible to BCTV (73, 284). Many varieties of squash were screened by exposure to viruliferous *C. tenellus* (74). Areas for growing garden beans in the western USA were circumscribed by the curly top disease (bean blight). Losses were heavy in years of intense *C. tenellus* infestation (53, 183). Laboratory tests and field observations showed quite clearly that the leafhopper prefers garden bean varieties to field bean varieties (150). As a result, field beans, or such resistant varieties as were found every year, were recommended to growers in areas adjacent to the vector's winter breeding sites (233).

Tomatoes grown near the natural winter breeding site of *C. tenellus* suffered much damage, due very often to the curly top disease (285). Tomatoes are infected by several strains of BCTV (128, 134, 341, 343) and this should be taken into consideration in the production of resistant varieties. The search for resistance in tomatoes took a relatively long time until success was achieved (85). High levels of resistance were found among inedible green-fruited *Lycopersicon* species. The levels in the red-fruited species, however, were inconsistent (209).

Pepper production in Washington State is also occasionally hampered by BCTV (30). Many pepper varieties were screened in Washington to determine the resistance to the dominant BCTV strain. Some varieties performed better than others (337).

Much curiosity was aroused in the scientific world as to the nature of the resistance to BCTV and how it is related to the vector. Sugar beet resistance is undoubtedly connected with virus titer. The virus concentration in plants of the resistant varieties was much lower than that in varieties showing very mild symptoms (20). Leafhoppers acquire a much larger quantity of virus when feeding on susceptible plants, and their ability to transmit virus increases thereby. In addition, the leafhopper vectorial capacity resulting from feeding on susceptible plants is greater than from feeding on resistant plants. Lately, some highly virulent strains of BCTV have attacked sugar beet fields in western USA (23). In all sugar beet varieties, resistance to BCTV infection and BCTV injury increases with the age of the plants (129). Resistance appears sooner in varieties known as resistant than in the susceptible ones. The reaction of very young seedlings of resistant varieties to BCTV is more effective than that of seedlings of susceptible ones, even during their sensitive period (97, 137, 277, 307, 327). Experiments revealed that additional important factors, quite apart from the age of the plants, did affect the rate of infection in both types of sugar beet varieties, namely (a) the site of the inoculation feeding on the plant, (b) the number of viruliferous leafhoppers per plant, (c) the length of time leafhoppers have access on the plant, (d) the temperature at the time of inoculation, and (e) the light intensity during the process of inoculation (277). Analyzing the experimental results obtained, it appears that resistance in sugar beet relates mostly to (i) the speed with which the virus moves from the inoculation site to the cells where it multiplies, (ii) the rate of its multiplication and (iii) its subsequent distribution to other parts of the plant (97). The inoculation process which starts with large doses of virus seems to be faster than that which starts with small doses. It appears that chances of introducing large doses of virus into the resistant plants increase during periods of massive infestation by *C. tenellus*. More inoculation sites per plant are thus expected and their distribution among several leaves leads to a faster translocation of the virus.

A similar response was observed in tomatoes as regards the age effect of the plant and the site and number of inoculation feedings (339). Several resistant varieties were obtained by cross breeding resistant wild varieties with sensitive edible ones (32, 211, 305, 306, 325, 340). In the last few years, levels of resistance have been increased greatly (210, 213), but the nature, of the resistance remains somewhat obscure. The most common explanation is that the virus delivered by *C. tenellus* fails to establish infection in resistant tomatoes as often as in susceptible plants (327). This process seems to be connected to some degree with the feeding preference of the vector (320, 321, 326) but does not appear to be related to the recovery from infection during the early stages of infection (324). Other factors which may explain the resistance of some of the tomato varieties are: (i) selective immunity to specific virus strains; (ii) tolerance or masking of symptoms (322); (iii) recovery from symptoms; (iv) restriction of the virus to roots only; (v) resistance to systemic translocation; (vi) localization of the virus at the site of inoculation; and (vii) slow movement of the virus from the site of inoculation to the site of infection. The question is, what happens to resistant varieties when a large wave of leafhoppers enters the field (212)?

Very little is known about the resistance in beans. Generally speaking, the resistance is connected with the titer of the virus, a low titer being found in varieties showing mild symptoms of infection (140). Furthermore, it has been known for some time that resistance to BCTV in some snap bean varieties may break down at high temperatures, regardless of the tolerance exhibited (302, 303).

Sugar Beet Curly Top East of the Rocky Mountains (USA)

Curly top disease has been shown to have strong links with the beet leaf-hopper. It is therefore expected that, when *C. tenellus* moves from its permanent sites in the southwestern and western states of the USA to the northern and eastern states, it will carry BCTV with it. Following a thorough study conducted on *C. tenellus* in western USA (69), all the leafhopper's major and minor breeding areas in winter, spring and summer were listed. The general direction of its annual migration from one area to another was mapped, six such areas were found in southwestern and western arid and semi-arid states of the USA (86). It is commonly accepted that the beet leafhopper is a desert insect (248). This explains why much of the BCTV damage is confined to the arid areas of the USA, usually west of the Rocky Mountains. Outbreaks of the beet leafhopper outside these borders were observed usually following several years of drought. The presence of *C. tenellus* in Kansas, east of the Rockies, was first reported in 1920 (186). In 1925, *C. tenellus* was found on sea purslane plants along the Atlantic coast of Florida (81). Large numbers of this leaf-hopper were identified in Illinois in 1937 (82) and again in 1953 (80), also in Kansas and, apparently, in a few more eastern states (84). Damage from BCTV in central and eastern states was recorded even before the beet leaf-hopper had been identified there (10, 57, 296, 314, 331). In 1953, BCTV caused heavy damage to sugar beet in some of these states (90, 136, 304). On occasion, the disease was observed in only a few plants (123, 275, 329, 338). In 1958,

BCTV was recorded once again following another leafhopper invasion to the east (276, 334). The small scale of BCTV infection in states like Virginia and Maryland seems to have resulted from the interference of some factors affecting the drift of the main wave of leafhoppers toward these areas. Efforts to detect *C. tenellus* in these states have failed (276, 334).

Control Measures

Several measures to limit BCTV damage have been prescribed in a general way by scientists carrying out research on this virus (25, 94, 181).

DIRECT CONTROL OF LEAFHOPPERS

The first attempts at direct control of the vector by means of contact insecticides gave results far from encouraging (25, 93, 251). In the early stages, even systemic insecticides lacked uniformity. Nevertheless, minor successes were recorded (154). As applications of systemic insecticides improved, the field experiments began to show consistently positive results (125, 227, 251). Often, the size of the existing leafhopper population was ignored (106, 201). During a large invasion of leafhoppers, there were indications that the applications of these systemic compounds in the field had met with limited success (180, 252) and that these materials had been effective on susceptible varieties of plants (232). It is known that the vector needs a very short IAP to introduce the virus into the plants. Chances that the insecticides will kill the vector before the virus is transmitted are very small. In addition, the amount of virus carried by *C. tenellus* from its breeding sites is large, especially during years of leafhopper outbreaks, at which time the spread of the virus cannot be prevented.

On the basis of the above evidence, leafhopper control was suggested as practicable during the autumn, when these insects leave cultivated fields for their winter hosts and reach their well known winter breeding sites. In winter, leafhopper populations are as a rule relatively small. The leafhoppers acquire the BCTV from source plants at their breeding sites, usually at the nymphal stage, when their control is relatively easy. Since 1931, sugar companies in several western states of the USA have been conducting a control program against the beet leafhopper outside cultivated areas during the autumn and winter (64). These insecticide applications have been sufficient to contain the spring population of the leafhoppers and to reduce damage by the BCTV to sugar beets (5, 64, 65, 66, 92). However, this program has proved inadequate in certain years and in some localities, and the outbreak of the spring population was not prevented due apparently to unexpected climatic conditions. During recent years, fields invaded by the vector have not been in a healthy condition (222).

The incidence of BCTV in table tomatoes growing in fields adjacent to winter breeding sites of *C. tenellus*, ranges from 30% to 100% every year. In similar areas, where the control program operates, incidence of the disease has ranged from zero to a maximum of 30% (4).

The control program must be the object of constant monitoring, as conditions vary. For example, in this program insecticides were applied in autumn on *Salsola iberica* plants, the most important "hold over" plants for *C. tenellus*. However, this host favored by the leafhopper is gradually being replaced by another species of *Salsola*, *S. paulsenii*, not the insect's favorite, thereby causing the latter to migrate to other species of plants for a transition period, until the winter host plants start growing. Then an alternative program has to be devised (196). As the timing of the insecticide application on the winter host is crucial, this application has to be carried out before the spring populations reach full maturity (31). Insecticide applications, after the planting of vegetations during the late autumn, ward off early infections in that season and minimize it in the next (223).

DESTRUCTION OF THE VECTOR'S FEEDING AND BREEDING HOST PLANTS

Mustard species, filaree, Russian thistle and annual saltbush are the favored breeding hosts, and their destruction in spring may be of help in the war against *C. tenellus* (2). This is impossible under ordinary circumstances, as breeding sites are very large and are located in the foothills. However, within the framework of a rangeland improvement project funded by the U.S. Congress (31) and executed in several states, this concept has also been taken into consideration (255). The areas were seeded with selected plant species to replace the insect breeding weeds. Unfortunately, the project was not completed and its success was limited. Several breeding areas in the desert, never plowed before, were irrigated and host species were destroyed. On the other hand, many areas previously irrigated were abandoned due to long periods of drought, thereby enabling the development of new breeding sites. Overstocking, overgrazing, burnings and other farming procedures may also cause breeding sites to expand (5, 90, 112, 185, 256).

SEEDING AND PLANTING PRACTICES

A correct seeding practice may reduce BCTV damage: (i) Considering that young infected plants die soon after infection, direct seedings and delayed thinnings give a better chance for development to the nearby plants which have escaped infection (94, 298); (ii) Direct seeding produces good stands of plants, and under favorable agricultural conditions the stands may create dense canopies with much shade and humidity, both anathema to *C. tenellus* (25, 49, 59, 105, 268, 298); (iii) The deliberate establishment of a beet-free period and the eradication of weed hosts in the cultivated areas may reduce the development of large summer populations of the vector and delay the vector's mass appearance in the next season; as a result, BCTV incidence is reduced (110); (iv) Keeping fields clean of weed hosts and destroying them before they can produce seeds seems to be the most practical way of controlling the spread of the virus (59); (v) The timing of seeding or planting is very important (153, 308). Early planting of sugar beets in California (December to February) is recommended to avoid leafhopper feeding on very young and sensitive plants (55, 96, 137, 230, 307). In the foggy areas of California, it is preferable to delay planting until after the spring migration of the insects (266, 281). Cantaloups are usually seeded in Arizona in February and are at the runner stage at the time of spring migration of *C. tenellus* (71). In Idaho, garden beans should be planted before insect migration to avoid attack of young seedlings by numerous leafhoppers (183, 228). This does not apply to areas of bean cultivation in central Washington (45).

RESISTANT VARIETIES

The use of resistant varieties of sugar beets and tomatoes enables planting in fields adjacent to the leafhoppers' breeding grounds (51, 52, 54). Breeders recommend the continuation of the development of new resistant varieties because more potent virus strains are likely to appear from time to time (105, 137, 217). Resistance in tomatoes seems to be adequate for the time being (213, 219) although it can break under heavy attack of viruliferous leafhoppers carrying some new strains. In tomatoes, resistance appears to be linked closely to the vector's feeding habits (320, 321, 326). Feeding preference is important also from the viewpoint of direct damage to tomatoes. Tomato varieties are known to suffer only from leafhopper feeding, apparently because of toxin secretion into the plants (124).

Resistant commercial bean varieties were released for use in western USA early in the 1940s (233, 234). A few tolerant varieties of snap beans were recommended over others, provided that (a) proximity to leafhopper sources was avoided, (b) the correct time of planting was chosen and (c) sprinkler irrigation was not applied (301).

In order to reduce damage from BCTV, an early idea was to use plant varieties that recover from infection. It is known that most tomato plants will die very soon after infection. However, in the late summer and autumn, when plants are in an advanced stage of growth, a late BCTV infection inflicts less damage because the plants are able to recover in greater numbers (48, 190, 191). Breeders look for varieties with a high recovery rate. This is determined in nature by type of plant, age of plant, site of infection, virus strain, environmental conditions, etc. (16, 343). Unfortunately, commercial varieties are not known to have this recovery characteristic. Recovery in Turkish tobacco is more common than in tomatoes (27), especially when infection starts either very early in growth or is located in old tissues (257). This property in tobacco can be transferred to tomatoes by grafting, but this was not found to be economical (342). In sugar beets, the recovery rate is very low (15).

It becomes more and more obvious that, in order to reduce damage from BCTV to an economic level, integrated pest management control must be enforced (214).

The Stubborn Disease Organism and its Transmission with *Circulifer* Species

The Stubborn Disease of Citrus

The stubborn disease has been known to affect citrus in California since 1915 (104). In spite of the fact that the disease has been known in the Mediterranean area since 1931 (265), studies of the disease in Mediterranean countries were started only in the 1950s (61). The disease causes severe damage to millions of citrus trees in western USA and to innumerable trees in the Mediterranean countries. The cause of the disease is a MLO (155, 188), and its identification in 1970 gave strong impetus to the study of the problem. In 1971 this MLO was cultivated "in vitro" in liquid media simultaneously in two different countries (122, 270). The cultured mycoplasma was identified as *Spiroplasma citri* (271), a helical motile prokaryote (75, 79). Koch's postulate was fulfilled by injecting cultured *S. citri* into leafhoppers. The leafhoppers transmitted the organism into white clover plants producing little leaf symptoms from which *S. citri* was recultured "in vitro" (75). The organism was transmitted by *Euscelis plebjus*, an experimental leafhopper vector (208), a few years before the natural vector of *S. citri* was discovered. It was also proved that the injection of *S. citri* into three different non-vector species led to its multiplication in their bodies (344).

The Natural Transmission of S. citri

A breakthrough in the search for the natural vector of *S. citri* occurred when the organism from field-collected *C. tenellus* was repeatedly cultured (187, 246). The final proof of a natural vector was obtained by transmitting *S. citri* with field collected *C. tenellus* (244). It is now accepted that the main vector of *S. citri* in nature is *C. tenellus*, while a few other leafhopper species in the USA are vectors of minor importance (143, 239, 242, 245). This discovery led to the idea among citrus pathologists in the Mediterranean countries that the beet leafhopper may also be a natural vector of the causal agent of stubborn disease in the region. *S. citri* was detected in seven out of 41 different species of leafhoppers collected in stubborn infected citrus groves in Morocco. Among these seven species, one *Circulifer* species alone was found: (*Circulifer*) *Neoaliturus haematoceps*, collected in Morocco, Turkey, Syria and Corsica (37, 38, 114, 236). In a survey carried out in the Mediterranean area by Frazier in 1951, the surprising discovery was made that *C. haematoceps*, and not *C. tenellus*, is the dominant leafhopper of *Circulifer* in all countries except Egypt. *C. haematoceps* was collected on a great variety of host plants, while *C. tenellus* was found in limited numbers and in areas adjacent to lakes or close to the sea. Cruciferous plants seemed to be their most important hosts (118, 120). (*C.*) *N. haematoceps* was also the main leafhopper species trapped in yellow sticky traps near citrus groves in Morocco. This is why (*C.*) *N. haemetoceps* rather than *C. tenellus* has been presumed, since about 1980, to be the natural vector of *S. citri* in the region (236). In 1986, following injection of cultured *S. citri* into leafhoppers, or AAPs of the leafhoppers, on infected periwinkle plants, the first definite proof was obtained of the vectorial capability of this leafhopper species (36, 113, 114). Periwinkle plants were infected with *S. citri* by individual (*C.*) *N. haematoceps* collected in the fields in Corsica (44). In 1985–6, during the search in Israel for the vectors of *S. citri*, a few *Circulifer* (*Neoaliturus*) specimens with different morphological aspects were collected on various host plants (170). Four populations, belonging to this collection, were kept in separate cultures. Two of the cultures comprised leafhoppers identified as the *C. tenellus* complex of species (168), one as the *C. haematoceps* complex and one as the *C. opacipennis* complex (171, 172). During experiments conducted with these four species of *Circulifer*, all the leafhoppers had their AAPs on *Matthiola incana* plants infected with *S. citri* (a red grapefruit isolate). However, only two populations of the *C. tenellus* complex were able to transmit the organism to the healthy seedlings of various plants including oranges (168, 173, 174, 264).

Strains and Isolates of S. citri

Five strains of *S. citri* have already been listed and approved by the International Committee for Systematic Bacteriology in sub-group 1–1: the Morocco R8A2T (27556) citrus strain, the California C189 (27665) orange strain, the Israel NCPPB (27565) orange strain (336), the A2-103 (33723) orange strain (1) and the Illinois HR-101 (33451) horseradish brittle root strain (260). The last *S. citri* strain does not affect horseradish in the fields of California (260) in spite of the fact that *S. citri* is very common on weeds in the field (143). In 1987, the transmission of a citrus-infecting isolate causing brittle root symptoms was documented for the first time in the laboratory (317). The Morocco strain differs slightly from the California C 189 strain in some biochemical aspects (157). Four approved strains are no longer used in transmission tests because most *S. citri* isolates lose their capability to be transmitted by the leafhopper vector after several passages through artificial media (194). Exceptionally, isolate A2-103 in England retained its pathogenicity after 170 passages (1).

Different isolates of *S. citri* are used for transmission tests. Reports from California indicate the use of isolates such as MV 101 and MH 135 originating from naturally infected periwinkle plants (148, 193), Moreno and Westside originating from naturally infected oranges (245), and Cir IB and C3B originating from field-collected *C. tenellus* leafhoppers (194, 263), etc. Same isolates may even show differences in their antigenic reaction (335). In France, a *C. haematoceps* isolate originating from Syria was used (114). Four isolates are being tested in Israel (264). An unusual, non-helical strain of *S. citri*, designated ASP-1, was used together with six other typical isolates in experiments conducted in England (333). The ASP-1 isolate exhibited erratic twitching and no rotary motion typical of all other isolates. In addition, this strange strain has a rather long log phase in culture. However, these defects have no effect on the capacity for transmission by means of injected experimental leafhopper vectors such as *Euscelis plebjus* and *Euscilidius varigatus* (333).

Strains which differed from the C189 in several aspects were obtained in California from field-grown plants infected with aster yellows (AY) or western-X (WX) MLOs. Following the separation between *S. citri* and the MLOs, transmission to test plants was of the *S. citri* type (262). *C. tenellus* was found to be an experimental vector for the WX MLO (258).

The Transmission of S. citri by C. tenellus in the USA

The relationship between *S. citri* and its vector, *C. tenellus*, and of the transmission process of the Californian isolates of *S. citri*, was investigated by Liu (193). Contrary to the BCTV, multiplication of *S. citri* in its vector was proved (193, 195). It was also found that *S. citri* multiplies in non-vector leafhoppers (332, 344) as well as in other non-vector insects (e.g. 169, 216). It seems that the titer of *S. citri* in the vector is initially reduced due to the digestion or destruction in the gut of an appreciable part of the acquired organisms, following which it dramatically increases to a maximum, usually after 15 days (193, 195). The minimum length of the LPs of *S. citri* in *C. tenellus* was found to be 10, 16 and 24 days respectively for leafhoppers injected, fed on source plants, or fed on membrane over *S. citri* solution. The minimum length of AAP was 6 h and the maximum, for maximum percentage of transmission, was 10 to 48 h. The minimum length of IAP was 2 h; maximum transmission was obtained following 48 h IAP (193).

Transmission with the brittle root (HR) isolate of *S. citri* by *C. tenellus* was more efficient, and single leafhoppers transmitted it from *Brassica rapa* plants at the rate of 64–89% (278). Similar tests, conducted with Californian isolates of *S. citri*, using *B. geniculata* plants as the source of infection, resulted in only 2–4% transmission (194, 205). The whole transmission process with respect to the HR isolate lasted less than that concerning California isolates. The minimum AAP was 45 minutes, the maximum percentage of transmission was obtained after AAPs of 12 h to 5 days. Minimum length of the LP in the vector was 7–9 days. A minimum of 5–15 min of IAP was sufficient for inoculation of test plants, while a 1- to 2-day IAP gave a maximum percentage of transmission. Males were found to be more efficient vectors than females (100).

Compared with *C. tenellus*, transmission of *S. citri* by the vector *Scaphytopius nitridus* was very poor; the latter required periods of over 5 days to acquire *S. citri* from infected plants (245). Transmission rates of the HR isolate of *S. citri* by the vector *Macrosteles fascifrons* were also very low (239). Work done to elucidate transmission of *S. citri* by the weak vector leafhopper, *M. fascifrons*, demonstrated the existence of barriers preventing transmission even from leafhoppers which had succeeded in acquiring the pathogen. The barriers seem to be located in the mesentron and postmesentron of the leafhopper (224). In spite of the fact that multiplication of *S. citri* in the vector was proved, no transovarial transmission has so far been observed in *C. tenellus* (194).

The translocation of *S. citri* within the plants has been investigated by the enzyme-linked immunosorbent assay (ELISA). In terminal shoots of side-grafted periwinkle plants, the organism was detected after 1 week; in their roots, the earliest it was detected was 14–26 days after grafting. Symptoms of infection appeared in these plants about 14 days after grafting (3). In the case of the HR isolate of *S. citri* which had been transmitted by *C. tenellus* to turnip plants, the organism was first detected in the roots 4 days after inoculation. In the young leaves *S. citri* was detected 8 days after inoculation. Very seldom, however, was the organism observed in the very old leaves. It was observed 5–9 days before the appearance of the symptoms (108). Titers of *S. citri* seemed to be much higher in turnip (108) than in grafted periwinkle plants (3).

Like MLOs, *S. citri* was found in the phloem tissue, particularly in the sieve tubes of the infected plants (155, 188, 215, 350). Only in one instance was *S. citri* reported in the nucleus of a parenchymal cell of the phloem of an infected *Chrysanthemum carinatum* plant (311).

The Transmission of S. citri in the Old World

The transmission ability of *Neoaliturus* (*Circulifer*) *haematoceps*, the suspected natural vector of *S. citri* in the Mediterranean region, has not yet been defined (114). Preliminary observations in Israel of a species of the *C. tenellus* complex conducted in the laboratory, gave LPs a minimum of 12 days in the vector. Relatively high rates of transmission were obtained with this species of leafhopper when it acquired *S. citri* from infected *Matthiola incana* plants (264).

Plants Infected by S. citri

Most of the *Citrus* species and varieties are naturally susceptible to *S. citri* infection. Soon after the discovery in 1976 that *C. tenellus* is a natural vector (244), many non-rutaceous weeds were found in the field infected with *S. citri* (243). Some of these weeds are now believed to be the pathogen reservoir for interseasonal transmission by *C. tenellus*. These plants are the vector's breeding or feeding hosts and the pathogen is acquired by the insect from them much more easily than from citrus trees (Oldfield, personal communication). In 1981, 20 plant species belonging to six botanic families, cultivated and non-cultivated, were known to be natural hosts of *S. citri* in the western USA. Other plants were experimentally infected with *S. citri* (243). Later on, a few more hosts were found (1). In mid-eastern USA, diseased horseradish plants in the field were found to be harboring *S. citri* (109, 260). The occurrence of BR disease in horseradish has been known in Illinois since 1936 (160). The pathogen was thought to be a BCTV strain, especially in view of the link existing between the disease and *C. tenellus* (329). Three brassicaceous weeds, common in the mideast, were experimentally inoculated with the BR strain of *S. citri* but were never found infected in the field (238). The HR isolate of *S. citri* have also been transmitted in the laboratory to five species of plants found naturally infected in California (107, 160).

Spiroplasma citri was also isolated from lettuce infected with the AY pathogen in fields of eastern USA (62, 176, 204, 262). *S. citri* was isolated in California from celery, aster, *Plantago* and periwinkle plants with AY symptoms. The symptoms exhibited in *Plantago* spp. plants by *S. citri* isolates transmitted by injected *Macrosteles severini* leafhoppers, were not typical of AY (262). *S. citri* was isolated from WX-infected peach and cherry trees (262). Of great interest is the discovery of *S. citri* in large numbers of cherry trees exhibiting no symptoms of infection. The isolation of *S. citri* from these trees took place in the spring, from April 30 to May 21. None of the isolates originating from these trees could be distinguished serologically from the citrus isolates of California (175). Similarly, very small numbers of helical and motile spiroplasmas were isolated from pear trees with pear decline (PD) symptoms. The spiroplasmas were isolated also from the pear psylla, the vector of PD. These isolates were serologically and culturally indistinguishable from *S. citri*. *S. citri* was isolated from pear trees from April to June and from the psylla in December (237, 261). We do not know as yet the role of infection of *S. citri* together with MLOs, and of *S. citri* alone, in "healthy" cherry trees within the context of the epidemiology of the stubborn disease of citrus. The same problem exists with an *Opuntia* isolate of *S. citri* (177, 178).

Epidemiology of the Stubborn Disease

The epidemiology of the stubborn disease in the USA is definitely associated with the behavior of the vector *C. tenellus* in the same manner as it is with that relating to the BCTV. Like BCTV, *S. citri* is disseminated in the western USA when there is a large migration of the leafhopper vector, especially after periods of drought (90, 107). In Illinois, epidemics of the BR disease of horseradish follow the same course, (107, 161, 254). Leafhoppers can carry both BCTV and *S. citri* during their long flight from the south western states to the eastern ones. Epidemics of the BR disease probably differ from those of the BCTV in Illinois because the HR isolate of *S. citri* has been identified as a local strain (317). The HR strain of *S. citri* is apparently kept for relatively long periods in perennials such as the horseradish, where it becomes a reservoir for inoculating the migrating *C. tenellus*. The strain has also been isolated from horseradish in Maryland (78). Its occurrence there is very rare and may be due either to using roots from infected sources, or to a few of the leafhoppers having reached there following migration from Illinois.

In the Old World, epidemiology of the stubborn disease is associated with more than one *Circulifer* species (35). The migratory nature of *N. (C.) haematoceps* has not yet been investigated nor has that of any other possible vector. In a disease spread over large areas of the Mediterranean a leaf-hopper such as *N. (C.) haematoceps* or a very closely related species, e.g., the *C. opacipennis* complex (35, 36), may be very important. However, another *Circulifer* species, such as one of the *C. tenellus* complex, has priority in spreading the disease in the deserts of the eastern Mediterranean and Asia (35). In confined areas, other *Circulifer* species are apparently responsible for the spread of the stubborn disease. In the Jordan Valley of Israel, an area with an acute stubborn infection, the species of *Circulifer* trapped most was a member of the *C. tenellus* complex (unpublished). In the hot, late summer, numerous adults and nymphs of another member of the *C. tenellus* complex could be captured in the coastal area of Israel on drying *Portulaca oleracea* plants (unpublished). Their role in the epidemiology of the disease in those areas is still under investigation.

Citrus trees are probably not the source of *S. citri* for the vectors in the USA and the Old World. All *Circulifer* species investigated are averse to feeding on any citrus variety (35, 44; G.N. Oldfield personal communication and personal observations in Israel). The importance of periwinkle plants in the epidemiology of stubborn disease is still an open question (39). *Circulifer* species are able to acquire *S. citri* from periwinkle plants when caged under laboratory conditions (114, 162). All *Circulifer* species tested in Israel rarely fed on periwinkle when having the choice to feed on a more suitable host (unpublished). Periwinkle plants are perennials but their lifespan, following infection by *S. citri*, is very short unless they are grown at moderate tempera-tures in the laboratory (38, 207) or are infected by a non-killing isolate of *S. citri* (unpublished). It has also been found that a mixed infection of *S. citri* and a MLO prevents the plant's death (148, 241, 247, 272), a common occurrence in the field (236). No thorough search has ever been conducted in the Old World for more suitable sources of *S. citri*, apart from periwinkle. Indeed, two cruciferous plants were once found in Israel harboring *S. citri* (63), but a regular occurrence, such as that in California (personal survey), was never observed in Israel. Undoubtedly, cruciferous plants are the source plants of the annual infection of *S. citri* in western USA. Source plants with masked infection cannot be ruled out as a source of infection in the Old World.

The Control of the Stubborn Disease

Following the discovery of Ishii et al. in 1967 that MLOs in plants are very sensitive to tetracycline antibiotics (159), numerous similar tests have been conducted on several of the spiroplasmas as well. Tetracycline treatments on spiroplasmas in the plants caused only temporary remission of symptoms (46, 156, 203, 346), but actually suppressed spiroplasma growth entirely in "in vitro" cultures (42, 271). A mere spray of tetracyclines on citrus trees proved ineffective (14) and applications of these materials inside the trees are very complicated and require special devices (345).

Stubborn disease attacks mostly young trees (47, 250). Preventive measures were considered, such as the treatment of young trees with systemic insecticides against the vector. These seemed to work in the laboratory but were ineffective in the field (163). It is assumed that, when the exact time of *S. citri* infection in nature is known, an intensive insecticide application at that particular time will reduce disease incidence.

A great many more infections occur in newly developed desert lands than in groves planted in areas with a long tradition of cultivation (35, 149). Land clearing operations under semi-desert and desert conditions and imbalanced vector-host relationships cause the leafhoppers to search for new green plants for its survival and they therefore feed on young citrus trees and inoculate them. Surrounding grove sections and citrus nurseries with trap plants favored by *C. tenellus* (i.e., sugar beets in the USA, or *Matthiola* plants in the Old World) may deflect the leafhoppers' attention from the young citrus trees. This may reduce infection, particularly in nurseries (149).

Reduction of stubborn infection in periwinkle plants in the USA was achieved in preliminary tests using whitewash treatment with kaolinite, montmorillonite or hydrated lime (348). It is assumed that, in nurseries, nets will protect citrus trees from leafhopper landing and infection in their first stage of growth.

The Beet Leafhopper-Transmitted Virescence Disease

The beet leafhopper-transmitted virescence disease was first discovered in 1974 in the laboratory when periwinkle seedlings were exposed to *C. tenellus* collected in the field in several California counties (247). Eight years later, the disease was discovered in Illinois (98). The damage caused by the disease to crops (radish in Idaho (300) and horseradish in Illinois) is not serious (98). Infected periwinkles were kept in the greenhouse for comparatively long periods without exhibiting wilt symptoms, as in the case of periwinkle infected by *S. citri* (247). BCTV was masked in several periwinkles showing virescence symptoms. Transfer of *C. tenellus* from these plants to sugarbeets resulted in BCTV infection alone of many of the test plants. MLOs were detected by an electron microscope in phloem cells of infected periwinkle (247). Most of the research on this MLO has been carried out by Oldfield (240) and Golino (141).

Transmission Characteristics of the Virescence Agent

The AAP of the beet leafhopper transmitted virescence agent (BLTVA) is comparatively short: 5 min was sufficient to initiate infection and after 1 h more than 20% of the insects had become inoculative. Maximum transmission (90%) was obtained by AAPs of 24–48 h. The minimum LP of the BLTVA in the leafhopper was 12 days. Maximum transmissions were obtained after 26–27 days from AAP an source plants. An IAP of 5 min was usually long enough to infect the plants. The highest rates of transmission were obtained by insects that had 2 days of IAP on test plants (144). The capacity to transmit the BLTVA is very high in comparison with that of a Californian isolate of *S. citri* and the BCTV (>60%, 2–4% and 33%, respectively) (144, 194, 282). The Illinois BLTVA is also transmitted very effectively from horseradish by *C. tenellus* (99).

Experimental Hosts for the BLTVA

In 1989, 43 of 60 plant species from 20 botanic families became infected with BLTVA agent, after exposure to inoculative *C. tenellus* in the laboratory. None of the monocots that were tested showed disease symptoms, and back-passages from these plants were not successful (144, 145). The BLTVA was also transmitted by means of *C. tenellus* to *Arabidopsis thaliana* plants. These are called botanical *Drosophila*, as the species is known for its many classified mutants available for research. The plant can also be used as a model system for MLO infection (146). The BLTVA was transmitted in Illinois to a few additional hosts (99). A polyclonal antiserum for this MLO is now available in California (142) and can be used in ELISA for the study of the relationships between the BLTVA and other MLOs and for the detection of the organism in suspected diseased plants and infected insects.

Safflower Phyllody and the Vector Neoaliturus Fenestratus

Since the early 1960s, a severe disease causing bushy growth with short internodes and small leaves has been observed in summer in experimental fields of safflower (*Carthamus tinctorius*) in Israel (167). The disease has been noted also in other uncultivated *Carthamus* species growing all over Israel in summer (unpublished). The cicadellid *Neoaliturus fenestratus* was found to transmit the agent of the disease both in nature and in the laboratory. Electron microscopy of infected material revealed MLOs in phloem sieve cells (167). *N. fenestratus* had been known earlier as a *Circulifer* species *C. fenestratus* (Herrich-Schaffer) (249).

Characteristics of the Transmission of the Phyllody Agent

The AAPs for the transmission of the safflower phyllody agent (SPA) have not been studied. Its LP in the vector was found to be 20–25 days. The vector transmitted SPA almost until it died. SPA was transmitted fairly efficiently, and 50–60% of the leafhoppers, which had been exposed to infected sources, became inoculative. The SPA was also found to shorten the vector's life and lower its rate of reproduction (259).

Plant Hosts for N. fenestratus

This leafhopper is very common in the Mediterranean region and Europe (H. Gunthert, personal communication), occurring as it does on several host species, mainly of the Compositae. However, it can be found also on species of other botanical families (118, 192).

Experimentally Infected Plants

Experiments carried out with SPA-inoculative leafhoppers showed that at least four more species of the Compositae are sensitive to infection. In addition, transmission of the SPA to periwinkle plants was observed. Ten other species, some of which were Compositae, produced no symptoms after exposure to inoculative *N. fenestratus*. Backpassage from these plants to safflower seedlings was negative (167). The list of plants infected by the SPA in Israel is shorter than that of plants infected by the BLTVA in California (145). The link between these pathogens is not known. No species of *Circulifer/Neoaliturus* other than *N. fenestratus* have been tested in Israel as additional vectors of SPA (personal knowledge). Analysis and comparison of the relationship between the BLTVA MLO in the US and MLOs causing phyllody in safflower in Israel and phyllody in periwinkle in the Mediterranean are necessary, as all these appear to be transmitted by leafhopper members of the same or close genera.

Other Suspected Leafhopper-Transmitted Pathogens in the Middle East

In 1984, a new plant pathogenic spiroplasma, *S. phoeniceum*, was found infecting periwinkles of a Syrian *S. citri* infection zone (274). *S. phoeniceum* differed from *S. citri* by having only 60% homology in the DNA (273, 336). The vector of this new spiroplasma is still unknown and, quite probably, plants become infected via leafhoppers. Periwinkle trap plants were placed in various locations of citrus growing areas along the Mediterranean coast of Syria in order to detect *S. citri* infection. Results indicate infection by *S. citri*, *S. phoeniceum* or phyllody alone, or by a combination of any two of these organisms (34). Quite probably, one or more members of the *Neoaliturus/Circulifer* species are the vectors of *S. phoeniceum* and of the phyllody in periwinkle, as leafhoppers of these genera are known to move around a lot in citrus areas and to transmit disease agents (34, 98, 167, 170, 247). However, there is no evidence of this at the present.

Concluding Remarks

This review was prepared according to our knowledge of the various diseases. The beet curly top disease has been investigated for almost 90 years and the information gathered about it and the vector is vast. The control program against the vector of the BCTV is still being carried out in some of the western states of the USA, and resistant varieties in sensitive areas are still being recommended. However, the interest of virologists and entomologists in the disease has slackened in recent years. Of late, there has been a slight revival of interest in the molecular level of the virus.

The stubborn disease is more problematic than the curly top disease, as it occurs in trees and not in annual crops such as sugar beets and tomatoes. No simple diagnostic tools are available to the growers for its control. In addition, very little progress, if any, has been made in producing resistant varieties of citrus. In the Old World, more than one vector exists, which complicates the research on taxonomic problems as well.

Information on the other diseases described in this chapter is scarce and the diseases' importance is doubtful, except on a purely scientific level. Much remains to be learned concerning aspects of MLOs and *S. citri* transmission by means of *Circulifer/Neoaliturus* spp. in nature (142).

Summary

1. Description of the epidemiology of the BCTV disease in the United States, particularly in view of the possible influence on it of the ecology of the vector *Circulifer tenellus*.
2. Ways to reduce the disease spread by *Circulifer* spp.
3. Discussion of the problems concerning the spread of the stubborn disease of citrus in the USA and in the Middle East.
4. *Circulifer* species are natural vectors of some MLOs in the USA and in the Old World. It would be interesting to know if the diseases are genetically close to one another. New tools have been developed for investigating the phenomenon.
5. Taxonomic problems are still unresolved in the Old World. It is suggested that future research involve more sophisticated techniques.

Acknowledgments. I thank Rene Modians for his help in preparing this review. I am also very grateful to the editor for his patience and understanding in view of the delays incurred in the submission of this paper owing to the Gulf crisis.

References

1. Allen, R.M., and Donndelinger, C.R., 1982, Pathogenicity proved for isolates of *Spiroplasma citri* from six host species, *Phytopathology* **72**:1004 (Abstr.).
2. Annad, P.N., 1931, Beet leafhopper's annual migrations studied in desert breeding areas, *U.S. Dept. Agric. Yearb. Agric.* **1931**:114–116.
3. Archer, D.B., Townsend, R., and Markham, P.G., 1982, Detection of *Spiroplasma citri* in plants and insect hosts by ELISA, *Plant Pathol.* **31**:299–306.

4. Armitage, H.M., 1952, Controlling curly top virus in agricultural crops by reducing populations of overwintering beet leafhoppers, *J. Econ. Entomol.* **45**:432–435.

5. Armitage, H.M., 1957, Report on sugarbeet leafhopper-curly top virus control in California, *Calif. Dept. Agric. Bull.* **46**, 8 pp.

6. Baker, C.F., 1896, New Homoptera from New Mexico Agricultural Experiment Station, *Psyche* **7** Suppl. 1:24–26.

7. Ball, E.D., 1905, The beet leafhopper (*Eutettix tenella*), *Utah Agric. Exp. Stn. Ann. Rep. (1904–1905)* **16**, 24 pp.

8. Ball, E.D., 1907, The genus *Eutettix, Proc. Davenport Acad. Sci.* **12**:27–94.

9. Ball, E.D., 1909, The leafhoppers of sugar beet and their relation to the "curly leaf" condition, *U.S. Dept. Agric. Bur. Entomol. Bull.* **66**:33–52.

10. Ball, E.D., 1917, The beet leafhopper and the curly leaf disease that it transmits, *Utah Agric. Exp. Stn. Bull.* **155**:1–56.

11. Ballantyne. B., 1969, Transmisison of summer death of beans, *Aust. J. Sci.* **31**:433–434.

12. Ballantyne, B., 1970, Field reactions of bean varieties to summer death in 1970, *Plant Dis. Rep.* **54**:903–905.

13. Banttari, E.E., and Moore, M.B., 1962, Virus cause of blue dwarf of oats and its transmission to barley and flax, *Phytopathology* **52**:897–902.

14. Bar-Joseph, M., and Raccah, B., 1973, Little leaf (stubborn): identification of the causal agent and ideas for of control, *Hassadeh* (in Hebrew) **54**:1007–1010.

15. Benda, G.T.A., and Bennett, C.W., 1964, Effect of curly top virus on tobacco seedlings: infection without obvious symptoms, *Virology* **24**:97–101.

16. Benda, G.T.A., and Bennett, C.W., 1967, Recovery from curly-top disease in tomato seedlings, *Am. J. Bot.* **54**:1140–1142.

17. Bennett, C.W., 1935, Studies on properties of the curly top virus, *J. Agric. Res.* **50**:211–214.

18. Bennett. C.W., 1949. The Brazilian curly top of tomato and tobacco resembling North America and Argentine curly top of sugar beet, *J. Agric. Res.* **78**:675–693.

19. Bennett, C.W., 1957, Interactions of sugar-beet curly top virus and an unusual mutant, *Virology* **3**:322–342.

20. Bennett, C.W., 1957, Influence of different combinations of tops and roots of susceptible and resistant varieties of sugar beet on curly top symptoms and virus concentrations, *J. Soc. Sugar Beet Technol.* **9**:553–565.

21. Bennett, C.W., 1962, Curly top virus content of the beet leafhopper influenced by virus concentration in diseased plants, *Phytopathology* **52**:538–541.

22. Bennett, C.W., 1962, Acquisition and transmission of curly top virus by artificially fed beet leafhopper, *J. Soc. Sugar Beet Technol.* **11**:637–648.

23. Bennett, C.W., 1963, Highly virulent strains of curly top virus in western United States, *J. Soc. Sugar Beet Technol.* **12**:515–520.

24. Bennett, C.W., 1967, Apparent absence of cross-protection between strains of the curly top virus in the beet leafhopper, *Circulifer tenellus, Phytopathology* **57**:207–209.

25. Bennett, C.W., 1971, The curly top disease of sugar beet and other plants, Monograph # 7. The American Phytopathological Society, St. Paul, Minnesota, 81 pp.

26. Bennett, C.W., Carsner, E., Coons, G.H., and Brandes, E.W., 1946, The Argentine curly top of sugarbeet, *J. Agric. Res.* **72**:19–48.

27. Bennett, C.W., and Esau, K., 1936, Further studies on the relation of the curly top virus to plant tissues, *J. Agric. Res.* **54**:479–502.

28. Bennett, C.W., and Tanrisever, A., 1958, Curly top disease in Turkey and its relationship to curly top in North America, *J. Soc. Sugar Beet Technol.* **10**:189–211.

29. Bennett, C.W., and Wallace, H.E., 1938, Relation of the curly top virus to the vector, *Eutettix tenellus, J. Agric. Res.* **56**:31–51.

30. Bienz, D.R., and Thornton, R.E., 1975, Growers Guide for Vegetable Crops: Peppers, Washington State Agric. Ext. Multilith 2946.

31. Blickenstaff, C.C., and Traveller, D., 1979, Factors affecting curly top damage to sugar beets and beans in southern Idaho, 1919–77, *USDA, SEA, Agric. Rev. Manuals* 22 pp.

32. Blood, H.L., 1939, Searching for curly top resistant tomatoes in Utah and South America, *Utah State Agric. Coll. Ext. Circ. NS* **98**:21–23.

33. Boncquet, P.A., 1923, Discovery of curly leaf of sugar beets in the Argentine Republic, *Phytopathology* **13**:458–460.

34. Bove, J.M., 1986. Stubborn and its natural transmission in the Mediterranean area and the Near East, *FAO Plant Prot. Bull.* **34**:15–23.

35. Bové, J.M., Fos, A., Lallemand, J., Raie, A., Ali, Y., Ahmed, N., Saillard, C., and Vignault, J.C., 1988, Epidemiology of *Spiroplasma citri* in the Old World, *Proc. Tenth IOVC Conf.*, University of California, Riverside, pp. 295–299.

36. Bové, J.M., Fos, A., Lallemand, J., Raie, A., Saillard, C., and Vignault, J.C., 1988, Epidemiology of *Spiroplasma citri* in the Old World, *Tenth IOVC Conf.*, University of California, Riverside, p. 98 (Abstr.).

37. Bové, J.M., Moutous, G., Saillard, C., Fos, A., Bonfils, J., Vignault, J.C., Nhami, A., Abassi, M., Kabbage, K., Hafidi, B., Mouches, C., and Viennot-Bourgin, G., 1979, Mise en évidence de *Spiroplasma citri* l'agent causal de la maladie du "stubborn" des agrumes dans 7 cicadelles du Maroc, *C.R. Acad. Sci.* Paris, **288**:335–338.

38. Bové, J.M., Nhami, A., Saillard, C., Vignault, J.C., Mouches, C., Garnier, M., Moutous, G., Fos, A., Bonfils, J., Abassi, M., Kabbage, K., Hafidi, B., and Viennot-Bourgin, G., 1979, Présence au Maroc de *Spiroplasma citri* l'agent causal de la maladie du "stubborn" des agrumes, dans des Pervenches (*Vinca rosea* L.) implantées en bordure d'orangeraies malades, et contamination probable du Chiendent (*Cynodon dactylon* (L.) Pers.) par le spiroplasme, *C.R. Acad. Sci.* Paris, **288**:399–402.

39. Bové, J.M., Vignault, J.C., Garnier, M., Saillard, C., Garcia-Jurado, 0., Bove, C., and Nhami, A., 1978, Mise en évidence de *Spiroplasma citri*, l'agent causal de la maladie du "stubborn" des agrumes, dans des Pervenches (*Vinca rosea* L.) ornementales de la ville de Rabat, Maroc, *C.R. Acad. Sci.* Paris, **286**:57–60.

40. Bowyer, J.W., 1974, Tomato big bug, legume little leaf and lucerne witches' broom: three diseases associated with different mycoplasma-like organisms in Australia, *Aust. J. Agric. Sci.* **25**:449–457.

41. Bowyer, J.W., and Atherton, J.G., 1971, Summer death of French bean: new hosts of the pathogen, vector relationship, and evidence against mycoplasmal etiology, *Phytopathology* **61**:1451–1455.

42. Bowyer, J.W., and Calavan, E.C., 1974, Antibiotic sensitivity "in vitro" of the mycoplasma-like organism associated with citrus stubborn disease, *Phytopathology* **64**:346–349.

43. Briddon, R.W., Watts, J., Markham, P.G., and Stanley, J., 1989, The coat protein of beet curly top virus is essential for infectivity, *Virology* **172**:628–633.

44. Brun, P., Riolacci, S., Vogel, R., Fos, A., Vignault, J.C., Lallemand, J., and Bové, J.M., 1988, Epidemiology of *Spiroplasma citri* in Corsica, *Proc. 10th IOCV Conf.*, University of California, Riverside, pp. 300–303.

45. Burke, D.W., 1964, Time of planting in relation to disease incidence and yields of beans in central Washington. *Plant Dis. Rep.* **48**:7879–793.

46. Calavan, E.C., 1975, The control of greening and stubborn, two mycoplasma-like diseases in citrus, in Raychaudhuri, S.P., Varma, A., Bhargava, K.S., and Mehrotra, B.S. (eds): Advances in Mycology and Plant Pathology, New Delhi, India, pp. 325–332.

47. Calavan, E.C., Kaloostian, G.H., Oldfield, G.N., Nauer, E.M., and Gumpf, D.J., 1979, Natural spread of *Spiroplasma citri* by insect vectors and its implications for control of stubborn disease of citrus, *Proc. Int. Soc. Citriculture, 1979*, pp. 900–902.

48. Cannon, O.S., 1960, Curly top in tomatoes, *Utah Agric. Exp. Stn. Bull.* **424**, 12 pp.

49. Carsner, E., 1919, Susceptibility of various plants to curly-top of sugarbeet, *Phytopathology* **9**:413–421.

50. Carsner, E., 1925, A bean disease caused by the virus of sugarbeet curly-top, *Phytopathology* **15**:731–732 (Abstr.).

51. Carsner, E., 1926, Seasonal and regional variations in curly-top of sugarbeets, *Science* **62**:213–214.

52. Carsner, E., 1926, Resistance in sugar beets to curly-top, *U.S. Dept. Agric. Circ.* **388**, 7 pp.

53. Carsner, E., 1926, Susceptibility of the bean to the virus of sugar beet curly-top, *J. Agric. Res.* **33**:345–348.

54. Carsner, E., 1933, Curly-top resistance in sugar beets and test of the resistant variety U.S. No. 1, *U.S. Dept. Agric. Tech. Bull.* **360**, 68 pp.

55. Carsner, E., and Stahl. C.F., 1924, Studies on curly-top disease of the sugarbeet, *J. Agric. Res.* **28**:297–320.

56. Carter, W., 1927, A technic for use with homopterous vectors of plant disease with special reference to the sugar beet leafhopper, *Eutettix tenellus* (Baker), *J. Agric. Res.* **34**:449–451.

57. Carter, W., 1927, Extensions of the known range of Eutettix tenellus Baker and curly-top of sugarbeets, *J. Econ. Entomol.* **20**:714–717.

58. Carter, W., 1928, Transmission of the virus of curly-top of sugarbeets through different solutions, *Phytopathology* **18**:675–679.

59. Carter, W., 1929, Ecological studies of curly-top of sugarbeets, *Phytopathology* **19**:467–477.

60. Carter, W., 1930, Ecological studies of the beet leaf hopper, *U.S. Dept. Agric. Tech. Bull.* **206**, 115 pp.

61. Chapot, H., 1957, First studies on the stubborn disease of citrus in some Mediterranean countries, *Proc. 1st IOCV Conf.*, University of California, Riverside, pp. 109–117.

62. Christiansen, C., Freundt, E.A., and Maramorosch, K., 1980, Identity of cactus and lettuce spiroplasmas with *Spiroplasma citri* as determined by DNA-DNA hybridization, *Curr. Microbiol.* **4**:353–356.

63. Clark, M.F., Flegg, C.L., Bar-Joseph, M., and Rottem, S., 1978, The detection of

Spiroplasma citri by enzyme-linked immunosorbent assay (ELISA), *Phytopath. Z.* **92**:332–337.

64. Cook, W.C., 1933, Spraying for control of the beet leafhopper in central California in 1931, *Calif. Dept. Agric. Monthly Bull.* **22**:138–141.

65. Cook, W.C., 1934, Spraying wild host plants in California reduces beet leafhopper injury, *U.S. Dept. Agric. Yearb. Agric.* **1934**:332–334.

66. Cook, W.C., 1943. Evaluation of a field-control program directed against the beet leafhopper, *J. Econ. Entomol.* **36**:382–385.

67. Cook, W.C., 1945, The relation of spring movements of the beet leafhopper (*Eutettix tenellus* Baker) in central California to temperature accumulation, *Ann. Entomol. Soc. Am.* **38**:149–162.

68. Cook, W.C., 1946, The ability of the beet leafhopper to regain lost vitality when transferred from unfavorable to favorable host plants, *Ecology* **27**:37–46.

69. Cook, W.C., 1967, Life history, host plants, and migrations of the beet leafhopper in western United States, *U.S. Dept. Agric. Tech. Bull.* **1365**, 122 pp.

70. Costa, A.S., 1952, Further studies on tomato curly-top in Brazil, *Phytopathology* **42**:396–403.

71. Coudriet, D.L., and Tuttle, J.M., 1963, Seasonal flight of insect vectors of several plant viruses in southern Arizona, *J. Econ. Entomol.* **56**:865–868.

72. Crawford, R.F., 1927, Curly-top in New Mexico, *U.S. Dept. Agric. Off. Rec.* **6**:8.

73. Dana, B.F., 1934, Progress in investigations of curly-top of vegetables, *Ann. Rep. Oregon St. Hort. Soc.* **26**:95–99.

74. Dana, B.F., 1938, Resistance and suceptibility to curly-top in varieties of squash, *Cucurbita maxima*, *Phytopathology* **28**:649–656.

75. Daniels, M.J., Markham, P.G., Meddins, B.M., Plaskitt, A.K., Townsend, R., and Bar-Joseph, M., 1973, Axenic culture of a plant pathogenic spiroplasma, *Nature (London)*, **244**:523–524.

76. Davis, M.J., Tsai, J.H., Cox, R.L., McDaniel, L.L., and Harrison, N.A., 1987, DNA probes for detecting the maize-bushy-stunt mycoplasma-like organism (MBA-MLO), *Phytopathology* **77**:1769 (Abstr.).

77. Davis, M.J., Tsai, J.H., Cox, R.L., McDaniel, L.L., and Harrison, N.A., 1988, Cloning of chromosomal and extra chromosomal DNA of mycoplasma-like organisms that cause maize bushy stunt disease, *Mol. Plant Microbe Interact.* **1**:295–302.

78. Davis, R.E., and Fletcher, J., 1983, *Spiroplasma citri* in Maryland: isolation from field-grown plants of horseradish (*Armoracia rusticana*) with brittle root symptoms. *Plant Dis.* **67**:900–903.

79. Davis, R.E., and Worley, J.F., 1973, Spiroplasma: motile, helical microorganism associated with corn stunt disease, *Phytopathology* **63**:403–408.

80. Decker, G.C., 1953, Beet leafhopper (*Circulifer tenellus*), *U.S. Bur. Entomol. Plant Quart., Coop. Econ. Insect Rep.* Illinois, 3 (37):674.

81. DeLong, D.M., 1925, The occurrence of the beet leafhopper, *Eutettix tenella* Baker, in the eastern U.S., *J. Econ. Entomol.* **18**:637–638.

82. DeLong, D.M., and Kadow, K.J., 1937, Sugarbeet leafhopper, *Eutettix tenellus* Baker, appears in Illinois, *J. Econ. Entomol.* **30**:210.

83. Dorst, H.E., and Davis, E.W., 1937, Tracing long-distance movements of beet leafhopper in the desert, *J. Econ. Entomol.* **30**:948–954.

84. Douglass, J.R., 1954. Outbreak of beet leafhoppers north and east of the permanent breeding areas, *Proc. Am. Soc. Sugar Beet Technol.* **8**:185–193.

85. Douglass, J.R., and Cook, W.C., 1952, The beet leafhopper, *U.S. Dept. Agric. Yearb. Agric.* **1952**:545–550.

86. Douglass, J.R., and Cook, W.C., 1954, The beet leafhopper, *U.S. Dept. Agric. Circ.* **942**, 21 pp.

87. Douglass, J.R., and Hallock, H.C., 1957, Relative importance of various host plants of the beet leafhopper in southern Idaho, *U.S. Dept. Agric. Tech. Bull.* **1155**, 11 pp.

88. Douglass, J.R., Hallock, H.C., Fox, D.E., and Hofmaster, R.N., 1946, Movement of spring-generation beet leafhoppers into beet fields of south-central Idaho, *Proc. Am. Soc. Sugar Beet Technol.* **4**:289–297.

89. Douglass, J.R., Hallock, H.C., and Peay, W.E., 1944, A new weed host of the beet leafhopper, *J. Econ. Entomol.* **37**:714–715.

90. Douglass, J.R., Peay, W.E., and Cowger, J.I., 1956, Beet leafhopper and curly-top conditions in the southern Great Plain and adjacent areas, *J. Econ. Entomol.* **49**:95–99.

91. Douglass, J.R., Romney, V.E., and Hallock, H.C., 1950, Survival of the beet leafhopper in southern Idaho during the severe winter of 1948–49, *Proc. Am. Soc. Sugar Beet Technol.* **6**:494–498.

92. Douglass, J.R., Romney, V.E., and Jones, E.W., 1955, Beet leafhopper control in weed-host areas of Idaho to protect snapbean seed from curly-top, *U.S. Dept. Agric. Circ.* **960**, 13 pp.

93. Douglass, J.R., Wakeland, C., and Gillett, J.A., 1939, Field experiments for the control of the beet leafhopper in Idaho, 1936–37, *J. Econ. Entomol.* **32**:69–78.

94. Duffus, J.E., 1983, Epidemiology and control of curly top disease of sugar beet and other crops, in Plumb, R.T., and Thresh, J.M. (eds): Plant Virus Epidemiology, Blackwell Scientific Publications, Oxford, England, pp. 297–304.

95. Duffus, J.E., and Gold, A.H., 1973, Infectivity neutralization used in serological tests with partially purified beet curly top virus, *Phytopathology* **63**: 1107–1110.

96. Duffus, J.E., Milbrath, G.M., and Perry, R., 1982, Unique type of curly-top virus and its relationship with horseradish brittle root, *Plant Dis.* **66**:650–652.

97. Duffus, J.E., and Skoyen, I.O., 1977, Relationship of age of plants and resistance to a severe isolate of the beet curly-top virus, *Phytopathology* **67**:151–154.

98. Eastman, C., Schultz, G., Fletcher, J., and McGuire, M., 1982, Presence in horseradish of a virescence agent distinct from the causal agent of brittle root, *Phytopathology* **72**:1005 (Abstr.).

99. Eastman, C.E., Schultz, G.A., Fletcher, J., Hemmati, K., and Oldfield, G.N., 1984, Virescence of horseradish in Illinois, *Plant Dis.* **68**:968–971.

100. Eastman, C.E., Schultz, G.A., McGuire, M.R., Post, S.L., and Fletcher, J., 1988, Characteristics of transmission of a horseradish brittle root isolate of Spiroplasma citri by the beet leafhopper, *Circulifer tenellus* (Homoptera: Cicadellidae), *J. Econ. Entomol.* **81**:172–177.

101. Egbert, L.N., Egbert, L.D., and Mumford, D.L., 1976, Physical characteristics of sugarbeet curly-top virus, *Ann. Meet. Am. Soc. Microbiol.*, Atlantic City, New Jersey, (Abstr.).

102. Esau, K., 1977, Virus-like particles in nuclei of phloem cells in spinach leaves infected with the curly-top virus, *J. Ultrastruct. Res.* **61**:78–88.

103. Esau, K., and Hoefert, L.L., 1973, Particles and associated inclusions in sugarbeet infected with the curly-top virus, *Virology* **55**:454–464.

104. Fawcett, H.S., Perry, J.C., and Johnston, J.C., 1944, The stubborn disease of Citrus, *Calif. Citrogr.* **29**:146–147.
105. Finkner, R.E., 1976, Cultivar blends for buffering against curly top and leafspot diseases of sugarbeet, *J. Am. Soc. Sugar Beet Technol.* **19**:74–82.
106. Finkner, R.E., and Scott, P.R., 1972, Sugarbeet cultivar and systemic insecticide interrelationships in the control of curly top virus, *J. Am. Soc. Sugar Beet Technol.* **17**:97–104.
107. Fletcher, J., 1983, Brittle root of horseradish in Illinois and the distribution of *Spiroplasma citri* in the United States, *Phytopathology* **73**:354–3577.
108. Fletcher, J., and Eastman, C.E., 1984, Translocation and multiplication of *Spiroplasma citri* in turnip (*Brassica rapa*), *Curr. Microbiol.* **11**:289–292.
109. Fletcher, J., Schultz, G.A., Davis, R.E., Eastman, C.E., and Goodman, R.H., 1981, Brittle root disease of horseradish: evidence for an etiological role of *Spiroplasma citri*, *Phytopathology* **71**:1073–1080.
110. Flock, R.A., and Deal, A.S., 1959, A survey of beet leafhopper populations on sugar beets in the Imperial Valley, California, 1953–1958, *J. Econ. Entomol.* **52**:470–473.
111. Forbes, S.A., and Hart, C.A., 1900, The economic entomology of sugarbeet, *Ill. Agric. Exp. Stn. Bull.* **60**:397–532.
112. Ford, H.P., Campbell, J.E., Osgood, J.W., and Deal, A.S., 1966, Control weeds to reduce curly-top, *Univ. Calif. Agric. Ext. Serv., Imperial County Circ.* 110, 20 pp.
113. Fos, A., Lallemand, J., and Bové, J.M., 1988, Experimental transmission of *Spiroplasma citri* to periwinkle plants by the leafhopper *Neoaliturus haematoceps*, *Seventh Int. Congr.* IOM, Baden, Austria, p. 182 (Abstr.).
114. Fos, A., Bové, J.M., Lallemand, J., Saillard, C., Vignault, J.C., Ali, Y., Brun, P., and Vogel, R., 1986, La cicadelle *Neoaliturus haematoceps* (Mulsant & Rey) est vecteur de *Spiroplasma citri* en Méditerranée, *Ann. Microbiol.* (*Paris*), **137A**:97–107.
115. Fulton, R.A., and Romney, Van E., 1940, The chloroform-soluble components of beet leafhopper as an indication of the distance they move in the spring, *J. Agric. Res.* **61**:737–743.
116. Fox, D.E., 1938, Occurrence of the beet leafhopper and associated insects on secondary plant successions in southern Idaho, *U.S. Dept. Agric. Tech. Bull.* **607**, 43 pp.
117. Fox, D.E., Chamberlin, J.C., and Douglass, J.R., 1945, Factors affecting curly-top damage to sugarbeets in southern Idaho, *U.S. Dept. Agric. Tech. Bull.* **897**, 29 pp.
118. Frazier, N.W., 1953, A survey of the Mediterranean region for the beet leafhopper, *J. Econ. Entomol.* **46**:551–554.
119. Freitag, J.H., 1936, Negative evidence on multiplication of curly-top virus in the beet leafhopper, *Eutettix tenellus*, *Hilgardia* **10**:305–342.
120. Freitag, J.H., Frazier, N.W., and Huffaker, C.B., 1955, Crossbreeding beet leafhoppers from California and French Morocco, *J. Econ. Entomol.* **48**:341–342.
121. Freitag, J.H., and Severin, H.H.P., 1936, Ornamental flowering plants experimentally infected with curly-top, *Hilgardia* **10**:263–302.
122. Fudl-Allah, A.E.-S.A., Calavan, E.C., and Igwegbe, E.C.K., 1971, Culture of a mycoplasma-like organism associated with stubborn disease of citrus, *Phytopathology* **61**:1321 (Abstr.).

123. Fulton, R.W., 1955, Curly-top of tobacco in Wisconsin, *Plant Dis. Rep.* **39**:799–800.

124. Gardner, D.E., and Cannon, O.S., 1972, Curly-top viruliferous and nonviruliferous leafhopper feeding effects upon tomato seedlings, *Phytopathology* **62**:183–186.

125. Georghiou, G.P., Laird, E.F. Jr., and Van Maren, A.F., 1964, Systemic insecticides reduce the spread of curly-top virus of sugarbeets, *Calif. Agric.* **18**(6):12–14.

126. Gibson, K.E., 1971, The incidence of curly-top virus on sugar beets in Iran—1966 to 1969, *Plant Dis. Rep.* **55**:85–86.

127. Gibson, K.E., and Oliver, W.N., 1970, Comparison of nymphal and adult beet leafhoppers as vectors of the virus of curly-top, *J. Econ. Entomol.* **63**:1321.

128. Giddings, N.J., 1938, Studies of selected strains of curly-top virus, *J. Agric. Res.* **56**:883–894.

129. Giddings, N.J., 1942, Age of plants as a factor in resistance to curly-top of sugarbeets, *Proc. Am. Soc. Sugar Beet Technol.* **3**:452–459.

130. Giddings, N.J., 1944, Additional strains of the sugar beet curly-top virus, *J. Agric. Res.* **69**:149–157.

131. Giddings, N.J., 1947, Some studies of curly top of flax, *Phytopathology* **37**:844 (Abstr.).

132. Giddings, N.J., 1950, Some interrelationships of virus strains in sugarbeet curly-top, *Phytopathology* **40**:377–388.

133. Giddings, N.J., 1950, Combination and separation of curly-top virus strains, *Proc. Am. Soc. Sugar Beet Technol.* **6**:502–507.

134. Giddings, N.J., 1954, Two recently isolated strains of curly-top virus, *Phytopathology* **44**:123–125.

135. Giddings, N.J., 1954, Some studies of curly-top on potatoes, *Phytopathology* **44**:125–128.

136. Giddings, N.J., 1954, Curly-top moves East, *Proc. Am. Soc. Sugar Beet Technol.* **8**:194–196.

137. Giddings, N.J., 1954, Relative curly-top resistance of sugar beet varieties in the seedling stage, *Proc. Am. Soc. Sugar Beet Technol.* **8**:197–200.

138. Giddings, N.J., 1959, The stability of sugar beet curly-top virus strains, *J. Am. Soc. Sugar Beet Technol.* **10**:359–363.

139. Gillette, C.P., and Baker, C.F., 1895, A preliminary list of the Hemiptera of Colorado, *Colorado Agric. Exp. Stn. Bull.* **3**:100.

140. Gold, A.H., 1969, Mechanism of curly top resistance in beans, *Phytopathology* **59**:1028 (Abstr.).

141. Golino, D.A., 1987, Characterization of the beet leafhopper transmitted virescence agent, a mycoplasma like organism, Ph.D. Dissertation, University of California, Riverside, 98 pp.

142. Golino, D.A., Kirkpatrick, B.C., and Fisher, G.A., 1989, The production of a polyclonal antisera to the beet leafhopper transmitted virescence agent, *Phytopathology* **79**:1138 (Abstr.).

143. Golino, D.A., and Oldfield, G.N., 1989, Plant pathogenic spiroplasmas and their leafhopper vectors, in Harris, K.F. (ed): Advances in Disease Vector Research, Volume 6, Springer-Verlag, New York, pp. 267–299.

144. Golino, D.A., Oldfield, G.N., and Gumpf, D.J., 1987, Transmission characteristics of the beet leafhopper transmitted virescence agent, *Phytopathology* **77**:954–957.

145. Golino, D.A., Oldfield, G.N., and Gumpf, D.J., 1989, Experimental hosts of the beet leafhopper transmitted virescence agent, *Plant Dis.* **73**:850–854.
146. Golino, D.A., Shaw, M., and Rappaport, L., 1988, Infection of *Arabidopsis thaliana* (L.) Heynh. with a mycoplasma-like organism, the beet leafhopper transmitted virescence agent, *Arabidopsis Inf. Serv.* **26**:9–14.
147. Goroder, N.K.N., 1990, Flight behavior and physiology of the beet leafhopper *Circulifer tenellus* (Baker) (Homoptera: Cicadellidae) in California, Ph.D. Thesis, University of California, Berkeley, pp. 1–162.
148. Granett, A.L., Blue, R.C., Harjung, M.K., Calavan, E.C., and Gumpf, D.J., 1976, Occurrence of *Spiroplasma citri* in periwinkle in California, *Calif. Agric.* **30**(3):18–19.
149. Gumpf, D.J., and Calavan, E.E., 1981, Stubborn disease of citrus, in Maramorosch, K., and Raychaudhuri, S.P. (eds): Mycoplasma Diseases of Trees and Shrubs, Academic Press, New York, pp. 97–134.
150. Hallock, H.C., 1946, Beet leafhopper selection of bean varieties and its relation to curly-top, *J. Econ. Entomol.* **39**:319–325.
151. Hallock, H.C., and Douglass, J.R., 1956, Studies of four summer hosts of the beet leafhopper, *J. Econ. Entomol.* **49**:388–391.
152. Hills, O.A., 1937, The beet leafhopper in the Central Columbia River breeding area, *J. Agric. Res.* **55**:21–31.
153. Hills, O.A., and Brubaker, R.W., 1968, Comparison of the effect of beet seed production of spring and fall infestations of beet leafhopper carrying curly top virus, *J. Am. Soc. Sugar Beet Technol.* **15**:215–220.
154. Hills, O.A., Coudriet, D.L., and Brubaker, R.W., 1964, Phorate treatments against beet leafhopper on cantaloups for prevention of curly top, *J. Econ. Entomol.* **57**:85–89.
155. Igwegbe, E.C.K., and Calavan, E.C., 1970, Occurrence of mycoplasmalike bodies in phloem of stubborn infected citrus seedlings, *Phytopathology* **60**:1525–1526.
156. Igwegbe, E.C.K., and Calavan, E.C., 1973, Effect of tetracycline antibiotics on symptom development of stubborn disease and infectious variegation of citrus seedlings, *Phytopathology* **63**:1044–1048.
157. Igwegbe, E.C.K., Stevens, C., and Hollis, J.J. Jr., 1979, An "in vitro" comparison of some biochemical and biological properties of California and Morocco isolates of *Spiroplasma citri*, *Can. J. Microbiol.* **25**:1125–1132.
158. Ikegami, M., 1989, Computer analysis between nucleotide and amino acid sequences of bean golden mosaic virus and those of maize streak, wheat dwarf, chloris mosaic and beet curly top viruses, *Microbiol. Immunol.* **33**:863–869.
159. Ishiie, T., Doi, Y., Yora, K., and Asuyama, H., 1967, Suppressive effects of antibiotics of tetracycline group on symptom development of mulberry dwarf disease, *Ann. Phytopathol. Soc. Japan*, **33**:267–275.
160. Kado, K.J., and Anderson, H.W., 1936, Brittle root disease of horseradish in Illinois, *Plant Dis. Rep.* **18**:228.
161. Kado, K.J., and Anderson, H.W., 1940, A study of horseradish diseases and their control, *Ill. Agric. Exp. Stn. Bull.* **469**:531–583.
162. Kaloostian, G.H., Oldfield, G.N., Calavan, E.C., and Blue, R.L., 1976, Leafhoppers transmit citrus stubborn disease to weed hosts, *Calif. Agric.* **30**(9):4–5.
163. Kaloostian, G.H., Oldfield, G.N., Gough, D., and Calavan, E.C., 1979, Control of citrus stubborn vectors in the laboratory, *Citrograph* **65**(11):17–18, 25.

164. Kheyri, M., Alimoradi, I., and Davatchi, A., 1969, The leafhoppers of sugarbeet in Iran and their role in curly top virus disease, Sugarbeet Seed Inst., Agric. College. Karaj, Entomol. Res. Div., Teheran, 30 pp.

165. Kirkpatrick, B.C., 1989, Strategies for characterizing plant-pathogenic mycoplasma-like organisms and their effects on plants, in Kosuze, T., and Wester, E.W. (eds): Plant-Microbe Interactions, Volume 3, Macmillan, New York, pp. 241–293.

166. Kirkpatrick, B.C., Stenger, D.C., Morris, T.J., and Purcell, A.H., 1987, Cloning and detection of DNA from a nonculturable plant pathogenic mycoplasma-like organism, *Science* **238**:197–200.

167. Klein, M., 1970, Safflower phyllody—a mycoplasma disease of *Carthamus tinctorius* in Israel, *Plant Dis. Rep.* **54**:735–738.

168. Klein, M., and Almeida, L., 1990, Two members of the *Circulifer tenellus* complex exhibit differences in vectoring *Spiroplasma citri*, *Seventh Int. Auchenorrhyncha Congr.*, Wooster, Ohio, (Abstr.).

169. Klein, M., and Purcell, A.H., 1987, Response of *Galleria mellonella* (Lepidoptera: Pyralidae) and *Tenebrio molitor* (Coleoptera: Tenebrionidae) to *Spiroplasma citri* inoculation, *J. Invert. Pathol.* **50**:9–15.

170. Klein, M., and Raccah, B., 1986, Morphological variations in Neoaliturus spp. (Cicadellidae) from various host plants in Israel, *Proc. 2nd Int. Workshop on Leafhoppers and Planthoppers of Economic Importance*, Provo, Utah, p. 267 (Abstr.).

171. Klein, M., and Raccah, B., 1991, Separation of two leafhopper populations of the *Circulifer haematoceps* complex on different host plants in Israel, *Phytoparasitica* 19(2). (in press)

172. Klein, M., and Raccah, B., 1991, Morphological characterization of two populations of *Circulifer* (Homoptera: Cicadellidae) from Israel, *Isr. J. Entomol.* 24. (in press)

173. Klein, M., Rasooly, R., and Raccah, B., 1988, New findings on the transmission of *Spiroplasma citri*, the citrus stubborn disease agent in Israel, by a beet leafhopper from the Jordan Valley, *Hassadeh*, (Hebrew, with English summary) **68**:1734–1737.

174. Klein, M., Rasooly, R., and Raccah, B., 1988, Transmission of *Spiroplasma citri*, the agent of citrus stubborn, by a leafhopper of the *Circulifer tenellus* complex in Israel, *Proc. Int. Citrus Congr. Middle East* (Tel Aviv, Israel), Volume 2, p. 49.

175. Kloeper, J.W., and Garrot, D.G., 1983, Evidence for a mixed infection of spiroplasmas and nonhelical mycoplasmalike organisms in cherry with X-disease, *Phytopathology* **73**:357–360.

176. Kondo, F., Maramorosch, K., McIntosh, A.H., and Varney, E.H., 1977, Aster yellows spiroplasma: isolation and cultivation "in vitro", *Proc. Am. Phytopathol. Soc. Meet.* Volume 4, pp. 190–191.

177. Kondo, F., McIntosh, A.H., Padhi, S.B., and Maramorosch, K., 1976, A spiroplasma isolated from the ornamental cactus *Opuntia tuna monstrosa, Proc. Soc. Gen. Microbiol.* **3**:154.

178. Kondo, F., McIntosh, A.H., Padhi, S.B., and Maramorosch, K., 1976, Electron microscopy of a new plant pathogenic spiroplasma isolated from *Opuntia, Proc. 34th Ann. Meet. Electron Microscopic Soc. Am.* Miami Beach, Florida, p. 56 (Abstr.).

179. Lafleche, D., and Bové, J.M., 1970, Mycoplasmes dans les agrumes atteints de "Greening", de "Stubborn" ou de maladies similaires, *Fruits* **25**:455–465.
180. Landis, B.J., Powell, D.M., and Hagel, G.T., 1970. Attempt to suppress curly top and beet western yellows by control of the beet leafhopper and the green peach aphid with insecticide-treated sugarbeet seed, *J. Econ. Entamol.* **63**:493–496.
181. Lange, W.H., 1987, Insect pests of sugar beet, *Ann. Rev. Entomol.* **32**:344–360.
182. Larsen, R.C., and Duffus, J.E., 1984, A simplified procedure for the purification of curly top virus and the isolation of its monomer and dimer particles, *Phytopathology* **74**:114–118.
183. Larson, A.O., and Hallock, H.C., 1942, Time of planting susceptible beans in relation to curly top injury in south central Idaho, *J. Econ. Entomol.* **35**:365–369.
184. Lawson, F.R., Chamberlin, J.C., and York, G.T., 1951, Dissemination of the beet leafhopper in California, *U.S. Dept. Agric. Tech. Bull.* **1030**, 59 pp.
185. Lawson, F.R., and Piemeisel, R.L., 1943, The ecology of the principal summer weed hosts of the beet leafhopper in the San Joaquin Valley, California, *U.S. Dept. Agric. Tech. Bull.* 343, 37 pp.
186. Lawson, P.B., 1920, The Cicadellidae of Kansas, *Kansas Univ. Sci. Bull.* **12**:180.
187. Lee, I.M., Cartia, G., Calavan, E.C., and Kaloostian, G.H., 1973, Citrus stubborn disease organism cultured from beet leafhopper, *Calif. Agric.* **27**(11):14–15.
188. Lee, I.-M., and Davis, R.E., 1983, Phloem-limited prokaryotes in sieve elements isolated by enzyme treatment of diseased tissue, *Phytopathology* **73**:1540–1543.
189. Lee, I.M., and Davis, R.E., 1988, Detection and investigation of genetic relatedness among aster yellows and other mycoplasma-like organisms using clonal DNA and DNA probes, *Mol. Plant-Microbe Interact.* **1**:303–310.
190. Lesley, J.W., 1931, The resistance of varieties and new dwarf races of tomato to curly top (Western yellows blight or yellows), *Hilgardia* **6**:27–44.
191. Lesley, J.W., and Wallace, J.M., 1938, Acquired tolerance to curly top in the tomato, *Phytopathology* **28**:548–553.
192. Linnavuori, R., 1962, Hemiptera of Israel, II. *Ann. Zool. Soc. Zool.-Bot. Fenn Vanamo* **24**:1–108.
193. Liu, H.-Y., 1981, The transmission, multiplication and electron microscopic examination of *Spiroplasma citri* in its vector, *Circulifer tenellus*, Ph.D. Thesis, University of California, Riverside, pp. 1–101.
194. Liu, H.-Y., Gumpf, D.J., Oldfield, G.N., and Calavan, E.C., 1983. Transmission of *Spiroplasma citri* by *Circulifer tenellus*, *Phytopathology* **73**:582–585.
195. Liu, H.-Y., Gumpf, D.J., Oldfield, G.N., and Calavan, E.C., 1983, The relationship of *Spiroplasma citri* and *Circulifer tenellus*, *Phytopathology* **73**:585–590.
196. Magyarosy, A.C., 1978, A new look at curly top disease, *Calif. Agric.* **32**(9):13–14.
197. Magyarosy, A.C., 1980, Beet curly top virus transmission by artificially fed and injected beet leafhoppers (*Circulifer tenellus*), *Ann. Appl. Biol.* **96**:301–305.
198. Magyarosy, A.C., and Duffus, J.E., 1977, Beet curly top virulence increased, *Calif. Agric.* **31**(6):12–13.
199. Magyarosy, A.C., and Duffus, J.E., 1977, The occurrence of highly virulent strains of the beet curly top virus in California, *Plant Dis. Rep.* **61**:248–251.
200. Magyarosy, A.C., and Sylvester, E.S., 1979, The latent period of beet curly top virus in the beet leafhopper, *Circulifer tenellus*, mechanically injected with infectious phloem exudate, *Phytopathology* **69**:736–738.
201. Malm, N.R., and Finkner, R.E., 1968, The use of systemic insecticides to reduce

the incidence of curly top virus disease in sugarbeets, *J. Am. Soc. Sugar Beet Technol.* **15**:246–254.

202. Maramorosch, K., 1955. Mechanical transmission of curly top virus to its insect vector by needle inoculation, *Virology* 1:286–300.

203. Maramorosch, K., Klein, M., and Wolanski, B.C., 1972, Beitrag zur Atiologie des Hexenbasenkrankheit der Kaktee *Opuntia tuna* (*Tuna monstrosa*). *Experientia* **28**:362–383.

204. Maramorosch, K., and Kondo, F., 1978, Aster yellows spiroplasma: infectivity and association with a rodshaped virus, *Zentralbl. Bakteriol. Parasitenkd. Infekionskr Hyg.* Abt. 1, **241**:196.

205. Markham, P.G., and Oldfield, G.N., 1983, Transmission techniques with vectors of plant and insect mycoplasmas and spiroplasmas, in Tully, J.G., and Razin, S. (eds): Methods in Mycoplasmalogy, Volume 2, Academic Press, New York, pp. 261–267.

206. Markham, P.G., Pinner, M.S., and Boulton, M.I., 1984, *Dalbulus maidis* and *Cicadulina* species as vectors of diseases in maize, *Maize Virus Dis. Newsl.* 1:33–34.

207. Markham, P.G., and Townsend, R., 1974, Transmission of *Spiroplasma citri* to plants, *INSERM Colloq.* **33**:201–206.

208. Markham, P.G., and Townsend, R., 1979, Experimental vectors of spiroplasmas, in Maramorosch, K., and Harris, K.F. (eds): Leafhopper Vectors and Plant Disease Agents, Academic Press, New York, pp. 413.–445.

209. Martin, M.W., 1963, Responses of curly top-resistant *Lycopersicon* species to curly top exposure in different areas of the west, *Plant Dis. Rep.* **47**:121–125.

210. Martin, M.W., 1970, Developing tomatoes resistant to curly top virus, *Euphytica* **19**:243–252.

211. Martin, M.W., and Cannon, O.S., 1963, Controlling tomato curly top by using resistant varieties, *Utah Sci.* **24**:3–5, 25–26.

212. Martin, M.W., and Thomas, P.E., 1969, C_5, a new tomato breeding line resistant to curly top virus, *Phytopathology* **59**:1754–1755.

213. Martin, M.W., and Thomas, P.E., 1986, Levels, dependability, and usefulness of resistance to tomato curly top disease, *Plant Dis.* **70**:136–141.

214. Martin, M.W., and Thomas, P.E., 1986, Increased value of resistance to infection if used in integrated pest management control of tomato curly top, *Phytopathology* **76**:540–542.

215. McCoy, R.E., 1979, Mycoplasma and yellows diseases, in Whitcomb, R.E., and Tully, J.G. (eds): Mycoplasmas, Volume 3., Plant and Insect Mycoplasmas, Academic Press, New York, pp. 229–264.

216. McCoy, R.E., Davis, M.J., and Dowell, R.V., 1981, "In vivo" cultivation of spiroplasmas in larvae of the greater wax moth, *Phytopathology* **71**:403–411.

217. McFarlane, J.S., 1969, Breeding for resistance to curly top, *J. Int. Inst. Sugar Beet Res.* **4**:73–93.

218. McKay, M.B., and Dykstra, T.P., 1927, Curly top of squash, *Phyhtopathology* **17**:48 (Abstr.).

219. McKay. M.B., and Dystra, T.P., 1927, Sugar beet curly top, the cause of western yellow tomato blight, *Phytopathology* **17**:39 (Abstr.).

220. Medina, V., Nebbache, S., and Markham, P.G., 1990, The intracellular pathway of maize streak virus in its leafhopper vector, *Cicadulina mbila*, 3rd Int. Work-

shop on Leafhoppers and Planthoppers of Economic Importance, Wooster, Ohio, (Abstr.).

221. Meyerdirk, D.E., Oldfield, G.N., and Hessein, N.A., 1983, Bibliography of the beet leafhopper, *Circulifer tenellus* (Baker), and two of its transmitted plant pathogens, curly top virus and *Spiroplasma citri* Saglio "et al"., *Bibliogr. Entomol. Soc. Am.* **2**:17–15.

222. Morrison, A.L., 1969, Curly top virus control in California, *Calif. Sugar Beet* **1969**:28,30.

223. Morrison, A.L., 1972, Curly top control, *Calif. Tomato Growers* **15**:8.

224. Mowry, T.M., 1986, Mechanisms of the barriers to *Spiroplasma citri* infection of *Macrosteles fascifrons*, Ph.D. Thesis, Michigan State University, East Lansing, pp. 1–75.

225. Mumford, D.L., 1972, A new method of mechanically transmitting curly top virus, *Phytopathology* **62**:1217–1218.

226. Mumford, D.L., 1974, Purification of curly top virus, *Phytopathology* **64**:136–139.

227. Mumford, D.L., and Griffin, G.D., 1973, Evaluation of systemic pesticides in controlling sugar beet leafhopper, *J. Am. Soc. Sugar Beet Technol.* **17**:354–357.

228. Mumford, D.L., and Peay, W.E., 1970, Curly top epidemic in western Idaho, *J. Am. Soc. Sugar Beet Technol.* **16**:185–187.

229. Mumford, D.L., and Thornley, W.R., 1977, Location of curly top virus antigen in bean, sugarbeet, tobacco and tomato by fluorescent antibody staining, *Phytopathology* **67**:1313–1316.

230. Murphy, A.M., 1942, Production of heavy curly-top exposures in sugar-beet breeding fields, *Proc. Am. Soc. Sugar Beet Technol.* **3**:459–462.

231. Murphy, A.M., Bennett, C.W., and Owen, F.V., 1959, Varietal reaction of sugar beets to curly top virus strain 11 under field conditions, *J. Am. Soc. Sugar Beet Technol.* **10**:281–282.

232. Murphy, A.M., and Douglass, J.R., 1952, Effect of DDT on beet leafhoppers, curly top, and yields of sugar beet varieties, *Proc. Am. Soc. Sugar Beet Technol.* **7**:497–502.

233. Murphy, D.M., 1940, A great northern bean resistant to curly-top and common bean-mosaic viruses, *Phytopathology* **30**:779–784.

234. Murphy, D.M., 1940, Bean improvement and bean diseases in Idaho, *Idaho Agric. Exp. Stn. Bull.* **238**, 22 pp.

235. Nault, L.R., and Knoke, J.K., 1981, Maize vectors, in Gordon, D.T., and Scott, G.E. (eds): Virus and Viruslike Diseases in Maize in the United States, *South. Coop. Serv. Bull.* 247, pp. 1–218.

236. Nhami, A., Bové, J.M., Bové, C., Monsion, M., Garnier, M., Saillard, C., Motous, G., and Fos, A., 1980, Natural transmission of *Spiroplasma citri* to periwinkles in Morocco. *Proc. 8th IOCV Conf.* University of California, Riverside, pp. 153–161.

237. Nyland, G., and Raju, B.C., 1978, Isolation and culture of a spiroplasma from pear trees affected by pear decline, *Phytopathol. News* **12**:216.

238. O'Hayer, K.W., Schultz, G.A., Eastman, C.E., and Fletcher, J., 1984, Newly discovered plant hosts of *Spiroplasma citri*, *Plant Dis.* **68**:336–338.

239. O'Hayer, K.W., Schultz, G.A., Eastman, C.E., Fletcher, J., and Goodman, R.M., 1983, Transmission of *Spiroplasma citri* by the aster leafhopper *Macrosteles fascifrons* (Homoptera: Cicadellidae), *Ann. Appl. Biol.* **102**:311–318.

240. Oldfield, G.N., 1982, A virescence agent transmitted by *Circulifer tenellus* Baker:

aspects of its plant range and association with *Spiroplasma citri*, *Rev. Infect. Dis.* **4**:S248.

241. Oldfield, G.N., 1984, Field ecology of *Spiroplasma citri* in western North America, *Isr. J. Med. Sci.* **20**:998–1001.

242. Oldfield, G.N., 1987, Leafhopper vectors of the citrus stubborn disease spiroplasma, *Proc. 2nd Int. Workshop on Leafhoppers and Planthoppers of Economic Importance* CIE, London, United Kingdom, pp. 151–159.

243. Oldfield, G.N., and Calavan, E.C., 1982, *Spiroplasma citri* non-rutaceous hosts, in Bové, J.M., and Vogel, R. (eds): Description and Illustration of Virus and Virus-like Diseases of Citrus: A Collection of Color Slides. 2nd ed., SECTCO -IFRA, Paris, France, pp. 1–9.

244. Oldfield, G.N., Kaloostian, G.H., Pierce, H.D., Calavan, E.C., Granett, A.L., and Blue, R.L., 1976, Beet leafhopper transmits citrus stubborn disease, *Calif. Agric.* **30**(6):15.

245. Oldfield, G.N., Kaloostian, G.H., Pierce, H.D., Calavan, E.C., Granett, A.L., Blue, R.L., Rana, G.L., and Gumpf, D.J., 1977, Transmission of *Spiroplasma citri* from citrus to citrus by *Scaphytopius nitridus*, *Phytopathology* **67**:763–765.

246. Oldfield, G.N., Kaloostian, G.H., Pierce, H.D., Granett, A.L., and Calavan, E.C., 1975, Natural occurrence of *Spiroplasma citri* in leafhoppers in California, *Phytopathology* **65**:117 (Abstr.).

247. Oldfield, G.N., Kaloostian, G.H., Pierce, H.D., Granett, A.L., and Calavan, E.C., 1977. Beet leafhopper transmits virescence of periwinkle, *Calif. Agric.* **31**(6):15.

248. Oman, P.W., 1948, Notes on the beet leafhopper *Circulifer tenellus* (Baker), and its relatives (Homoptera: Cicadellidae), *J. Kansas Entomol. Soc.* **21**:10–14.

249. Oman, P.W., 1970, Taxonomy and nomenclature of the beet leafhopper *Circulifer tenellus* (Homoptera: Cicadellidae), *Ann. Entomol. Soc. Am.* **63**:507–512.

250. Pappo, S., and Bauman, I., 1969, A survey of the present status of little-leaf (stubborn) disease in Israel, *Proc. 1st Int. Citrus Symp.* University of California, Riverside, Volume 3, pp. 1439–1444.

251. Peay, W.E., 1959, Laboratory tests for control of the beet leafhopper on snap beans grown for seed, *J. Econ. Entomol.* **52**:700–703.

252. Peay, E.W., and Oliver, W.N., 1964, Curly top prevention by vector control on snap beans grown for seed, *J. Econ. Entomol.* **57**:3–5.

253. Perkes, R.E., 1970, *Circulifer tenellus* (Baker) (Homoptera: Cicadellidae): mating behavior, ecology of reproduction and attraction to monochromatic electromagenetic radiation, Ph.D. Thesis, University of California, Riverside, 120 pp.

254. Petty, H.B. Jr., 1955, The insect pests of horse-radish in southwestern Illinois, Ph.D. Thesis, University of Illinois, Urbana.

255. Piemeisel, R.L., 1942, A general appraisal of plant cover in relation to beet leafhoppers, forage production, and soil protection, *Proc. Am. Soc. Sugar Beet Technol.* **3**:462–464.

256. Piemeisel, R.L., and Carsner, E., 1951, Replacement control and biological control, *Science* **113**:14–15.

257. Price, W.C., 1943, Severity of curly top in tobacco affected by site of inoculation, *Phytopathology* **33**:586–601.

258. Purcell, A.H., and Gonot, K., 1981, Transmission of X-disease agent by injected beet leafhoppers, *Phytopathology* **71**:900 (Abstr.).

259. Raccah, B., and Klein, M., 1982, Transmission of the safflower phyllody mollicute by *Neoaliturus fenestratus*, *Phytopathology* **72**:230–232.

260. Raju, B.C., Nyland, G., Backus, E.A., and McLean, D.L., 1981, Association of a spiroplasma with brittle root of horseradish, *Phytopathology* **71**:1067–1072.
261. Raju, B.C., Nyland, G., and Purcell, A.H., 1983, Current status of the etiology of pear decline, *Phytopathology* **73**:350–353.
262. Raju, B.C., Purcell, A.H., and Nyland, G., 1984, Spiroplasmas from plants with aster yellows disease and X-disease: isolation and transmission by leafhoppers, *Phytopathology* **74**:925–931.
263. Rana, G.L., Kaloostian, G.H., Oldfield, G.N., Granett, A.L., Calavan, E.C., Pierce, H.D., Lee, I.M., and Gumpf, D.J., 1975, Acquisition of *Spiroplasma citri* through membranes by homopterous insects, *Phytopathology* **65**:1143–1145.
264. Rasooly, R., 1988, Isolation and transmission of the citrus stubborn disease pathogen, M.Sc Thesis, Hebrew University of Jerusalem. Faculty of Agriculture. Rehovot, Israel, (Hebrew, with English summary) 84 pp.
265. Reichert, I., and Perlberger, J., 1931, Little leaf disease of citrus trees and its causes, *Hadar* **4**:193–194.
266. Ritenour, G., Hills, F.J., and Lange, W.H., 1970, Effect of planting date and vector control on the suppression of curly top and yellows in sugarbeet, *J. Am. Soc. Sugar Beet Technol.* **16**:78–84.
267. Roberts, I.M., Robinson, D.J., and Harrison, B.D., 1984, Serological relationships and genome homologies among geminiviruses, *J. Gen. Virol.* **65**:1723–1730.
268. Romney, V.E., 1943, The beet leafhopper and its control in beets grown for seed in Arizona and New Mexico, *U.S. Dept. Agric. Tech. Bull.* **85**, 24 pp.
269. Rose, D.J.W., 1978, Epidemiology of maize streak disease, *Ann. Rev. Entomol.* **23**:259–282.
270. Saglio. P., Laflèche, D., Bonissol, C., and Bové, J.M., 1971, Isolement et culture "in vitro" des mycoplasmes associés au "Stubborn" des agrumes et leur observation au microscope électronique, *C.R. Acad. Sci. Paris,* **272**:1387–1390.
271. Saglio, P., L'Hospital, M., Laflèche, D., Dupont, G., Bové, J.M., Tulley, J.G., and Freundt, E.A., 1973, *Spiroplasma citri* gen. and sp.n.: a mycoplasma-like organism associated with "Stubborn" disease of citrus, *Int. Syst. Bacteriol.* **23**:191–204.
272. Saillard, C., Vignault, J.C., Fos, A., and Bové, J.M., 1984, *Spiroplasma citri*-induced lethal wilting of periwinkles is prevented by prior or simultaneous infection of the periwinkle by an MLO, *Ann. Microbiol. (Paris),* **135A**:163–168.
273. Saillard, C., Vignault, J.C., Bové, J.M., Raie, A., Tulley, J.G., Williamson, D.L., Fos, A., Garnier, M., Gadeau, A., Carle, P., and Whitcomb, R.F., 1987, *Spiroplasma phoeniceum* sp. nov., a new plant pathogenic species from Syria, *Int. J. Syst. Bacteriol.* **37**:106–115.
274. Saillard, C., Vignault, J.C., Gadeau, A., Carle, P., Garnier, M., Fos, A., Bové, J.M., Tully, J.G., and Whitcomb, R.F., 1984, Discovery of a new plant-pathogenic spiroplasma, *Isr. J. Med. Sci.* **20**:1013–1015.
275. Schneider, C.L., 1955, Incidence of curly top of sugar beets in Minnesota and Iowa in 1954, *Plant Dis. Rep.* **39**:453–454.
276. Schneider, C.L., 1959, Occurrence of curly top of sugar beets in Maryland in 1958, *Plant Dis. Rep.* **43**:681.
277. Schneider, C.L., Jafri, A.M., and Murphy, A.M., 1968, Greenhouse testing of sugar beet for resistance to curly top, *J. Am. Soc. Sugar Beet Technol.* **14**:727–734.
278. Schultz, G., McGuire, M., Eastman, C., O'Hayer, K., and Fletcher, J., 1982,

Properties of transmission of a brittle root isolate of *Spiroplasma citri* by the leafhopper *Circulifer tenellus, Phytopathology* **72**:1005 (Abstr.).

279. Severin, H.H.P., 1921, Minimum incubation periods of causative agent of curly leaf in beet leafhopper and sugar beet, *Phytopathology* **11**:424–429.

280. Severin, H.H.P., 1921, Summary of the life history of beet leafhopper (*Eutettix tenella* Baker), *J. Econ. Entomol.* **14**:433–436.

281. Severin, H.H.P., 1923, Investigations of beet leafhopper (*Eutettix tenella* Baker) in Salinas Valley of California, *J. Econ, Entomol.* **16**:479–485.

282. Severin, H.H.P., 1924, Curly leaf transmission experiments, *Phytopathology* **14**:88–93.

283. Severin, H.H.P., 1924, Causes of fluctuations in numbers of beet leafhopper (*Eutettix tenella* Baker) in a natural breeding area of the San Joaquin Valley of California, *J. Econ. Entomol.* **17**:639–645.

284. Severin, H.H.P., 1927, Crops naturally infected with sugar beet curly top, *Science* **67**:137–138.

285. Severin, H.H.P., 1928, Transmission of tomato yellows or curly top of the sugar beet, by *Eutettix tenella* (Baker), *Hilgardia* **3**:251–271.

286. Severin, H.H.P., 1929, Additional host plants of curly top, *Hilgardia* **3**:595–629.

287. Severin, H.H.P., 1931, Modes of curly top transmission of the beet leafhopper, *Eutettix tenella* (Baker), *Hilgardia* **6**:253–276.

288. Severin, H.H.P., 1933, Field observations of the beet leafhopper, *Eutettix tenellus* in California, *Hilgardia* **7**:282–360.

289. Severin, H.H.P., 1934, Weed host range and overwintering of curly-top virus, *Hilgardia* **8**:263–277.

290. Severin, H.H.P., 1947, Location of curly top virus in the beet leafhopper, *Eutettix tenellus, Hilgardia* **17**:545–551.

291. Severin, H.H.P., and Freitag, J.H., 1933, Some properties of the curly-top virus, *Hilgardia* **8**:1–48.

292. Severin, H.H.P., and Freitag, J.H., 1934, Ornamental flowering plants naturally infected with curly top and aster yellows viruses, *Hilgardia* **8**:233–260.

293. Severin, H.H.P., and Henderson, C.F., 1928, Beet leafhopper, *Eutettix tenellus* (Baker) does not occur in Argentine Republic, *J. Econ. Entomol.* **21**:542–544.

294. Severin, H.H.P., and Henderson, C.F., 1928, Some host plants of curly top, *Hilgardia* **3**:339–393.

295. Severin, H.H.P., and Schwing, E.A., 1926, The 1925 outbreak of the beet leafhopper (*Eutettix tenella* Baker) in California, *J. Econ. Entomol.* **19**:478–483.

296. Severin, H.H.P., and Severin, H.C., 1927, Curly top of sugar beets in South Dakota, *J. Econ. Entomol.* **20**:586–588.

297. Shapovalov, M., 1928, Yellows, a serious disease of tomatoes, *U.S. Dept. Agric. Misc. Publ.* **13**, 1–4 pp.

298. Shapovalov, M., 1940, Direct seeding of tomatoes for production and disease control, *Utah State Agric. Coll. Ext. Circ.* NS **105**:25–27.

299. Shaw, H.B., 1910, The curly-top of sugar beets, *U.S. Dept. Agric. Bur. Plant Industry Bull.* **181**:1–46.

300. Shaw, M.E., Golino, D.A., and Kirkpatrick, B.C., 1990, Infection of radish in Idaho by the beet leafhopper transmitted virescence agent, *Plant Dis.* **74**:252 (Abstr.).

301. Silbernagel, M.J., 1965, Differential tolerance to curly top in some snap bean varieties, *Plant Dis. Rep.* **49**:475–477.

302. Silbernagel, M.J., and Jafri, A.M., 1973, Effect of temperature on curly top resistance in *Phaseolus vulgaris*, *Phytopathology* **63**:1218 (Abstr.).
303. Silbernagel, M.J., and Jafri, A.M., 1974, Temperature effects on curly top resistance in *Phaseolus vulgaris*, *Phytopathology* **64**:825–827.
304. Sill, W.H. Jr., and Padi, S.M., 1954, 1953, epiphytotic of curly top of sugar beets in Kansas, *Plant Dis. Rep.* **38**:57.
305. Simpson, W.R., 1959, The Owyhee Tomato, *Idaho Agric. Exp. Stn. Bull.* **298**:1–3.
306. Simpson, W.R., 1962, Payette, a new curly top resistant dwarf tomato variety, *Univ. Idaho Agric. Exp. Stn. Bull.* **387**, pp. 1–7.
307. Skuderma, A.W., Cormany, C.E., and Hurst, L.A., 1933, Effects of time of planting and fertilizer mixtures on the curly top resistant sugar beet variety U.S. No. 1 in Idaho, *U.S. Dept. Agric. Circ.* **273**:1–16.
308. Smith, R.E., 1906, Tomato diseases in California, *Calif. Agric. Exp. Stn. Bull.* **175**:1–16.
309. Smith, R.E., and Bonequet, P.A., 1915, New light on curly top of the sugar beet, *Phytopathology* **5**:103–107.
310. Stahl, C.F., 1920, Studies on the life history and habits of the beet leafhopper, *J. Agric. Res.* **20**:245–252.
311. Stanarius, A., Muller, H.M., and Kleinhempel, H., 1979, Elektronenmikroskopischer Nachweis von *Spiroplasma citri* im Zellkern infizierter Wirtspflanzen, *Zentialbl. Bakteriol.* **134**:559–562.
312. Stanley, J., 1985, The molecular biology of geminiviruses, *Adv. Virus Res.* **30**:139–177.
313. Stanley, J., Markham, P.G., Callis, R.J., and Pinner, M.S., 1986, The nucleotide sequence of an infectious clone of the gemini virus beet curly to virus, *EMBO J.* **5**:1761–1767.
314. Starrett, R.C., 1929, A new host of sugar beet curly top, *Phytopathology* **19**:1031–1035.
315. Stenger, D.C., Carbonaro, D., and Duffus, J.E., 1990, Genomic characterization of phenotypic variant of beet curly top virus, *J. Gen. Virol.* **71**:2211–2215.
316. Stenger. D.C., and Duffus, J.E., 1989, Genomic characterization of beet curly top virus isolates, *Phytopathology* **179**:1158 (Abstr.).
317. Sullivan, D.A., Oldfield, G.N., Eastman, C.E., Fletcher, J., and Gumpf, D.J., 1987, Transmission of citrus-infecting strain of *Spiroplasma citri* to horseradish. *Plant Dis.* **71**:469.
318. Thomas, P.E., 1969, Thirty-eight new hosts of curly top virus, *Plant Dis. Pep.* **53**:548–549.
319. Thomas, P.E., 1970, Isolation and differentiation of five strains of curly top virus, *Phytopathology* **60**:844–848.
320. Thomas, P.E., 1972, Mode of expression of host preference by *Circulifer tenellus*, the vector of curly top virus, *J. Econ. Entomol.* **65**:119–123.
321. Thomas, P.E., and Boll, R.K., 1977, Effect of host preference on transmission of curly top virus to tomato by the beet leafhopper, *Phytopathology* **67**:903–905.
322. Thomas, P.E., and Boll, R.K., 1978, Tolerance to curly top virus in tomato, *Phytopathol. News* **12**:181 (Abstr.).
323. Thomas, P.E., and Bowyer, J.E., 1980, Properties of tobacco yellow dwarf and bean summer death viruses, *Phytopathology* **70**:214–217.
324. Thomas, P.E., and Martin, M.W., 1969, Association of recovery from curly top in tomatoes with susceptibility, *Phytopathology* **59**:1864–1867.

325. Thomas, P.E., and Martin, M.W., 1971, Apparent resistance to establishment of infection by curly top virus in tomato breeding lines, *Phytopathology* **61**:550–551.

326. Thomas, P.E., and Martin, M.W., 1971, Vector preference, a factor of resistance to curly top virus in certain tomato cultivars, *Phytopathology* **61**:1257–1260.

327. Thomas, P.E., and Martin, M.W., 1972, Characterization of a factor of resistance in curly top virus resistant tomatoes, *Phytopathology* **62**:954–958.

328. Thomas, P.E., and Mink, G.I., 1979, Beet curly top virus, CMI/AAB Description of Plant Viruses. No. 210, 6 pp.

329. Thornberry, H.H., and Takeshita, R.M., 1954, Sugar beet curly top virus and curly top disease in Illinois and their relation to horseradish brittle root, *Plant Dis. Rep.* **38**:3–5.

330. Thornley, W.R., and Mumford, D.L., 1979, Intracellular location of beet curly top virus antigen as revealed by fluorescent antibody staining, *Phytopathology* **69**:738–740.

331. Townsend, C.O., 1908, Curly top, a disease of the sugar beet, *U.S. Dept. Agric. Bur. Plant Industry Bull.* **122**, pp. 1–37.

332. Townsend, R., Markham, P.G., and Plaskitt, K.A., 1977, Multiplication and morphology of *Spiroplasma citri* in the leafhopper *Euscelis plabejus*, *Ann. Appl. Biol.* **87**:307–313.

333. Townsend, R., Markham, P.G., Plaskitt, K.A., and Daniels, M.J., 1977, Isolation and characterization of a non-helical strain of *Spiroplasma citri*, *J. Gen. Microbiol.* **100**:15–21.

334. Troutman, J.L., and Fenne, S.B., 1959, The occurrence of curly top in Virginia, *Plant Dis. Rep.* **43**:155–156.

335. Tsai, P., and Allen, R.M., 1980, Comparative serology of several isolates of *Spiroplasma citri*, Proc. 4th Meet. Int. Council on Lethal Yellowing, University of Florida, p. 16 (Abstr.).

336. Tully, J.G., Rose, D.L., Clark, E., Carle, P., Bové, J.M., Henegar, R.B., Whitcomb, R.F., Colflesh, D.E., and Williamson, D.L., 1987, Revised group classification of the genus *Spiroplasma* (Class Molicutes), with proposed new groups XII to XXIII, *Int. J. Syst. Bacteriol.* **37**:357–364.

337. Ungs, W.D., Woodbridge, C.G., and Csizinski, A.A., 1977, Screening peppers (*Capsicum annuum* L.) for resistance to curly top virus, *Hortic. Science* **12**(2):161–162.

338. Valleau, W.D., 1953, False broomrape and leaf curl: two new diseases in Kentucky, *Plant Dis. Rep.* **37**:538–539.

339. Vest, H.G. Jr., 1964, The relation of age and site of inoculation of tomato plants and their reaction to curly top virus, M.Sc. Thesis, Utah State University, Logan, pp 1–40.

340. Virgin, W.J., 1940, The Chilean tomato, *Lycopersicon chilensis*, found resistant to curly top, *Phytopathology* **30**:280 (Abstr.).

341. Wallace, J.M., 1942, Virus strains in relation to acquired immunity from curly top in tomato, *Phytopathology* **32**:18–19 (Abstr.).

342. Wallace, J.M., 1944, Acquired immunity from curly top in tobacco and tomato, *J. Agric. Res.* **69**:187–214.

343. Wallace, J.M., and Lesley, J.W., 1944, Recovery from curly top in the tomato in relation to strains of the virus, *Phytopathology* **34**:116–123.

344. Whitcomb, R.F., Tully, J.G., Bové, J.M., and Saglio, P., 1973, Spiroplasmas and acholeplasmas: multiplication in insects, *Science* New York, **182**:1251–1253.
345. Williams, D.S., and McCoy, R.E., 1983, Treatment of woody plants with antibiotics, in Tully, J.G., and Razin, S. (eds): Methods in Mycoplasmology, Volume 2, Diagnostic Mycoplasmology, Academic Press, New York, pp. 275–286.
346. Wolanski, B.C., Klein, M., and Maramorosch, K., 1971, Electron microscopic studies on the effect of tetracycline-HC1 on the mycoplasmalike bodies in corn stunt and aster yellows infected plants, *Phytopathology* **61**:917 (Abstr.).
347. Yokomi, R.K., 1969, Phenological studies of *Circulifer tenelllus* (Baker) (Homoptera: Cicadellidae) in the San Joaquin Valley of California, Ph.D. Thesis, University of California, Davis, 92 pp.
348. Yokomi, R.K., Bar-Joseph, M., Oldfield, G.N., and Gumpf, D.J., 1981, A preliminary report of reduced infection by *Spiroplasma citri* and virescence in whitewash-treated periwinkle, *Phytopathology* **71**:914 (Abstr.).
349. Young, D.A. Jr., and Frazier, N.W., 1954, A study of the leafhopper genus *Circulifer* Zakhavatkin (Homoptera, Cicadellidae), *Hilgardia* **23**:25–52.
350. Zelcer, A., Bar-Joseph, M., and Loebenstein, G., 1971, Mycoplasma-like bodies associated with the little leaf disease of citrus, *Isr. J. Agric. Res.* **21**:137–142.

6
Thrips-Tomato Spotted Wilt Virus Interactions: Morphological, Behavioral and Cellular Components Influencing Thrips Transmission

Diane E. Ullman, John J. Cho, Ronald F.L. Mau, Wayne B. Hunter, Daphne M. Westcot, and Diana M. Custer

Introduction

Epidemics of insect-transmitted plant viruses in agricultural ecosystems require the interaction of 3 basic components: the host plant of the virus, the insect vector and the plant pathogenic virus. While this triad sounds quite straight forward, the relationships and interactions occurring between and among the basic triad components and the environment are complex and dynamic, frequently defying complete understanding by scientists and agricultural practitioners worldwide. The economically devastating tomato spotted wilt virus (TSWV) epidemics that occurred globally in recent years have inspired a multidisciplinary, interactive research approach aimed at unraveling the relationship between TSWV, its thrips vectors and the epidemiology of this important plant disease (23, 24, 75, 92). Our goal in this chapter is to integrate information regarding thrips morphology, feeding behavior, biology on different plant hosts and cellular thrips/TSWV interactions into a

Diane E. Ullman, Department of Entomology, University of Hawaii, 3050 Maile Way, Honolulu, Hawaii 96822, USA.

John J. Cho, Department of Plant Pathology, Hawaiian Institute of Tropical Research and Human Resource-Maui Research Station, P.O. Box 269, Kula Hawaii 96790, USA.

Ronald F.L. Mau, Department of Entomology, University of Hawaii, 3050 Maile Way, Honolulu, Hawaii 96822, USA.

Wayne B. Hunter, Department of Entomology, University of Hawaii, 3050 Maile Way, Honolulu, Hawaii 96822, USA.

Daphne M. Westcot, Department of Entomology, University of Hawaii, 3050 Maile Way, Honolulu, Hawaii 96822, USA.

Diana M. Custer, Department of Plant Pathology, Hawaiian Institute of Tropical Research and Human Resources-Maui Research Station, P.O. Box 269, Kula, Hawaii 96790, USA.

We dedicate this chapter to K. Sakimura in recognition of his tremendous contributions to our understanding of thrips biology, systematics, and relationships with tomato spotted wilt virus.

pivotal foundation for understanding TSWV epidemiology and control. As an introduction, we present the essential elements of TSWV epidemics: the thrips, the virus and the interactions of these two entities with plant hosts.

Thrips (Order: Thysanoptera) are minute insects that are widespread throughout the world in habitats ranging from forests, grasslands and scrub to cultivated crops and gardens (65). Many of the known thrips species are recognized as economically important insects causing direct damage to plants as a result of their feeding, as well as indirect damage as vectors of plant pathogens (18, 23, 75, 96, 100, 104, 126). Tomato spotted wilt virus (TSWV) (96, 104), tobacco streak virus (TSV) (42, 108) and maize chlorotic mottle virus (MCMV) (Jiang, Meinke, Wright, Wilkinson and Campbell unpublished data) are the only plant viruses known to be transmitted by thrips, although recent evidence demonstrates that thrips transmission is involved in the spread of prunus necrotic ringspot virus and prune dwarf virus (43, G. Mink, pers. comm.).

Among these plant viruses, TSWV is the only virus thrips transmit in a persistent, circulative fashion (104). Unlike TSWV, the viruses causing tobacco streak, prunus necrotic ringspot and prune dwarf apparently have a noncirculative relationship with their thrips vectors. The suspected mechanism underlying transmission of these viruses revolves around the hypothesis that infected pollen particles contact feeding wounds caused by thrips stylets thus introducing virus particles (42, 43, 108, G. Mink, pers. comm.). The mechanism underlying inoculation of maize chlorotic mottle is currently unknown; however, this virus may also have a noncirculative relationship with its thrips vector (Jiang, Meinke, Wright, Wilkinson and Campbell, unpublished data).

The thrips/TSWV relationship is unique among plant virus/vector associations, as only larval thrips can acquire the virus (54, 65, 67, 94, 102, 104, 105). A latent period is required before transmission can occur, although the length of the latent period is highly variable, even within a single thrips species (94, 95, 104, 105, 107). Occasionally larval thrips can transmit TSWV before they pupate, but adults that have acquired the virus as larvae more commonly transmit TSWV (104, 105). Certainly, as the dispersal stage of the insect, viruliferous adults are most important epidemiologically (23, 75, 104). Although thrips have been recognized as vectors of TSWV for more than sixty years (94), much remains to be learned regarding the specific interactions underlying thrips inoculation of plants with TSWV.

Like many insect vector/plant virus associations (71), the TSWV/thrips relationship is very specific, with only six among the 5,000 known thrips species serving as competent vectors of TSWV (4, 89, 90, 94–96, 102–107, 117, 119). Confirmed vector species include the western flower thrips (WFT), *Frankliniella occidentalis* (Pergande) (1, 18, 22, 75, 90), the tobacco thrips (TT), *F. fusca* (Hinds) (105), the common blossom or cotton bud thrips (CBT), *F. schultzei* (Trybom) (4, 22, 106), the onion thrips (OT), *Thrips tabaci* Lindeman (23, 67, 94, 102, 105), *T. setosus* Moulton (104) and the chilli thrips (CT),

Scirtothrips dorsalis Hood (4). A seventh thrips species, the melon thrips (MT), *T. palmi* Karny has recently been implicated as vector of a watermelon isolate of TSWV in Taiwan and groundnuts isolate in India (56, 75, 97). Most of these thrips species disperse readily over large areas, are highly polyphagous feeders and can also reproduce on a variety of plant species (65, 75, 129, 131). The wide range of plants on which adult WFT may feed is perhaps best illustrated by a faunistic survey taken in California, in which this ubiquitous insect was present on nearly every plant species sampled (16).

Like its thrips vector species, TSWV has an extensive host range of more than 200 plant species from 33 dicotyledonous and five monocotyledonous families (9, 19–21, 56). At least two serologically distinct isolates of TSWV are known, the lettuce isolate (TSWV-L), so called because of its original isolation from lettuce (41) and the impatiens isolate (TSWV-I) detected in a wide variety of flower crops throughout the United States (64). Members of the two serogroups are distinguished primarily by serologically distinct nucleocapsid (N) proteins and paracrystalline arrays of cytoplasmic filaments present in infected cells (64, 121). Advances in the understanding of TSWV molecular biology may identify other important differences (29, 64, 70, Kim and German, pers. comm.). Many isolates in the TSWV-L serogroup have been detected in various geographic locations, some of which are serologically distinct using specific antisera to viral structural proteins (26, 124). The existence of distinct serotypes has been proposed and certainly warrants further investigation (26). Defective isolates of TSWV also form readily when the virus is passed mechanically several times (53, 92, 98, 122). Considerable evidence exists suggesting that thrips vector species transmit different TSWV-L isolates with varying efficiencies (75, 89, 90). Hence, the components determining specificity between TSWV, its thrips vectors, and the efficiency with which various vector species transmit TSWV, probably involves complex interactions occurring between populations of different thrips vector species, TSWV isolates and the many plant hosts that these two entities share.

TSWV particles are spherical with a diameter of 80–110 nm (9). Particles consist of three 'pseudo' circular nucleocapsid structures represented by at least four structural proteins, surrounded by a lipid envelope with distinctive surface projections consisting of at least two membrane glycoproteins (9, 79). The TSWV genome consists of three linear, single-stranded RNA molecules denoted S RNA, M RNA and L RNA (27, 79). These characteristics set TSWV apart from any other known plant virus in terms of morphology and genome structure and supported its classification as the type member of a monotypic tomato spotted wilt group of plant viruses (51, 92). Recently, reclassification of TSWV as a new member of the Bunyaviridae, a family of animal viruses with which TSWV shares many characterics, including particle morphology, genome structure and cytoplasmic site of particle maturation has been proposed (27, 31, 79). More than 300 viruses, mostly arthropod-transmitted, are classified into the family Bunyaviridae, including serious human pathogens such as Rift Valley fever, Crimean-Congo haemorrhagic fever and California encephalitis (31). Although further investigations regarding TSWV molecular biology and vector/virus relationships are needed to clarify relationships between TSWV and members of the Bunyaviridae, TSWV reclassification has been accepted by the scientific community, making TSWV the first phytobunyavirus (32) (See addendum for additional information).

Much can be gained by comparing current findings regarding thrips/TSWV/plant interactions on a cellular and molecular level to relatively well studied viruses among the Bunyaviridae. Indeed, some of our discussion herein centers on important analogies between TSWV and other members of the Bunyaviridae. While these analogies can serve to direct future experimental approaches and enhance our understanding of TSWV biology, overemphasis of what appear to be analogous characteristics and/or functions may also veil differences existing in the thrips/TSWV/plant system and should be considered with caution.

Thrips Anatomy, Morphology and Behavior

Thrips have long been thought to rasp the plant surface disrupting or removing epidermal cells after which they randomly suck up exuding plant fluids (77, 125). Consequently, thrips have been classified as rasping-sucking insects and categorized in epidermal feeding guilds (33, 77, 91, 114, 125). Research focusing on the morphology and feeding of diverse thrips species across the Tubiliferan and Terebrantian suborders during the past 20 years, strongly support the contrasting viewpoint that thrips likely feed in a manner more analogous to piercing-sucking insects in which they puncture the epidermis and ingest the cytoplasm from an array of plant cells beyond the epidermal layer (17, 44, 45, 48–50, 82, 118, 119, 127, 128). Through presentation of morphological and behavioral data in the following sections of this chapter, we provide further support for the hypothesis that thrips are indeed piercing-sucking insects with more complex and diverse feeding behavior than previously thought. Thrips internal morphology and anatomy is presented to furnish an orientation for later discussion of TSWV location, fate and pathway in thrips cells.

Feeding Structures

Thrips are unique among the Insecta in that the mouthcone and the feeding structures contained therein are asymmetrical (8, 49, 65, 82, 128). Scanning electron microscopy of the WFT preparing to probe a plant surface, reveals the opisthorhynchous condition of the mouthcone which is typical of the Thysanoptera, as well as the paired maxillary and labial palps on the mouthcone (Figure 6.1). In an analysis of mouthpart morphology and movement during feeding, Hunter and Ullman (1989) (49) demonstrated that the mouthcones of the WFT and CBT are composed of paired paraglossae and house a

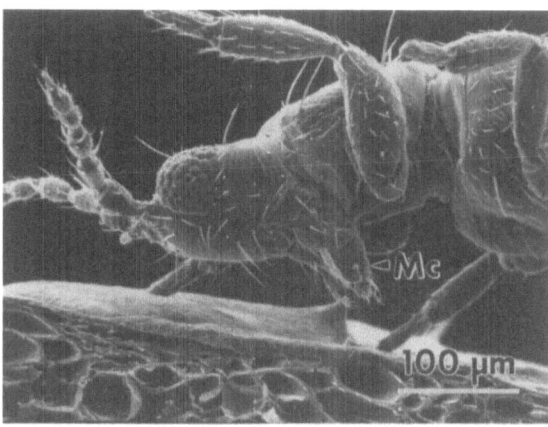

FIGURE 6.1. The typical preprobing behavior of the western flower thrips, *Frankliniella occidentalis*, is captured in this scanning electron micrograph showing an adult thrips engaged in antennal waving and scraping of one proleg across the plant surface. Following this behavior, thrips frequently draw the proleg across the antennae and then repeat the sequence. The opisthorhynchous condition of the mouthcone (mc) is shown, as well as the internal aspect of the plant cells.

labral pad, a single, left mandibular stylet, and a pair of maxillary stylets (Figures 6.1–6.4). Each of the paired paraglossae possesses 10 sensory pegs (Figure 6.2, 3b) of 3 distinct morphologically different types: sensilla basiconica lacking a cuticular collar (Figure 6.3b: 1,3), s. basiconica with cuticular collar (Figure 6.3b: 2, 4, 5, 7, 8) and s. trichoidea (Figure 6.3b: 6, 9, 10) (*sensu* Chisholm and Lewis 1984 (17)). Individual axons leading to each of the 10 sensory pegs are housed in the paraglossae and can be seen when the lateral surface is etched away (49). When the paraglossae are closed, as they are prior to probing, the sensory pegs cover the tip of the mouthcone and may come into direct contact with the plant when the mouthcone is pressed against the leaf surface (Figure 6.3b). Similar sensory pegs have been observed on the paraglossae of many thrips species in the suborder Terebrantia, thus appearing to represent the general condition of the group (17, 49, 128, Hunter and Ullman, unpublished data).

Whether the paraglossal sensory pegs are chemosensory or mechanosensory and what role they play in the chain of sensory events leading thrips to ingest from particular plants or plant cells has not yet been clearly elucidated. However, recent evidence from transmission electron microscopy (TEM) of thin sections suggest that some paraglossal sensory pegs are mechanosensory while others may have both mechano- and chemosensory functions (Hunter and Ullman, unpublished data). Understanding the sensory role of these structures in thrips host finding, host choice and feeding site selection is a topic of current interest and research (Hunter and Ullman unpublished data,

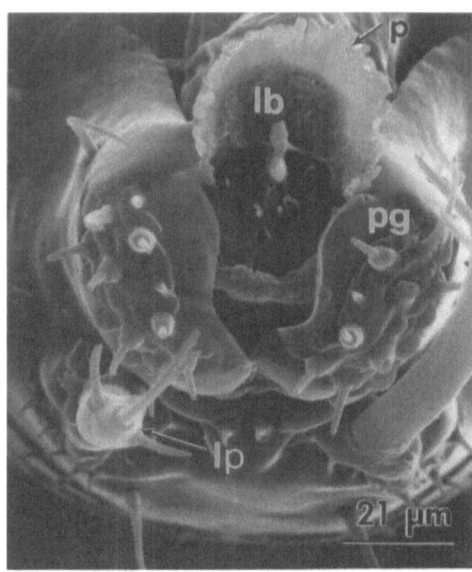

FIGURE 6.2. Tip of the mouthcone of the western flower thrips, *Frankliniella occiden-talis*, with the labral pad (lb) extruding from the paraglossae (pg) as they separate. Dorsal edge of labral pad is fringed with many small palpi (p) of possibly sensory function. Labial palps can also be seen (lp). Modified from Hunter and Ullman (49) with permission from Pergamon Press.

Kumar, Ullman and Cho unpublished data). Certainly, a strong foundation in understanding the sensory basis for various thrips behaviors will go a long way toward directing our search for plant germplasm with characters that are nonattractive to thrips or not preferred for feeding (115, 116). In addition, a comprehension of how specific feeding events lead to thrips inoculation of TSWV, will contribute to our understanding of epidemiology and may lead to development of plant varieties with resistance to thrips transmission of TSWV.

When the insect begins to probe, the paraglossae separate and extrusion of the labral pad is initiated (Figure 6.2). The labral pad is fringed with many small palpi (Figure 6.3b) and when fully extruded is omega shaped with the ends meeting ventrally. During feeding, the labral pad is pressed against the plant surface, probably stabilizing the mouthcone, providing support for the stylets and preventing loss of plant fluids during feeding. When the labral pad is withdrawn following a probe, surface waxes on the plant surface are often removed (17, 49) and probably ingested. Sensory cues regarding plant surface chemistry received during this process may be important in the sensory processes leading to prolonged feeding from a plant host. Removal of plant surface waxes during the feeding process may also contribute to increased transpirational loss and decreased photosynthesis resulting from thrips feed-

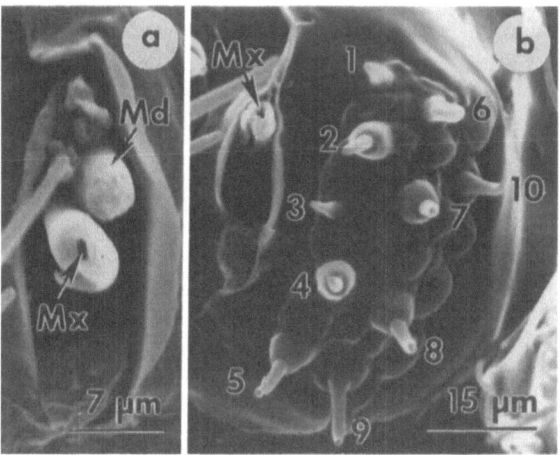

FIGURE 6.3. Tip of the mouthcone of the western flower thrips, *Frankliniella occidentalis*. (a) Single mandibular stylet (Md) is closed apically and used to puncture leaf cells creating an entry point for insertion of paired maxillary stylets (mx) that form a feeding tube. Arrow denotes subapical opening of feeding tube. (b) 10 sensory pegs shown on left paraglossa. Three morphologically distinct types can be seen, basiconica lacking a distinct cuticular collar (1,3), basiconica with a distinct cuticular collar (2, 4, 5, 7, 8), and sensilla trichoidea (6, 9, 10). (Modified from Hunter and Ullman (49) with permission from Pergamon Press.)

ing, as well as the typical silvery appearance of thrips feeding damage (8, 100, 125, 126). Chisholm and Lewis (1984) (17) proposed that the labral pad palpi may serve a sensory function; however, innervation of the palpi has not yet been observed.

As probing of the plant begins, the stylets emerge from the center of the labral pad and open paraglossae (Figures 6.2, 6.3a, 6.3b). The single, left mandible is closed apically (Figure 6.3a) and internally innervated by 3 central and 2 lateral dendrites of mechanosensory function (Figure 6.4b). Because the mandibular stylet is fused to the exoskeleton (49, 84), it can only be extruded through compression of the mouthcone. The mandible serves primarily to pierce plant cells after which it is followed by insertion of the maxillae which are under muscular control (17, 49, 84). At their origin high in the mouthcone, the maxillary stylets are widely separated from one another lying on either side and posterior to the mandible (Figure 6.4b). The maxillary stylets converge distally in the mouthcone until they meet and interlock in a tongue and groove fashion approximately 12 microns from the mouthcone tip (Figures 6.3a, 6.4a, 6.4a: inset). The maxillae open subapically and thus, serve as a tube for ingesting plant fluids (Figure 6.4a, 6.4a: inset) (49, 82). Both maxillary stylets are innervated and attached proximally to muscles.

By analogy to other piercing-sucking insects (5, 6), these data support the

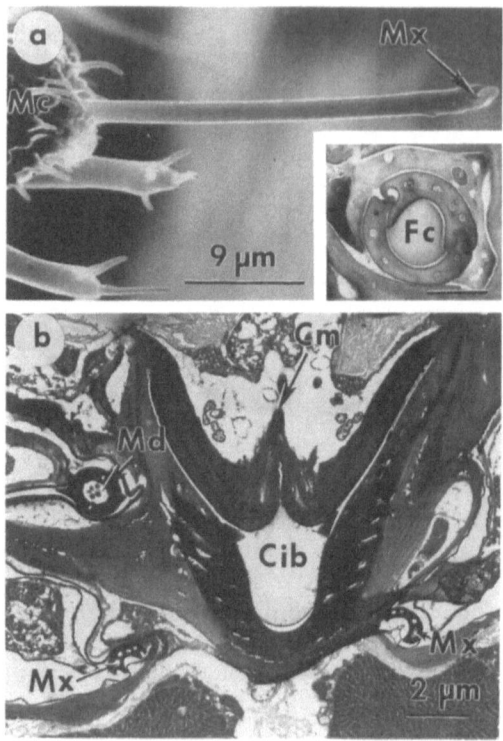

FIGURE 6.4. (a) Lateral view of the mouthcone (Mc) of the western flower thrips, *Frankliniella occidentalis*, with maxillary stylets (Mx) extended 30 μm beyond the mouthcone tip. Arrow denotes subapical opening to the feeding tube. Inset shows a transmission electron micrograph (TEM) of a cross section of the maxillary stylets. The manner in which the stylets interlock and form a food canal (Fc) can be seen, as well as the dendritic canals and dendrites present in each stylet. Modified from Hunter and Ullman (49) with permission from Pergamon Press. (b) TEM of a cross section of the proximal portion of the mouthcone showing the cibarium (Cib), the point at which cibarial muscles attach (Cm) and the orientation of the maxillary (Mx) and mandibular (Md) stylets relative to the cibarium and one another. Arrow denotes dendritic canal.

idea that thrips receive considerable mechanosensory information regarding stylet placement and movement and can direct mandibular and maxillary stylet movements. The maxillary stylets can be extended at least 30 μm (Figure 6.4a), suggesting that thrips may feed from cells at least 30 μm within plant tissue, a depth that could include vascular cells of small veins in some plants (49). Although vascular feeding has never been demonstrated, feeding often occurs from mesophyll cells well beneath the epidermal leaf layer (17, 48). Thrips possess a relatively large cibarial pump (Figure 6.4b) indicating that these small insects can readily pump fluids from plant cells. Innervation of the stylets, in conjunction with the thrips ability to insert the maxillary stylets relatively deeply in plant tissue, suggest thrips have far more control over the cells from which they feed and the amount of fluid they ingest from any given plant tissue than previously thought (Hunter and Ullman, unpublished data, 49, 50).

Observations of mouthcone and feeding structure morphology of 5 thrips species strongly support the hypothesis that a single mandible closed apically, paired maxillary stylets forming a feeding tube and paraglossal sensory structures are traits shared by most members of the Thysanoptera (49, Hunter and Ullman unpublished data). Current literature addressing morphology of several Terebrantian and Tubiliferan species lend support to this conclusion (17, 44, 45, 48). Rasping structures have not been observed on the thrips mouthcone, while all the structures required for piercing plant cells and sucking up plant cell cytoplasm can be found associated with every thrips species that has been studied. Direct observation of thrips feeding (17, 48, 49, 50), in conjunction with histological evidence (17, 48), clearly supports our earlier contention that the feeding strategy of the Thysanoptera should be reclassified to piercing-sucking (17, 48, 49, 118).

More than an exercise in semantics, reclassification of the way these tiny insects feed profoundly influences the way direct thrips feeding damage on leaves and fruit and assessment of concomitant yield losses are viewed (17, 48, 100, 113, 125, 126). Furthermore, our understanding of the thrips feeding repetoire will dramatically impact hypotheses regarding acquisition and transmission of plant viruses, particularly TSWV. Several recent Entomology texts have already categorized the Thysanoptera as piercing-sucking insects (10, 34), although the notion of thrips rasping epidermal layers away and sucking up exuding sap persists in others (33, 91, 125). Future considerations of thrips ecological niches, feeding behavior, feeding damage and impact on yield loss will hopefully be evaluated in light of the current understanding of Thysanopteran feeding modes. Much may then be learned by comparing thrips host finding, feeding site selection and feeding behavior to better studied systems among guilds of piercing-sucking insects (6, 48, 49, 71).

Thrips Feeding Behavior

Many thrips species feed on pollen and flower structures, as well as on vegetative portions of the plant (65). Our discussion will focus primarily on vegetative feeding, although most of the concepts we discuss probably have relevance to feeding on other plant parts.

The first behavior observed when thrips arrive upon a surface is antennal dragging and a scraping movement in which the insect draws its proleg across the plant surface, reaching beyond the head and other proleg, then drawing the leg across the antennae. This behavior may represent gathering of important plant cues leading to host selection, as sensory structures are present on the antennae (84, Hunter and Ullman, unpublished data). An adult WFT engaged in this behavior on the surface of a green bean pod has been captured with scanning electron microscopy (SEM) and is shown in Figure 6.1. Following these tarsal and antennal movements, the mouthcone is placed on the plant surface and the mouthcone and stylet movements described earlier under anatomy of feeding structures begin. Observation of WFT feeding on young, translucent lettuce leaves allowed visualization of stylet insertion and movement, as well as movement of plant fluids into the feeding tube (49). Examination of the feeding of *Limothrips cerealium*, the WFT and the soybean thrips, *Sericothrips variabilis* (Beach) with light microscopy and C14 inulin, has been combined with light, scanning and transmission electron microscope observation of plant cells fed upon by the various thrips species (17, 48, 49). These investigations reveal that thrips pierce and empty individual plant leaf cells of their cytoplasm (17, 48, 49), rather than rasping the leaf surface and rupturing many epidemal cells (33, 91, 125). Whole plant cell organelles, such as chloroplasts, are frequently ingested (17) and can be seen in thin sections of the thrips midgut following feeding (120). Leaf cells beneath a pierced epidermal cell usually are emptied completely, resulting in collapsed mesophyll and epidermal cells (17, 48).

Recently, a system for electronically monitoring the feeding of the WFT was used to further characterize thrips feeding (50, 60). Figure 6.14 shows an adult WFT attached with silver conductive paint to fine gold wire in a typical feeding posture during electronic monitoring. A complete description of these experiments and the current state of the art in electronic monitoring of insect feeding can be found in The Proceedings of an Informal Conference on Electronic Monitoring of Insect Feeding to Honor D.L. McLean and M.G. Kinsey (50). Evidence from electronic monitoring suggests that the WFT has at least 2 modes of feeding in which they frequently engage when ingesting leaf cell contents. First, thrips make many probes of short duration during which they apparently salivate into and empty the contents of single or small groups of plant cells probably just under the epidermal surface. Secondly and less often, probes of much longer duration are made which consist of a short period of apparent salivation followed by what appears to be long term ingestion, possibly from vascular cells. Histological evidence is still required to determine the depth at which these various types of probes occur in plant tissue. Sakimura (105) suggested, based on light microscope observations of thrips feeding damage that thrips make shallow and deep probes and that TSWV inoculation was more common following shallow probes. The above mentioned results of electronic monitoring suggest, that, hypothetical shallow and deep modes of thrips feeding (104, 105) should be reconsidered relative to transmission of TSWV and comprehending sensory mechanisms governing host and feeding site selection. No doubt an understanding of varying feeding modes will greatly advance our understanding of how thrips acquire and transmit TSWV and warrants future attention.

Distribution of feeding on leaf surfaces of different plant species has not been well documented for thrips species that transmit TSWV. However, many other thrips species have been shown to have characteristic leaf surface distributions (65). This aspect of thrips feeding may be very important in understanding differential acquisition of TSWV from various plant hosts, as TSWV distribution within plant tissues is known to vary among plant species. Furthermore, TSWV is frequently unevenly distributed in plants giving rise to ring spots, streaks and to characteristically "lopsided" symptomology (23, 35, 52, 61). Whether or not virus infection alters selection of feeding sites by thrips vector species is not known; however, the contention that thrips use plant cues to discriminate feeding sites on the plant surface and possibly chemistry of other plant fluids to determine duration of ingestion and mode of feeding suggests such a hypothesis warrants investigation. Furthermore, feeding site selection on various plant hosts with varying distributions of TSWV in infected tissues may be a critical element of thrips acquisition and transmission efficiency.

Internal Anatomy and Morphology

The following section on internal anatomy and morphology provides an orientation for tracing the movement of ingested plant fluids and TSWV particles through thrips tissues. To ingest the contents of plant cells, the thrips cibarial pump (Figure 6.4b) expands and plant fluids are drawn up into the food canal formed by the maxillae (Figure 6.4a) to the hypopharynx and then into the cibarial cavity. When the cibarial pump collapses, ingested plant fluids are pushed up through the circumoesophageal passage into the oesophagus (Figures 6.5a, 6.5b) (83–85, 118). The oesophagus is a flexible structure that apparently expands and contracts during the feeding process and is probably important in the process of drawing fluids up out of the cibarial pump (118, 119). The thrips midgut is looped, but to our knowledge not morphologically or functionally differentiated (Figures 6.5a, 6.5b). Ingested plant components leave the oesophagus and enter the midgut, which is formed by a single layer of columnar epithelial cells that serve as an interface between ingested plant material and the hemolymph. Internally, the midgut is lined with many microvilli that extend into the large gut lumen they encircle (Figures 6.20, 6.21) and mediate passage of macromolecules, such as virus particles, across the apical plasmalemma and into the columnar epithelial cell cytoplasm (13, 38, 72, 74, 101). The basal surface of the columnar epithelial cells of the midgut is formed by the basement membrane, the external surface of which faces the hemocoel where it is bathed by the hemolymph (Figure 6.6). Together, the apical plasmalemma and basement membrane represent 2 barriers through which ingested TSWV particles must pass to enter the insect hemocoel and infect epidemiologically important target organs, such as the salivary glands.

The pyloric valve denotes the juncture between the midgut and the hindgut (Figures 6.7a, 6.7b). Numerous membranous infoldings occur where the malphighian tubules meet the valve region just anterior to the hindgut forming

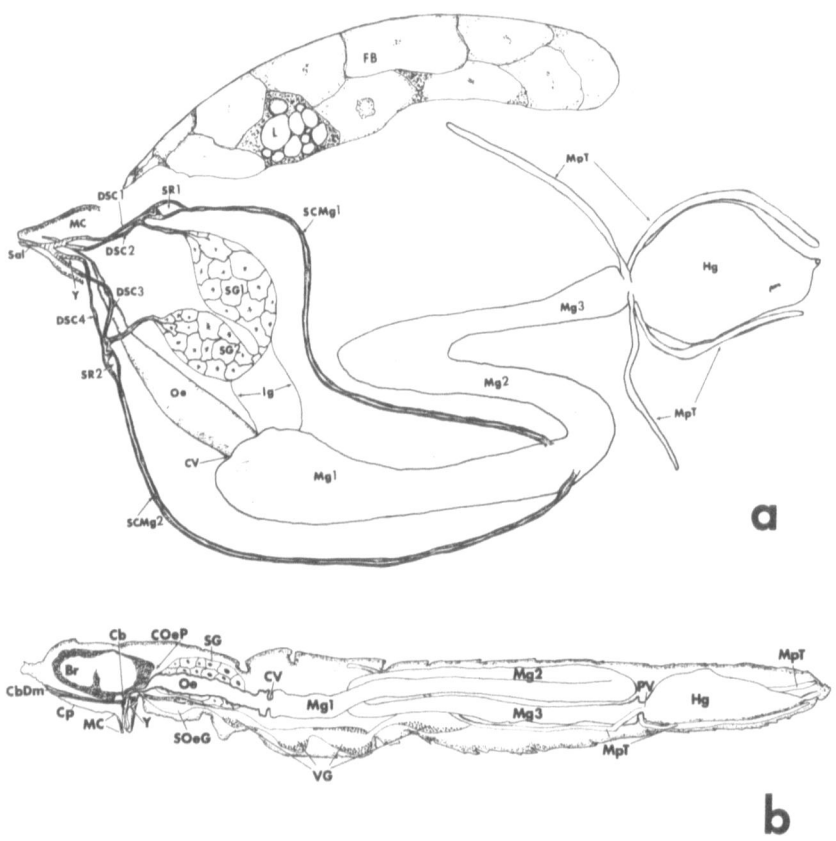

FIGURE 6.5. (a) Diagrammatic view of *Frankliniella occidentalis* organs in dissection. Note the large free floating fat body (FB). In situ this organ originates at the circumoesophageal passage (COeP) above the brain, intertwines with the salivary glands (SG1, SG2) and is pressed against the length of the midgut (Mg1, Mg2, Mg3). Orientation of salivary glands (SG1, SG2) and associated ducts (SCMg1, SCMg2, SR1, SR2) relative to the midgut (Mg1-Mg3) and deferent salivary canals (DSC1-DSC4) in the mouthcone (MC) are shown. (b) Diagrammatic sagittal/parasagittal view of adult *F. occidentalis*, showing relative orientation of internal organs in situ. Ducts between salivary glands and midgut and deferent salivary canals are not shown (see 6.5a), however, the Y-shaped salivary ducts leading to salivarium from their juncture with the deferent salivary canals can be seen (see (a). Br = brain, COeP = circumoesophageal passage, CV = cardiac valve, DSC1-DSC4 = deferent salivary canals, Hg = hindgut, MC = mouthcone, Mg1 = anterior midgut, Mg2 = central midgut, Mg3 = posterior midgut, MpT = malpighian tubules, Oe = oesphagus, Ov = Ovarioles, SCMg1, SCMg2 = salivary canals fastened to midgut, SG1, SG2 = salivary gland, SOeG = suboesophageal ganglion, Y = Y-shaped salivary ducts leading to salivarium. (Modified from Ullman, et al (118) with permission from Pergamon Press.)

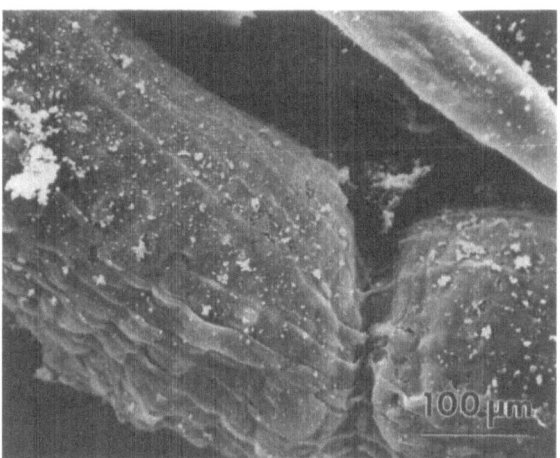

FIGURE 6.6. External aspect of the midgut columnar epithelial cells of the western flower thrips, *Frankliniella occidentalis*. The basement membrance of these cells, the external surface of which is shown here, represents the interface between ingested plant material entering the epithelial cell cytoplasm and the hemocoel. Note columnar morphology of the cells. Modified from Ullman, et al (118) with permission from Pergamon Press.

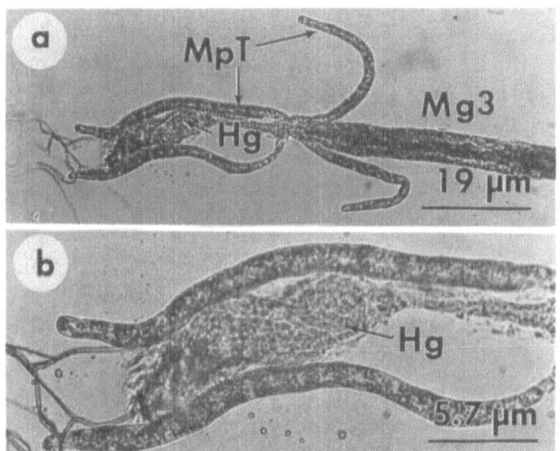

FIGURE 6.7. Light micrographs (LM) of the midgut (Mg3), hindgut (Hg) and malpighian tubules (MpT) of a dissected adult western flower thrips, *Frankliniella occidentalis*. (a) Typical orientation, with 2 malpighian tubules (MpT) free floating anteriorly, and 2 adhered to the hindgut (Hg) posteriorly is represented. (b) At higher magnification, posteriorly oriented tubules are shown as they leave their juncture at the pyloric valve and adhere for part of their length with the hindgut (Hg).

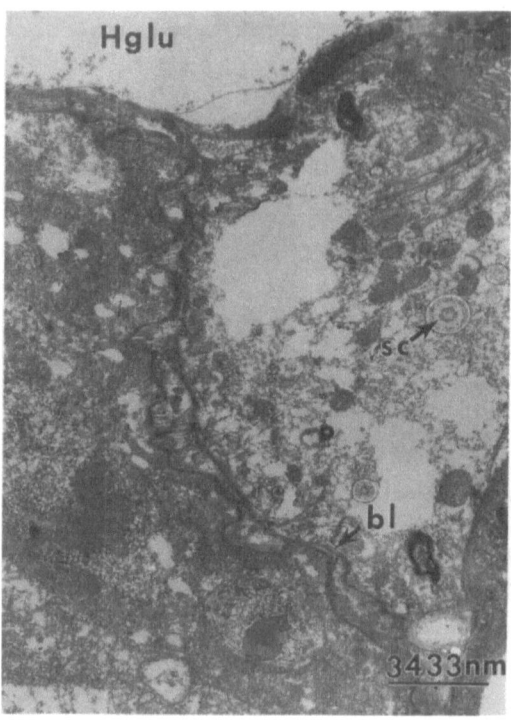

FIGURE 6.8. The malpighian tubules (MpT) arise at the pyloric valve area in the western flower thrips, *Frankliniella occidentalis*. The basal labyrinth (bl), characteristic at this juncture, hindgut lumen (Hglu) and spherocrystals (sc) that are common in primary cells of malphighian tubule are shown. (Modified from Ullman, et al (118) with permission from Pergamon Press.)

the basal labyrinth (Figure 6.8). Many spherocrystals, suggesting mineral sequestration, can be seen (Figure 6.8). Similar mineral spherocrystals are common in the malpighian tubules of insects (73). Contrary to previous observations of WFT internal morphology (118), 4, not 3 malpighian tubules arise at this juncture (Figures 6.7a, 6.7b). Thus, the presence of 4 malpighian tubules probably represents the common condition of all members of the Thysanopteran order (86, 93, 109). Two of the tubules are free floating and appear to lie in an anterior orientation in the insect. The other 2 malpighian tubules are always observed lying in a posterior orientation closely aligned with the basal membrane of the hindgut (Figures 6.7a, 6.7b, 6.9). In dissection, the posteriorly oriented malpighian tubules are free floating as they leave the pyloric valve area, after which they adhere to the hindgut (Figures 6.7a, 6.7b, 6.9). In the area of adhesion, distinct hindgut and malpighian tubule basal lamina can be seen in cross section (Figure 6.9). All 4 malpighian tubules are relatively simple in structure consisting of a single layer of cells with a micro-

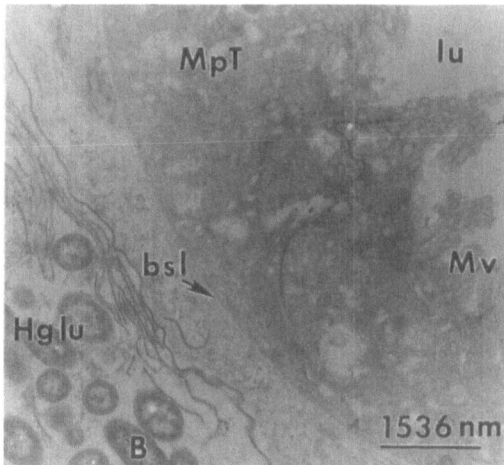

FIGURE 6.9. Cross section through a posteriorly oriented malpighian tubule (MpT) of the western flower thrips, *Frankliniella occidentalis*, in the region at which the tubule adheres to the hindgut. Distinctly separate basal lamina (bsl) can be seen in the area in which the tubule and hindgut epithelial cell adhere. Bacteria (B) observed in the hindgut lumen (Hglu) can be seen. The simple structure of the malpighian tubule, consisting of a single layer of epithelial cells surrounding a microvilli (Mv) lined lumen (lu) is shown. (Modified from Ullman, et al (118) with permission from Pergamon Press.)

villi lined lumen (Figure 6.9). Direct studies of thrips malipighian tubule function have not been done to our knowledge. However, by analogy to other insect systems, although simple in structure and association to the hindgut, functionally, the malpighian tubules must play an important role in osmoregulation for this tiny insect (101).

Like the midgut, the hindgut is lined with numerous microvilli. A rich microflora of rod shaped bacteria inhabit the hindgut lumen (Figure 6.10) (118). These bacterial organisms are present in all individuals we have sectioned regardless of age or diet. It is possible they are symbiotic organisms and play an important, although unstudied, role in thrips survival. Symbiotic bacteria have been previously reported in the Thysanoptera, although the reported descriptions do not contain enough detail to facilitate comparison to the bacteria we observe (12).

A most unusual characteristic of the internal anatomy of the WFT and at least 4 other thrips species we have investigated is a pair of ducts associated with the salivary glands anteriorly and adhered to the midgut posteriorly (Figures 6.11a, 6.12a) (118, 119). Light microscopy and TEM observation reveal a duct of striated cuticle within portions of these structures (Figures 6.11b, 6.12a) and microvilli lining other portions (30, 118). The cuticle lined ducts end at the midgut interface where both structures are clearly adhered

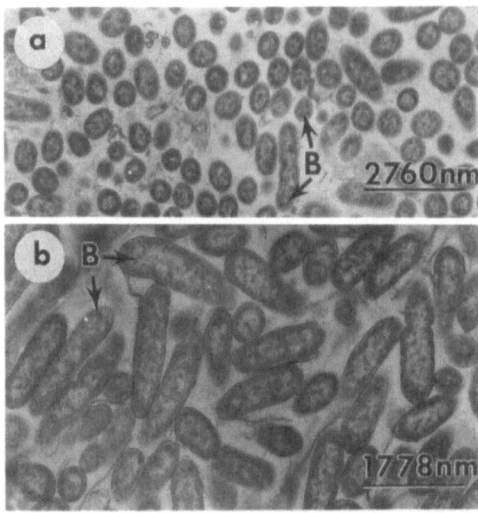

FIGURE 6.10. Hindgut lumen of the western flower thrips, *Frankliniella occidentalis*, showing resident bacteria (B). (a) Characteristic rod-shape and abundance of bacteria is shown. (b) At higher magnification the bacterial cell wall can be seen.

(Figure 6.11). TEM observations of thin sections of the area at which the ducts interface with the midgut epithelia, show that the duct ends at this juncture; however, the tissue surrounding the ducts appears to merge with the midgut epithelial cells for a minute distance, suggesting possible functional connections. Whether this adhesion provides a functional link between the midgut and the salivary glands is not known; however, the morphology of these structures and their relationship to the salivary glands and deferent ducts led us to classify them as salivary canals forming a union between the midgut, lobed salivary glands and the outside environment (118). Pesson (93) observed a similar pair of structures in *Aptinothrips rufus* and describes them as tubular salivary glands. He did not describe internal ducts, nor did he discuss their association with the midgut and ducts leading out to the salivarium. Recently, Del Bene, et al (30) suggest that the internal morphology of the structures supports their classification as tubular salivary glands. Their study differs from ours in that they were unable to observe a union between the structures and the midgut epithelium and suggested that the adhesion represented a blind ending held by ligaments or collagen. Final classification of these unusual structures as salivary canals, tubular salivary glands or some combination thereof will depend upon future studies of function. Certainly, the fascinating possibility that these structures produce and or carry enzymes or possibly virus particles between the midgut, lobed salivary glands and the outside environment deserves further attention. In the meantime, we continue

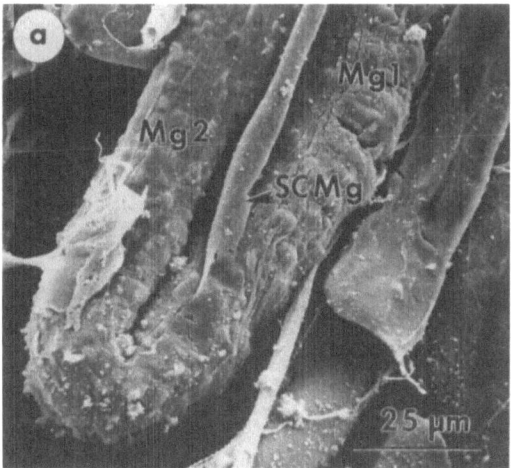

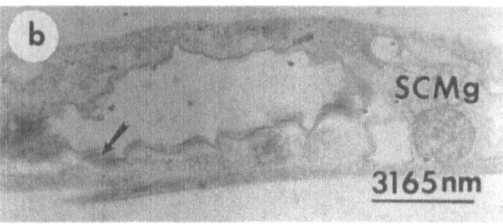

FIGURE 6.11. Midgut and salivary ducts fastened to the midgut of the western flower thrips, *Frankliniella occidentalis*. (a) External aspect of the region between the anterior and intermediate loops of the midgut (Mg1 and Mg2, respectively) where one of 2 ducts (SCMg1) leading from the midgut to the salivary glands is adhered. (b) Cross section of one of the 2 ducts (SCMg) showing striated cuticle lining a portion of duct lumen (arrow) and mitochondria in cellular portion of the duct. (Modified from Ullman, et al (118) with permission from Pergamon Press.)

to refer to the structures as salivary canals fastened to the midgut (SCMg1, SCMg2). Anteriorly, the ducts meet an apparent reservoir (Figures 6.5a, 6.12a, SR1, SR2). Three sets of ducts leave each reservoir, 1 pair connecting to the lobed salivary gland and the other 2 leading to Y-shaped ducts that eventually converge and empty into the salivarium in the mouthcone (Figures 6.5a, 6.5b, 6.12a, DSC1-4).

The lobed salivary gland lies within the prothoracic segment sometimes extending into the mesothoracic segment (Figure 6.5b). The gland has 2 lobes joined by a narrow piece of tissue and is held in place posteriorly by fine ligaments that attach to the anterior midgut (Mg1) just below the cardiac valve (Figures 6.5a, 6.12a). The gland is readily dissected from live thrips

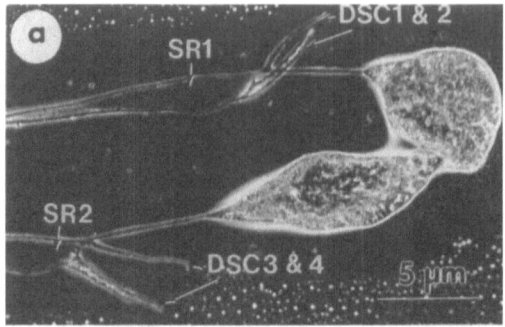

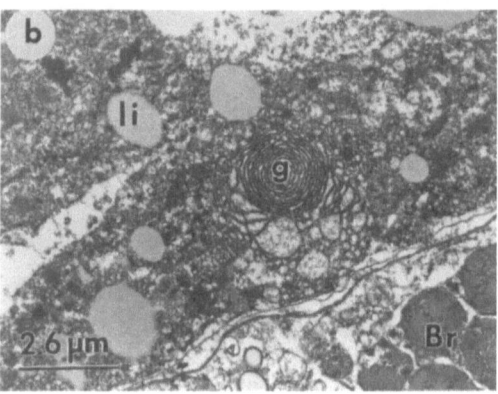

FIGURE 6.12. (a) Salivary gland dissected from an adult western flower thrips, *Frankliniella occidentalis*. Light micrograph shows the salivary reservoirs (SR1, SR2) that are present at the juncture between the salivary canals fastened to the midgut (SCMg in Figures 6.5a, 6.5b, and 6.11), the deferent salivary canals (DSC1-DSC4) and the canals leading to and from the salivary gland. The deferent salivary canals lead to the Y-shaped salivary ducts that ultimately empty into the salivarium in the mouthcone (see Figure 6.5a). The external aspect of the large loosely aggregated cells making up the lobed salivary gland can be seen. (b) Transmission electron micrograph (TEM) showing the internal aspects of a salivary gland cell where it presses against the brain (Br). These cells are rich in lipids (li) and Golgi bodies (g). (Lower figure modified from Ullman, et al (118) with permission from Pergamon Press.)

(Figure 6.12a) and the lobes are composed of large, loosely aggregated cells that are rich in lipid droplets, golgi, coated vesicles and numerous lamellar infoldings (Figure 6.12a, 6.12b). The lumen of each lobe is lined with microvilli (Figure 6.21) and empties into a canal that apparently leads to a deferent salivary canal and possibly to the salivary reservoirs (Figures 6.5a, 6.12a, DSC1 & 3, SR1 & 2). The salivary glands are rich in golgi bodies and coated vesicles, organelles that are potentially important in TSWV replication. By analogy to other members of the Bunyaviridae that replicate in the golgi body and are transported intracellularly via coated vesicles, the salivary glands should be considered as a potential TSWV replication site (118, 119).

The reproductive systems of those Thysanopterans that have been described appear to be uniformly similar (93, 118). Females have a pair of ovaries, one lying ventral to the midgut and the other to one side (Figures 6.5b, 6.13a). Each ovary consists of 4 ovarioles suspended anteriorly from a terminal filament and merging posteriorly into a lateral oviduct that eventually connects to the median oviduct and the vagina (Figure 6.13b). Males have a pair of accessory glands that lie ventral to the hindgut and a pair of testes with ducts leading to a common ejaculatory duct (86, 93, 118).

The fat bodies of insects are highly variable in their external and internal morphology and their location within the insect hemocoel (68, 101). Recently, we have observed a large structure in thick and thin sections, as well as in dissections of live WFT, that we tentatively classify as a fat body (Figure 6.5a).

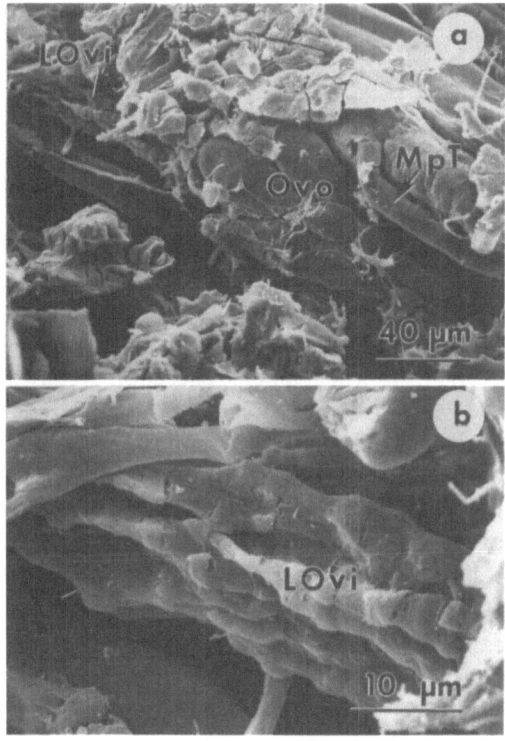

FIGURE 6.13. External aspects of one ovary and set of oviducts of an adult western flower thrips, *Frankliniella occidentalis*. (a) Three of 4 ovarioles (Ovo) can be seen where they press against the junction between the midgut and hindgut. Ovarioles merge posteriorly into a lateral oviduct (LOvi). One malpighian tubule (MpT) is shown arising from the junction between the midgut and hindgut. (Modified from Ullman, et al (118) with permission from Pergamon Press.)

Like fat bodies in many other insects, this organ is a sac-like structure surrounded by a membrane that is apparently free floating in the hemocoel and contains tremendous quantities of lipids, many vacuoles, membranous swirls and infoldings. Anteriorly, this putative fat body is very narrow and pressed against the circumoesophageal passage and the brain (Figures 6.5a, 6.5b). Posteriorly, the organ rapidly widens and is intertwined with the lobed salivary glands and loops of the midgut. At certain times in WFT development, the fat body occupies a large portion of the thorax and abdomen. In numerous dissections of live WFT of varying age, we have observed that size, appearance and lipid content of this organ changes with development and age of the insect, further supporting its classification as a fat body. Currently, evidence is accumulating that may more definitively classify this previously undescribed organ as a fat body (Ullman and Westcot, unpublished data) and to elucidate its role in TSWV fate in thrips vector species.

Thrips Acquisition and Transmission of TSWV

Importance of Thrips Development in the Epidemiological Significance of Plant Hosts

When developing and implementing strategies for managing the spread of TSWV in agricultural or ornamental crops, it is critical to remember that acquisition of the virus can only occur during the larval stages of the insect (18, 20, 23, 24, 75, 88, 104–106, 119). Thrips larvae are not winged and are not readily dispersed by the wind, while adults are winged and disperse readily (65, 75, 104). Thus, female adults must select plant hosts on which their larval offspring develop and from which a new generation of fully developed adults ultimately disperse.

The epidemiological significance of this important aspect of thrips biology and TSWV relationships is illustrated in Figure 6.15. Dispersing viruliferous adult thrips may encounter and feed upon many plant species. Some of these plant species may not be hosts of TSWV, thus, feeding of viruliferous adult thrips will not result in infection. Among these non-hosts of TSWV, some species are hosts of the thrips and will support thrips development. Adults may oviposit on these plants, with eclosing larvae completing their development and dispersing from these plants. In this situation, none of the dispersing adults will be viruliferous. Thus, thrips populations developing on plants immune to TSWV do not later contribute to primary infections in crops, nor can they be directly involved in secondary spread. Essentially, these insects represent a dead end for the disease cycle (See Figure 6.14).

The only contribution non-viruliferous adults may make to TSWV epidemics is through oviposition on TSWV infected plant hosts that may increase the number of infectious adult thrips in the next generation. When management strategies are being designed for crops that support thrips popu-

FIGURE 6.14. A western flower thrips adult, *Frankliniella occidentalis*, beginning to probe the leaf surface during electronic monitoring. Arrow denotes the gold wire fastened to the dorsum of the insect to facilitate electronic monitoring.

lations, such as lettuce, this contribution to thrips populations within the crop may be significant and warrant control. Build up of viruliferous thrips populations in these crops is particularly significant when growers plant sequentially, because viruliferous thrips populations developing in one planting cycle heavily influence disease incidence in future planting cycles (132).

In contrast, when viruliferous adult thrips encounter plants susceptible to TSWV, infection may occur. Among the many plant species falling into this category, many will not support thrips egg and larval development (75, 88, Bautista and Mau, unpublished data). Hence, these plants also represent a dead end for the TSWV disease cycle and are probably not epidemiologically significant. If the plant species is a crop plant, reduced or no secondary spread can be expected from foci within the crop. Consequently, control measures should be aimed at limiting primary infection. This can be done by controlling epidemiologically significant alternate crops and weed hosts, limiting dispersal of viruliferous thrips into the crop and/or using plant varieties that are unattractive to thrips. Breeding plant varieties that incorporate characteristics that reduce thrips landing due to nonattractive leaf and flower colors, chemistry or structure are worthy of considerable attention (115, 116, Kumar, Ullman and Cho, unpublished data).

Plant species of greatest importance to spread of TSWV are those that viruliferous adult thrips feed upon, infect and that subsequently support egg and larval development. These plant species serve as tremendous reservoirs of TSWV and viruliferous thrips, thus, acting as foci for primary infection in crops and newly emerging weeds. Within cropping systems, these plants are important foci for secondary and in some cases, tertiary virus spread.

The epidemiological concepts, graphically depicted in Figure 6.15, are

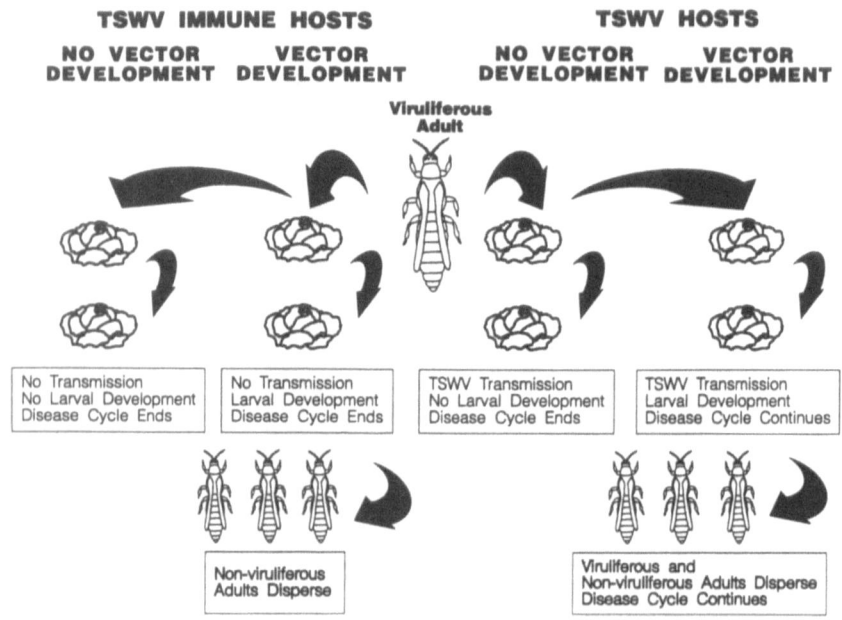

FIGURE 6.15. Graphic representation of how thrips biology and development on various plant species can impact the TSWV disease cycle. Larval thrips do not disperse readily and are the only stage of the insect that can acquire TSWV. Hence, the plant species of greatest epidemiological significance are those that are susceptible to inoculation with TSWV by dispersing adult thrips and also subsequently support thrips development. Viruliferous and non-viruliferous thrips will disperse from such plants serving to introduce TSWV foci into crop plantings or to spread TSWV from previously existing foci. On the other end of the spectrum are plants that do not host either TSWV or thrips vector species and are, hence, of no epidemiological significance. The intermediate conditions on this continuum are also represented.

pivotal to development of TSWV and thrips management strategies in varying geographic locations and crops. Success of integrating a variety of management tactics based on these concepts has been demonstrated in Hawaii, where crop rotation with epidemiologically "dead end" plants, and spatial and temporal avoidance of epidemiologically important plants reduced virus incidence on some farms by 50% (23). In Hawaii, a large number of noncrop plant species are TSWV hosts; however, many species may not support WFT development, thus reducing their epidemiological significance and importance as targets of control programs (18–23, 75, 129, 131). In order to develop a directed and effective management program for Hawaiian growers, much effort has been focused on developing an understanding of thrips biology on varying TSWV plant hosts, as well as elucidating variability of transmission efficiencies following acquisition from different plant hosts (75, Bautista and Mau, unpublished data).

In other geographic regions where TSWV is an important economic problem, far fewer noncrop plant species may be of epidemiological importance (80). Thus, management approaches in these regions may target manipulation of crop plantings with less need to consider weed species as alternative virus sources. Clearly, the thrips/TSWV relationships illustrated in Figure 6.15 greatly influence management. Furthermore, quarantine decisions regarding TSWV spread through movement of plant transplants and cuttings from one geographic region to another should be based on an understanding of thrips biology on the plant material in question.

WFT Biology on Plant Hosts of TSWV

Much of our discussion revolves around the WFT, primarily because this species has been considered the most significant TSWV vector species in Hawaii and elsewhere due to its abundance (75). Other vector species may be of greater importance, depending on species compositions and virus isolate(s) predominating in epidemics in various geographical regions (75, 80, 88, 96, 97). It should be noted that little is known about relative efficiency and propensity (*sensu* Irwin and Ruesink (55)) of different thrips vector species. Certainly, further attention should be focused on developing an understanding of vector intensities of different thrips species, their relationships with different virus isolates and subsequent impacts on TSWV epidemiology. Preliminary information on this topic is presented by Mau, et al (75).

As part of an effort to target control of plant species of greatest epidemiological significance in the Hawaiian islands, suitability of plant hosts based on oviposition and feeding preference, as well as thrips development have been extensively studied (75, Bautista and Mau unpublished data). Among the many plant hosts tested, vegetative stages of *Fagopyrum esculentum* (buckwheat), *Amaranthus hybridus* L. (green amaranth), *Sonchus oleraceus* (sowthistle), *Arcticum lappa* L. (burdock), *Brassica campestris* L. (cabbage), *Malva parviflora* L. (cheeseweed), *Verbesina encelioides* (Cav.) Benth. & Hook. (golden crown-beard), *Datura stramonium* L. (jimson weed), *Lactuca sativa* var. *longifolia* (lettuce) were shown to support thrips development and are thus, of considerable potential importance in contributing to TSWV epidemics. When flowering stages were examined, cheeseweed and golden crown-beard increased in their suitability as thrips hosts, thus, amplifying the potential importance of these species in contributing viruliferous thrips to dispersing populations (Bautista and Mau, unpublished data). Shifts in thrips host preference and plant suitability during flowering have been previously documented and are not surprising considering the predilection of WFT for flowers (129, 131). TSWV has been detected in flowering plant parts, although to our knowledge nothing is known regarding thrips acquisition of TSWV from infected flowers. Certainly, this is an important data gap that may change the way epidemiologically important TSWV plant hosts are viewed.

The suitability of a number of other plant species were also examined.

Vegetative stages of *Nicandra physalodes* (L.) Gaertn. (apple of Peru), *Emilia sonchifolia* (L.) (Flora's paint brush), *Tropaeolum majus* (nasturtium), *Galinsoga quadriradiata* (Raf.) Blake (Peruvian daisy), *Limonium latifolium* (Sm.) Ktze. (statice), *Lycopersicon esculentum* Mill. (tomato), *Nicotiana tabacum* L. (tobacco) and *Verbena litoralis* (verbena) were relatively unsuitable or did not support larval WFT development. Thus, these plant species are considered less important epidemiologically in the Hawaiian islands (75, Bautista and Mau unpublished data). These data provide an important foundation for developing regional strategies for TSWV management where the WFT is the primary vector species (ie., directed weed control), as well as influencing the formulation of appropriate strategies for different crops (ie., lettuce, a highly suitable thrips host vs. crops that are relatively unsuitable thrips hosts). There is a need for similar information to be developed for other vector species and TSWV plant hosts that may be of importance in various geographical regions where TSWV epidemics occur.

Role of Plant Host in TSWV Acquisition Efficiency

The epidemiological importance of thrips biology on different TSWV infected plant hosts has been discussed earlier in this chapter and is illustrated in Figure 6.15. Another component influencing the relative epidemiological significance of plant species that host both thrips and TSWV, is the efficiency with which thrips acquire virus from these plant species. The importance of plant host species to efficiency of TSWV transmission by the WFT has been demonstrated in Ontario greenhouses (1–3, 14, 15).

Preliminary evidence from recent investigations with the WFT demonstrate that acquisition of TSWV by larval thrips varies with plant host species as indicated by enzyme-linked immunosorbent assay (ELISA) of individual larval thrips following acquisition feeding (Figure 6.16). When larvae fed on infected jimson weed (*Datura* in Figure 6.16), virus could be detected with ELISA in 75–100% of the individuals tested. This rate of acquisition was seldom achieved in larval populations fed on infected burdock (*Arcticum*). Furthermore, the high TSWV titers acquired by larvae feeding from infected jimson weed (optical density readings (OD) at 405 nm of 0.9–1.1 in individual larvae) were never achieved when acquisition feeding occurred on burdock.

There are several possible explanations for these results most of which revolve around the suitability of a plant to support virus replication and/or the internal distribution of virus particles in systemically infected plant tissues. First, jimson weed may support higher virus titers than burdock. Secondly, TSWV may be more evenly and widely distributed in jimson weed cells giving WFT larvae greater opportunity to access virus particles during feeding. Finally, thrips may prefer feeding on jimson weed, thus ingesting more infected plant material from this plant host.

With regard to the first explanation, high virus titers can be detected in

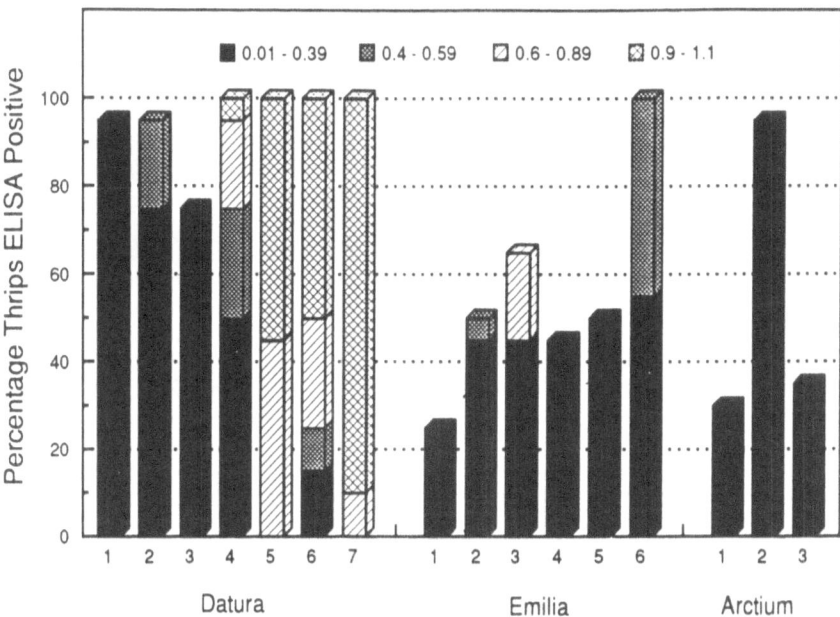

FIGURE 6.16. Stacked bar histogram showing the results of enzyme-linked immuno-sorbent assay (ELISA) of individual western flower thrips larvae (*Frankliniella occidentalis*), following acquisition feeding from three different TSWV plant host species, *Datura stramonium*, *Emilia sonchifolia* and *Articum lappa*. The percentage of thrips in which TSWV was detected is indicated on the vertical axis, while hatching in bars indicates the optical density (OD) at 405 nm of individual thrips submitted to ELISA testing (see upper portion of graph for legend). The highest TSWV titers, as indicated by OD at 405 nm, and greatest percentage of ELISA positive larvae were found following acquisition feeding on *Datura stramonium*.

both plant species when macerated leaf tissue is tested by ELISA (19–22). Distribution of virus within the leaves of these plant species docs vary, as indicated by differences in symptomology and preliminary evidence from direct immunoblotting of leaf tissue (47, 66, 119, Ullman and Cho, unpublished data). With direct immunoblotting, TSWV was very evenly distributed among cells of systemically infected jimson weed leaves. In contrast, TSWV distribution was patchy in the leaves of burdock. Thus, distribution of TSWV in jimson weed appears to provide increased opportunity for virus ingestion to larval thrips populations and may partially explain the high virus titers detected. Feeding preference may also play a role, as jimson weed, the plant from which larvae acquired TSWV most efficiently, is also highly suitable for thrips feeding, oviposition and development (75).

How important is the virus titer acquired by larval thrips to subsequent numbers of viruliferous adults in the population? Preliminary evidence from experiments in which larvae and adults developing from the same larval

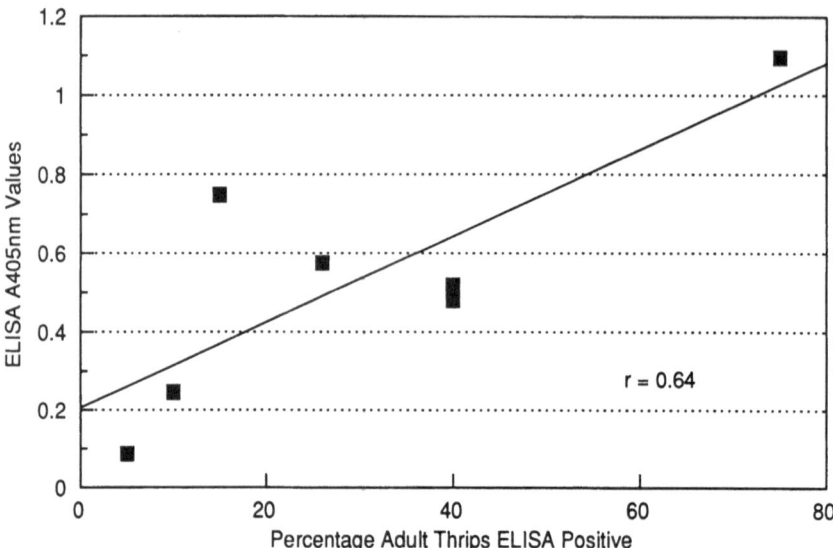

FIGURE 6.17. Regression of the percentage of adult thrips in which tomato spotted wilt virus (TSWV) could be detected with enzyme-linked immunosorbent assay (ELISA) against the OD at 405 nm of individual larvae from which they were reared following TSWV acquisition feeding. There is a low, but possibly important correlation between high and/or low virus titers following larval acquisiton and the percentage of ELISA positive adults reared from these cohorts.

cohort were subsampled and tested with ELISA, show that there is some correlation between OD at 405 nm and the percentage of adults thrips in which TSWV can be detected (Figure 6.17). When these data are considered in conjunction with what is known about host suitability and abundance of WFT in Hawaii, jimson weed clearly emerges as a very significant non-crop virus source (24, 75). While burdock is a good host for thrips development, thrips acquisition from this plant species does not appear to be very efficient. Nevertheless, this plant species may be of some epidemiological significance, because, inefficiency in acquisition could be offset by the large number of WFT that can develop in burdock plantings (burdock is an important local crop in the Hawaiian islands, while considered a weed species elsewhere).

Under conditions of forced thrips feeding in the laboratory, Flora's paintbrush (*E. sonchifolia*), a plant that does not support WFT larval development and hence, is not epidemiologically important in nature, served as a better acquisition host than burdock (Figure 6.16). While Flora's paintbrush is not important to TSWV epidemiology in regions where the WFT is the primary vector species, these data are experimentally helpful as this plant species is an excellent host for long term maintenance of thrips-transmissible isolates of TSWV (authors, personal observation).

Impact of Plant Host and Feeding Behavior on Thrips Inoculation of TSWV

In many insect transmitted plant virus systems, vector feeding behaviors that are optimal for virus inoculation differ from those that are optimal for virus acquisition. While the above mentioned data demonstrates the importance of host preference and sustained ingestion of infected plant material to thrips acquisition of TSWV, little information is available regarding varying inoculation efficiencies when acquisition hosts vary or when feeding preference on the inoculation host varies. Sakimura (105) recorded the acquisition and inoculation efficiencies of the OT (onion thrips) and the TT (tobacco thrips) on various plant hosts. His data suggest that interactions occurring at both the acquisition and inoculation thrips/plant interface contribute to altered inoculation efficiencies. While Sakimura did not investigate or discuss the possible mechanisms underlying these results, his work clearly demonstrated that both the TT and OT transmitted TSWV with the highest efficiency, 42% and 61%, respectively, when acquisition and inoculation occurred on Flora's paintbrush. In contrast, when the thrips acquired TSWV from Flora's paintbrush, inoculation efficiency on aster was reduced to 22% and 35%, respectively, half of that observed when Flora's paintbrush was used for both acquisition and transmission. Acquisition from aster and inoculation on Flora's paintbrush resulted in 24% and 26% transmission, respectively. When aster was used for both acquisition and inoculation the OT did not transmit TSWV at all and the TT transmitted to only 12% of the test plants. Although proposing an explanation for these results now would be speculative, Sakimura's data represents one of the most complete demonstrations of how varying acquisition and inoculation plant host species influence thrips transmission efficiencies.

More recently, the influence of feeding preference on WFT transmission of TSWV to varying chrysanthemum clones was investigated (3, 15). These data demonstrated that the size of leaf scars (a means of quantifying thrips feeding damage) were poorly correlated with TSWV inoculation of chrysanthemum clones. This result may have occurred because susceptibility of chrysanthemum clones to TSWV is epidemiologically more significant than the feeding activity of the WFT (15). An alternative explanation that may warrant consideration is that thrips inoculate TSWV more efficiently to less preferred chrysanthemum clones on which they feed poorly. While initially such an idea may seem counter intuitive, it is possible that an inverse relationship between feeding duration by thrips and TSWV inoculation to plants exists. Feeding events occurring on lesser preferred hosts, in brief feeding bouts, may be more important to successful inoculation than the duration or quantity of feeding that occurs. Support for this interpretation is provided by the findings of Sakimura (105) which indicate that TSWV inoculation was most efficient when thrips feeding scars were extremely small. The idea that plant hosts that are nonpreferred by thrips could be better inoculation hosts warrants further investigation, particularly in light of how such a concept might alter the goals of plant breeders aiming to produce plant varieties with resistance to thrips transmission of TSWV.

Variation in TSWV Isolates and Vector/Virus Relationships

As mentioned earlier in this chapter, recent evidence suggests that several TSWV isolates and serotypes exist (26, 56, 64, 87, 92, 98, 117, 124) (see addendum). Specific relationships between TSWV isolates and thrips vector species are not yet well understood. However, there has long been reason to believe that vector specificity to different TSWV isolates exists and that thrips vector species do not acquire and transmit TSWV isolates with equal efficiency (75, 89, 90, 104–106).

Paliwal (90) demonstrated that the TT transmitted two Canadian isolates of TSWV more efficiently than did the WFT. Furthermore, the OT did not transmit two Canadian isolates of TSWV. Additional evidence of vector specificity to different TSWV isolates was provided by the research of Amin, et al (4), that demonstrated differing inoculation efficiencies for the CBT and the CT when fed on varying peanut isolates of TSWV. Recent preliminary data suggest that the severe TSWV-L isolate used in our laboratory studies is readily acquired by the WFT, but not by the OT or CBT (Mau and Cho, pers. communication). These results contrast early data (102–106), that suggest Hawaiian epidemics were once dominated by an isolate readily acquired by the OT (102, 105). Subsampling of thrips species following TSWV acquisition suggest that transtadial passage of the above mentioned lettuce isolate occurs only in the WFT (Cho and Mau, pers. comm.).

By analogy to other insect/virus systems, changes in viral components, such as membrane glycoproteins or nucleocapsid proteins, may underscore specificity between TSWV isolates and thrips vector species (13, 71). Among the Bunyaviridae, membrane glycoproteins play a critical role in receptor mediated endocytosis and cell fusion critical to the infection pathway in humans, as well as arthropod vectors (31, 39, 69). Investigations of defective TSWV isolates that have lost thrips transmissibility indicate that viral membrane glycoproteins are altered or absent, thus, highlighting the potential importance of these structural proteins in thrips vector specificity (92, 122). The role of TSWV membrane glycoproteins in mediation of TSWV acquisition of various TSWV isolates is a topic of interest and future research (Ullman, German and Sherwood, unpublished data).

Cellular Events Preventing TSWV Acquisition by Adult Thrips

As illustrated and discussed earlier in this chapter, the thrips/TSWV relationship is unique among insect transmitted plant viruses in that adult thrips can only transmit TSWV when acquisition occurred during larval feeding upon

TSWV-infected plants (12, 14, 22, 23). Once acquired by thrips larvae, TSWV is circulative in the insect, passed transtadially, possibly replicates in thrips cells and is transmitted persistently by adult thrips (4, 7, 18, 22, 24, 37, 75, 102–106). In contrast, adult thrips that have not fed on infected plants during larval instars can not become viruliferous even if they are allowed lengthy feeding on TSWV infected plants (24, 25, 75, 102–106, 118-120). This unique insect/virus relationship has been recognized for many years; however, until recently, the underlying mechanisms have eluded entomologists and plant pathologists alike (25, 94, 102–106).

Early researchers explored the potential role of physiological factors in this phenomenon, such as, differences in larval and adult midgut pH that may affect virus stability or dissemination and the potential of the adult midgut serving as a barrier to viropexis and dissemination of TSWV into the thrips hemocoel (25, 104). The internal morphology of larval and adult thrips, discussed earlier in this chapter demonstrated that there was no obvious anatomical explanation for this unique relationship (118). Hence, our research focused on possible behavioral and/or cellular explanations.

Enzyme-linked immunosorbent assay (ELISA) of individual thrips showed that nonvector thrips species, such as *Hercinothrips femoralis* and vector species, such the WFT, ingested TSWV when given feeding access to infected plants as larvae or as adults (Figure 6.18). Hence, simple differences in the feeding behavior of larval and adult thrips can not entirely explain vector specificities or the inability of adult thrips to acquire TSWV. The data presented in Figure 6.18 also reaffirm that, when WFT larvae are fed on infected plants, a percentage of the population acquires TSWV and the virus is transtadially passed through the non-feeding pupal stages to the adult where it apparently persists throughout the life of the insect. In contrast, TSWV was only retained for a short time in adult WFT given their first access to infected plants as adults (Figure 6.18).

When these adult cohorts were removed from infected plants and held on noninfected plant material a percentage of the population retained the virus longer than one would expect if retention was based simply on excreting ingested virus particles. Following 4 days of postacquisition feeding on noninfected plant material, TSWV could still be detected with ELISA in 7% of the adults tested (Figure 6.18). These results demonstrate that ELISA testing of adult thrips trapped in nature may lead to erroneous conclusions regarding viruliferous thrips populations. It is apparent that TSWV would be detected in any insect that recently fed on an infected plant, although only those that acquired the virus as larvae could inoculate a plant and be of epidemiological importance.

These data led us to wonder if TSWV entered adult WFT cells and in which cells it might be retained. To address this question, we fed larval and adult thrips on TSWV infected jimson weed (*D. stramonium*) for varying times. After feeding, the insects were fed on sucrose for 1 hour and dissected so the head ends contained the salivary glands, a piece of the fat body and the feeding

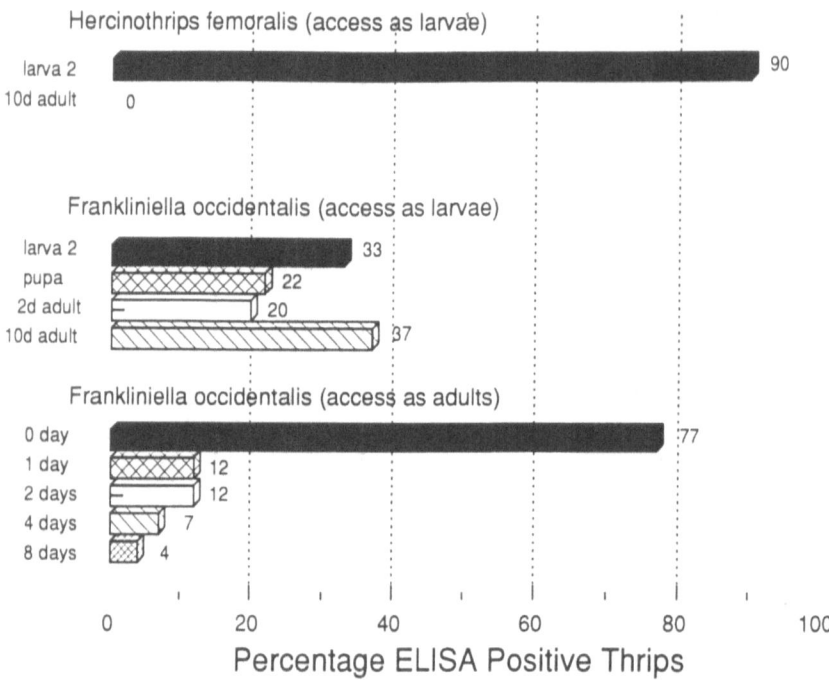

FIGURE 6.18. Horizontal histogram showing the percentage of individual thrips in which TSWV was detected with ELISA following acquisition access on TSWV infected plants and various subsequent post acquisition feeding periods on noninfected plant material. A nonvector thrips species, *Hercinothrips femoralis* given feeding access to TSWV infected plants is compared to a vector thrips species, *Frankliniella occidentalis* given feeding access to TSWV infected plants during larval or adult stages.

structures while the posterior end (labeled tail in Figure 6.19) contained the oesophagus, the majority of the fat body, the digestive tract and all the other internal organs (see Figures 6.5a, 6.5b and earlier discussion of internal anatomy for orientation). After dissection, thrips pieces were tested with ELISA for TSWV presence (22, 24). TSWV could be detected in dissected larval heads and posterior parts after only 1 hour of acquisition feeding (Figure 6.19). In contrast, TSWV was not detected in adult heads even after 24 hours of acquisition feeding, although virus was detected in the posterior ends following only 1 hour of acquisition feeding. Observation with TEM of thin sections of WFT from these cohorts revealed that the midgut and hindgut lumens of WFT adults contained many TSWV particles after 24 hours of acquisition feeding (Figures 6.20, 6.21, 6.23) (119, 120).

Virions could be seen lining the lumen and cytoplasm near the apical plasmalemma of the midgut microvilli of adults following acquisition feeding from TSWV infected plants (Figure 6.21). Virus particles were also seen in

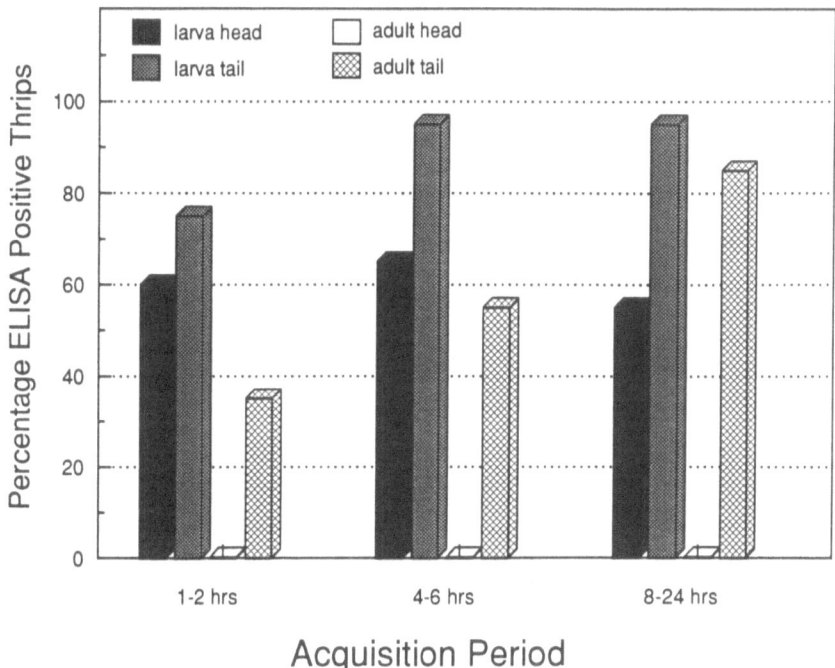

FIGURE 6.19. Histogram showing percentage of anterior (head) or posterior (abdomen or tail end) *Frankliniella occidentalis* dissections in which TSWV could be detected with ELISA following various acquisition feeding periods on TSWV infected plants.

possible vesicles near the apical plasmalemma and numerous TSWV particles could be seen in the cytoplasm of the midgut epithelial cells (Figure 6.22) (120). While the hindgut lumen was frequently observed full of virus particles, TSWV was not seen fusing to the microvillar membrane nor were the virions seen in the cytoplasm of hindgut epithelial cells (Figure 6.23). Perhaps of greatest importance, TSWV was confined to the digestive tract and midgut epithelia of these WFT adults. TSWV particles were not observed in the hemocoel of any WFT given feeding access to infected plants as an adult.

In contrast, virus particles could be detected with ELISA in the heads of larvae after only 1 hour of acquisition feeding. Virions were also observed with TEM in the hemocoel of larvae prepared for viewing after 24 hours of acquisition feeding. When viruliferous larvae were allowed to develop to adulthood, virus particles could be seen in the salivary glands (Figure 6.24) and TSWV was transmitted to 68% of the plants we inoculated using thrips from these cohorts. From these data, it appears that the midgut serves as a barrier to TSWV acquisition by adult thrips as previously postulated (25, 104, 119, 120).

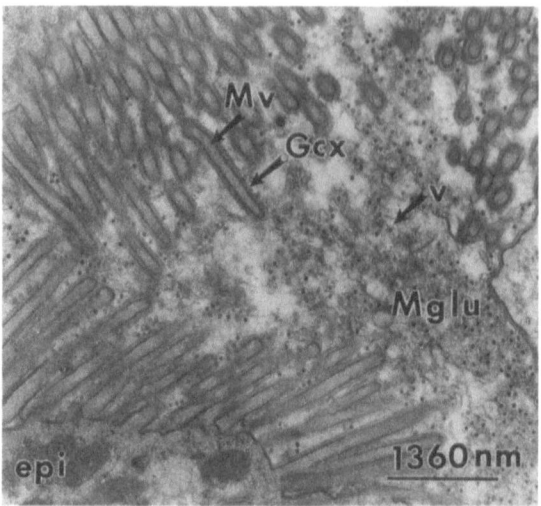

FIGURE 6.20. The midgut of an adult *Frankliniella occidentalis* following acquisition feeding on a tomato spotted wilt virus (TSWV) infected plant. Many TSWV particles can be seen in the midgut lumen (Mglu) and pressed against the microvilli (Mv) lining the midgut epithelial cells (epi). An unusual glycocalyx (Gcx) common to *F. occidentalis* can be seen on the apical plasmalemma of the microvilli (Mv). The function of the glycocalyx is unknown.

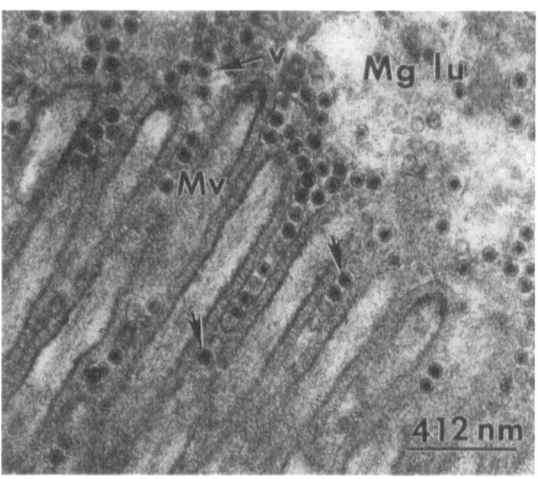

FIGURE 6.21. At high magnification, tomato spotted wilt virus (TSWV) particles can be seen in the midgut of an adult *Frankliniella occidentalis* following acquisition feeding on a TSWV infected plant. Virions (v) are abundant in the midgut lumen (Mglu), and are closely aligned (see arrow) with the apical plasmalemma of the midgut microvilli (Mv).

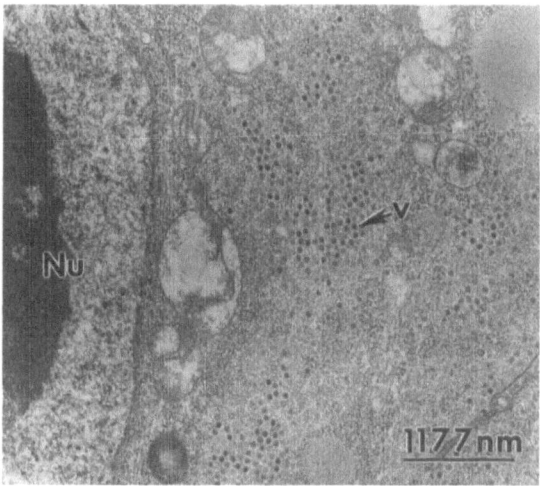

FIGURE 6.22. Tomato spotted wilt virus (TSWV) particles (v) accumulating in the cytoplasm of a midgut epithelial cell of an adult *Frankliniella occidentalis* following acquisition feeding on a TSWV infected plant. As in plant cells, virions have not been observed in the nuclei (Nu).

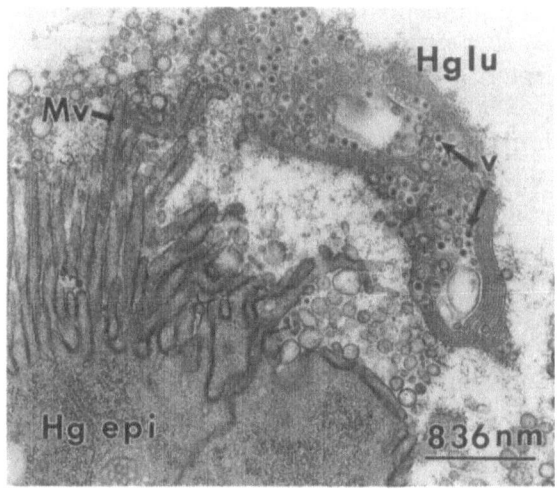

FIGURE 6.23. Tomato spotted wilt virus (TSWV) particles in the hindgut lumen (Hglu) of an adult *Frankliniella occidentalis* following acquisition feeding on a TSWV infected plant. TSWV particles were not observed closely aligned with the microvilli (Mv) lining the hindgut lumen, nor were virus particles observed in hindgut epithelial cells (Hg epi).

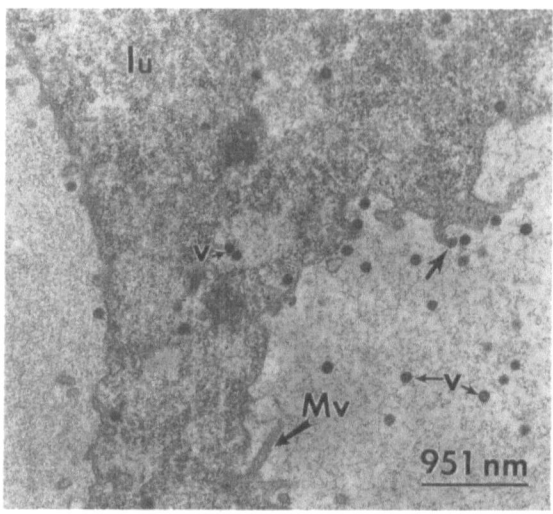

FIGURE 6.24. Tomato spotted wilt virus (TSWV) particles (v) in the cytoplasm (c) and lumen (lu) of the salivary gland of a 10 day old adult *Frankliniella occidentalis* following larval acquisition feeding on a TSWV infected plant. TSWV particles (v) can be seen apparently budding off the microvilli (Mv) lining the lumen of the salivary gland.

Specific biochemical and molecular explanations for the above mentioned cellular data have not yet been tested. Enzyme conditions or pH in the midgut cytoplasm may alter virus stability and/or membrane glycoproteins, or, appropriate receptors may be lacking at the midgut basement membrane of adult thrips preventing virus dissemination to epidemiologically important sites in the thrips hemococoel. By analogy to LaCrosse virus, a mosquito-transmitted bunyavirus, pH and enzymatic processing of viral membrane glycoproteins could play a vital role in cellular events leading to virus stability, fusion and dissemination (31, 39, 40, 69).

The eventual loss of virus following adult acquisition access feeding and the visual aspects of TSWV particles seen in adult midgut epithelial cells with TEM, strongly suggest that virus entering these cells is sequestered and decomposed by some cellular process. Previous researchers attempted feeding adult thrips on TSWV infected plants and then puncturing their midguts to see if virus could then be transmitted (25, 104). In addition, potential pH differences between larval and adult thrips have been investigated (25). The recent data just discussed suggest that these early hypotheses warrant renewed investigation using more sensitive technologies that are currently available. Further consideration of the importance of thrips physiology in understanding of how TSWV membrane glycoproteins interact in receptor mediated events governing thrips acquisition of TSWV are topics of ongoing research (Ullman, German, Sherwood and Westcot, unpublished data).

Importance of Virus Replication in Thrips Cells

The possibility that TSWV may replicate in its thrips vectors has long been a topic of interest. Sakimura (104) concluded that although TSWV could persist for much of the adult thrips life, the virus probably did not replicate in its vector and should be considered a circulative, nonpropagative virus. In recent years, many lines of evidence have been presented that suggest TSWV does replicate in thrips vector species (24, 37, 119, Robb, unpublished data), although fully definitive and direct evidence is still lacking. Preliminary data from ELISA of insects from a single cohort over time suggest that virus titers increase with increasing thrips age, indicating that replication is occurring (24). Thin sections of 10 day old adult WFT that acquired TSWV as larvae, reveal paracrystalline arrays, viroplasm and dense masses (Figure 6.25) (119), some of which closely resemble those observed in plant tissues where TSWV is replicating (62, 87, 92, 121). These cytoplasmic structures have never been observed in non-infected control thrips; however, definitive association between cytoplasmic inclusions in thrips cells and TSWV replication awaits completion of experiments involving labelling with virus specific antibodies (119, Ullman, Cho and Westcot, unpublished data).

Nucleic acid dot blot procedures have been developed for detection of TSWV RNA in plant and thrips tissues by cDNA clones (97). Recently, cDNA probes have been developed that specifically detect genomic and complementary L RNA strands. Presence of both strands occurs only when the virus is replicating, thus, strand specific cDNA probes provide a means for investigat-

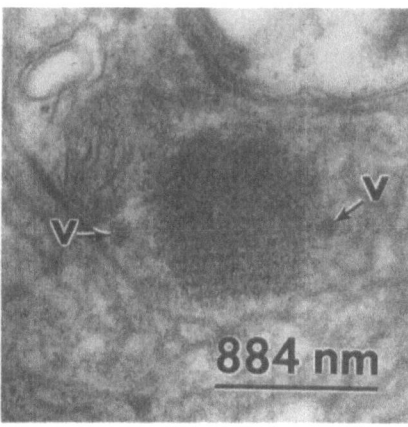

FIGURE 6.25. A paracrystalline array and potentially associated tomato spotted wilt virus (TSWV) particles (v) in the fat body of an adult *Frankliniella occidentalis* following larval acquisition feeding on a TSWV infected plant. Paracrystalline arrays were found only in tissue of thrips given TSWV acquisition feeding during larval stages and were not observed in noninfected control thrips. Definitive association of these arrays with TSWV presence and replication awaits labelling with TSWV-specific antibodies.

ing viral replication in plant and thrips tissues (37). In preliminary tests using these TSWV L RNA specific cDNA probes, both viral strands were detected in larval thrips (37). Although additional testing is required, these data add significantly to accumulating evidence suggesting that TSWV replication is occurring in thrips tissues (37).

Further evidence suggesting that TSWV may replicate in thrips tissues was recently provided by life history analyses that indicate WFT survival and longevity is dramatically decreased on TSWV-infected leaves (Robb unpublished data). While these investigations suggest that TSWV is pathogenic to thrips, effects due to differences in nutritional quality between healthy and infected leaves can not be eliminated as an alternative explanation. Similar results have been obtained in our laboratories (authors, unpublished data) and investigations using purified TSWV preparations and an in vitro feeding system (63) to eliminate plant variables and search for possible cytopathologies in thrips tissues are ongoing (Hu, Wang, Ullman, Cho and Westcot, unpublished data).

Pathogenicity of TSWV to its thrips vectors is of considerable interest from both evolutionary and practical outlooks. If TSWV is pathogenic to thrips, this aspect of the vector/virus relationship would set TSWV apart from the rest of the Bunyaviridae that replicate in their arthropod vectors, but do not cause decreased longevity or cytopathology (31). Moreover, pathogenicity of TSWV to thrips raises further questions regarding the origin and evolution of this unusual virus. Did TSWV originate as an insect virus and later coevolve to infect members of the Plant Kingdom? This is a question of considerable interest to evolutionary biologists, particularly in light of TSWV's new classification as the only phytopathogen among the Bunyaviridae (32). In practical and applied terms, TSWV pathogenicity to thrips could influence the abundance of viruliferous thrips in nature, thus, impacting TSWV epidemiology.

Conclusion

The integration of information on thrips biology, morphology, behavior and cellular thrips/TSWV interactions presented in this chapter provides a foundation for building comprehensive management strategies appropriate for specific crops and regions. Clearly, plant species vary in their suitability as hosts for both TSWV and thrips vector species. These plant host/thrips relationships directly impact viruliferous thrips populations and TSWV epidemiology in different regions and individual cropping systems. Ultimately, transmission efficiency is impacted by thrips feeding preferences on both acquisition and inoculation hosts, as well as interactions between thrips and TSWV isolates on cellular and molecular levels. For example, conditions influencing TSWV membrane glycoproteins probably influence receptor mediated events and potentially, virus replication in thrips cells. In turn, these events govern acquisition, the latent period, thrips longevity and efficiency of TSWV inoculation to healthy plants.

Tremendous advances have been made in our understanding of thrips biology, feeding behavior, biochemical and molecular characteristics of TSWV, as well as the interactions occurring between these two entities and their plant hosts. Clearly, much remains to be learned before we can completely comprehend how thrips/TSWV/plant interactions relate to TSWV spread in agricultural and ornamental cropping systems. Undoubtably, the multidisciplinary research efforts that are now occurring on a global scale, will result in an explosion of information on a broad front. The final result of these advances will likely be a dramatic increase in our understanding of this unusual plant virus and its relationship to its insect vectors, accompanied by introduction of novel and environmentally sound management strategies.

Addendum

The *Tospovirus genus* has been established in the family Bunyaviridae with TSWV as the type member. In addition, research on serologically distinct isolates, formerly designated as strains (ie. TSWV-L, lettuce strain and TSWV-I, impatiens strain), revealed differences sufficiently distinct to warrant establishment of at least two separate viruses, now called TSWV (formerly TSWV-L) and impatiens necrotic spot virus (INSV, formerly TSWV-I), within the *Tospovirus* genus. A complete review of these changes and *Tospovirus* diagnosis, molecular biology, and phylogeny can be found in German, T.L., Ullman D.E., and J.W. Moyer, 1992. *Annu. Rev. Phytopathol.* **30**:315–348.

Acknowledgment. This research was supported in part by the United States Department of Agriculture under CSRS Special Grant #88-34135-3593 managed by the Pacific Basin Advisory Group (PBAG), Biomedical Research Support Grant, University of Hawaii-Manoa, and a Research Center in Minority Institutions Grant RR-03061, National Institutes of Health. We thank David Fisher, Tina Weatherby, and Marilyn Dunlap for technical assistance and use of the Pacific Biomedical Research Center's Biological Electron Microscope Facility. This is Journal series #3696 of the Hawaiian Institute of Tropical Agriculture and Human Resources.

References

1. Allen, W.R., and Broadbent, A.B., 1986, Transmission of tomato spotted wilt virus in Ontario greenhouses by the western flower thrips, *Frankliniella occidentalis* (Pergande), *Can. J. Plant Pathol.* **8**:33–38.

2. Allen, W.R., and Matteoni, J.A., 1988, Cyclamen ringspot: Epidemics in Ontario greenhouses caused by the tomato spotted wilt virus, *Can. J. Plant Pathol.* **10**:41–46.

3. Allen, W.R., Matteoni, J.A., and Broadbent, A.B., 1991, Factors relating to epidemiology and symptomatology in florist's chrysanthemum infected with the tomato spotted wilt virus, in Hsu, H., and Lawson, R.H. (eds): *Virus-Thrips-Plant Interaction of Tomato Spotted Wilt Virus, Proceedings of a USDA Workshop*, United States Department of Agriculture, Agricultural Research Service, ARS-87, pp. 28–45.

4. Amin, P.W., Reddy, D.V.R., and Ghanekar, A.M., 1981, Transmission of tomato spotted Wilt virus, causal agent of bud necrosis of peanut, by *Scirtothrips dorsalis* and *Frankliniella schultzei, Plant Dis.* **65**:663–665.

5. Backus, E.A., and McLean, D.L., 1985, Behavioral evidence that the precibarial sensilla of leafhoppers are chemosensory and function in host discrimination, *Entomol. Exp. Appl.* **37**:219–228.

6. Backus, E.A., 1985, The anatomical and sensory mechanisms of feeding behavior, in Nault, L.S., and Rodriguez, J.G. (eds): The Leafhoppers and Planthoppers, John Wiley and Sons, pp. 163–19.

7. Bald, J.G., and Samuel, G., 1931, Investigation on "spotted wilt" of tomatoes. II. Australia, Commonwealth Council Sci. *Ind. Research Bull. No. 54.*

8. Bedford, H.W., 1921, The cotton thrips (*Heliothrips indicus* Bagn.) in the Sudan, *Bull. Wellcome Trop. Res. Labs. Entomol.* Section 18.

9. Best, R.J., 1968, Tomato spotted wilt virus, in Smith, K.M., and Lauffer, M.A. (eds): Advances in Virus Research, Volume 13, Academic Press, New York, pp. 65–145.

10. Borrer, D.J., Triplehorn, C.A., and Johnson, N.F., 1989, An Introduction to the Study of Insects, Saunders College Publishing, Philadelphia, 875 pp.

11. Bournier, A., and Bournier, J.P., 1987, L'introduction en France d'un nouveau ravageur: *Frankliniella occidentalis, Phytoma* **388**:14–17.

12. Bournier, A., 1961, Sur l'existence et l'evolution d'un mycetome au cours de l'embryologenese de *Caudothrips buffai* Karny, *Verh. XI Intern. Kongr. Entomol.* 1960, **1**:352.

13. Briddon, R.W., Pinner, M.S., Stanley, J., and Markham, P.G., 1990, Geminivirus coat protein gene replacement alters insect specificity, *Virology* **177**:85–94.

14. Broadbent, A.B., Allen, W.R., and Footitt, R.G., 1987, The association of *Frankliniella occidentalis* (Pergande) (Thysanoptera:Thripidae) with greenhouse crops and the tomato spotted wilt virus in Ontario, *Can. Ent.* **119**:501–503.

15. Broadbent, A.B., Matteoni, J.A., and Allen, W.R., 1990, Feeding preferences of the western flower thrips, *Frankliniella occidentalis* (Pergande) (Thysanoptera: Thripidae), and incidence of tomato spotted wilt virus among cultivars of florist's chrysanthemum, *Can. Ent.* **122**:1111–1117.

16. Bryan, D.E., and Smith, R.F., 1956, The *Frankliniella occidentalis* (Pergande) complex in California (Thysanoptera:Thripidae), University of California Publ. Ent., **10**:359–410).

17. Chisholm, I.F., and Lewis, T., 1984, A new look at thrips (Thysanoptera) mouthparts, their action and effects of feeding on plant tissue, *Bull. Entomol. Res.* **74**:663–675.

18. Cho, J.J., Mitchell, W.C., Yudin, L., and Takayama, L., 1984, Ecology and

epidemiology of tomato spotted wilt virus (TSWV) and its vector, *Frankliniella occidentalis, Phytopathology* **74**:866, (Abstr.).

19. Cho, J.J., Mau, R.F.L., Gonsalves, D., and Mitchell, W.C., 1986, Reservoir weed hosts of tomato spotted wilt virus, *Plant Dis.* **70**:1014–1017.

20. Cho, J.J., Mitchell, W.C., Mau, R.F.L., and Sakimura, K., 1987, Epidemiology of tomato spotted wilt virus on crisphead lettuce in Hawaii, *Plant Dis.* **71**:505–508.

21. Cho, J.J., Mau, R.F.L., Mitchell, W.C., Gonsalves, D., and Yudin, L., 1987, Host list of tomato spotted wilt virus (TSWV) susceptible plants, University of Hawaii, College of Tropical Agriculture and Human Resources, Research-Extension Series 078, 12 pp.

22. Cho, J.J., Mau, R.F.L., Hamasaki, R.T., and Gonsalves, D., 1988, Detection of tomato spotted wilt virus in individual thrips by enzyme linked immunosorbent assay, *Phytopathology* **78**:1348–1352.

23. Cho, J.J, Mau, R.F.L., German, T.L., Hartmann, R.W., Yudin, L.S., Gonsalves, D., and Provvidenti, R., 1989, A multidisciplinary approach for tomato spotted wilt virus (TSWV) management in Hawaii, *Plant Dis.* **73**:375–383.

24. Cho, J.J., Mau, R.F.L., Ullman, D.E., and Custer, D.M., 1991, Serological detection of tomato spotted wilt virus within thrips, in Hsu, H., and Lawson, R.H. (eds): Virus-Thrips-Plant Interaction of Tomato Spotted Wilt Virus, Proceedings of a USDA Workshop, United States Department of Agriculture, Agricultural Research Service, ARS-87, pp. 144–152.

25. Day, M.F., and Irzykiewicz, H., 1954, Physiological studies on thrips in relation to transmission of tomato spotted wilt virus, *Australian J. Biol. Sci.* **7**:274–281.

26. de Avila, A.C., Huguenot, C., Resende, R. de O., Kitajima, E.W., Goldbach, R.W., and Peters, D., 1990, Serological differentiation of 20 isolates of tomato spotted wilt virus, *J. gen. Virol.* **71**:2801–2807.

27. de Haan, P., Wagemakers, L., Goldbach, R., and Peters, D., 1989a, Tomato spotted wilt virus, a new member of the Bunyaviridae? in Kolakofsky, D., and Mahy, B.W.J. (eds): Genetics and Pathogenicity of Negative Strand Viruses, Elsevier, Amsterdam, pp. 287–290.

28. de Haan, P., Wagemakers, L., Peters, D., and Goldbach, R., 1989b, Molecular cloning and terminal sequence determination of the S and M RNAs of tomato spotted wilt virus, *J. gen. Virology* **70**:3469–3473.

29. de Haan, P., Wagemakers, L., Peters, D., and Goldbach, R., 1990, The S RNA segment of tomato spotted wilt virus has an ambisense character, *J. gen. Virol.* **71**:1001–1007.

30. Del Bene, G., Dallai, R., and Marchini, D., 1991, Ultrastructure of the midgut and the adhering tubular salivary glands of *Frankliniella occidentalis* (Pergande) (Thysanoptera:Thripidae), *Intl. J. Insect Morphol. Embryol.* **20**:15–24.

31. Elliott, R.M., 1990a, Molecular biology of the bunyaviridae, *J. Gen. Virology* **71**:501–522.

32. Elliott, R.M., 1990b, Bunyaviridae, in *Abstr. Intl. Congress of Virology*, Berlin, p. 499.

33. Elzinga, R.J., 1987, Fundamentals of Entomology, 3rd Edition, Prentice-Hall, New Jersey, 456 pp.

34. Evans, H.E., 1984, Insect Biology: A Textbook of Entomology, Addison-Wesley, Massachusetts, 436 pp.

35. Francki, R.I.B., and Grivell, C.J., 1970, An electron microscope study of the

distribution of tomato spotted wilt virus in systemically infected *Datura stamonium* leaves, *Virology* **42**:969–978.

36. Francki, R.I.B., and Hatta, T., 1981, Tomato spotted wilt virus, in Kurstak, E. (ed): Handbook of Plant Virus Infection and Comparative Diagnosis, Elsevier Biomedical Press, North Holland, pp. 491–512.

37. German, T.L., Hu, Y., Cho, J.J., and Ullman, D.E., 1991, Detection of tomato spotted wilt virus RNA in plant and thrips using strand-specific probes, in Hsu, H., and Lawson, R.H. (eds): *Virus-Thrips-Plant Interaction of Tomato Spotted Wilt Virus, Proceedings of a USDA Workshop,* United States Department of Agriculture, Agricultural Research Service, ARS-87, pp. 137–143.

38. Goldstein, J.L., Brown, M.S., Anderson, R.G.W., Russell, D.W., and Schneider, W.J., 1985, Receptor-mediated endocytosis: Concepts emerging from the LDL receptor system, *Ann. Rev. Cell Biol.* **1**:1–39.

39. Gonzalez-Scarano, F., 1985, La crosse virus Gl glycoprotein undergoes a conformational change at the pH of fusion, *Virology* **140**:209–216.

40. Gonzalez-Scarano, F., Pobjecky, N., and Nathanson, N., 1984, La crosse bunyavirus can mediate pH-dependent fusion from without, *Virology* **132**:222–225.

41. Gonsalves, D. and Trujillo, E.E., 1986, Tomato spotted wilt virus in papaya and detection of the virus with ELISA, *Plant Dis.* **70**:501–506.

42. Greber, R.S., Klose, M.J., Teakle, D.S., and Milne, J.R., 1991, High incidence of tobacco streak virus in tobacco and its transmission by *Microcephalothrips abdominalis* and pollen from *Ageratum houstonianum, Plant Dis.* **75**:450–452.

43. Greber, R.S., Klose, M.J., Milne, J.R., and Teakle, D.S., 1991, Transmission of prunus necrotic ringspot virus using plum pollen and thrips, *Anals. Appl. Biol.* **118**:589–591.

44. Heming, B.S., 1978, Structure and function of mouthparts in larvae of *Haplothrips verbasci* (Osborn) (Thysanoptera, Tubilifera, Phlaeothripidae), *J. Morphol.* **156**:1–37.

45. Heming, B.S., 1985, Thrips (Thysanoptera) in Alberta, *Agric. For. Bull.* **8**:19–24.

46. Horton, J.R., 1918, The citrus thrips, *Bull. U.S. Dept. Agric.* p. 616.

47. Hsu, H.T., and Lawson, R.H., 1991, Detection of tomato spotted wilt virus by enzyme-linked immunosorbent assay, dot-blot immunoassay and direct tissue blotting, in Hsu, H., and Lawson, R.H. (eds): *Virus-Thrips-Plant Interaction of Tomato Spotted Wilt Virus, Proceedings of a USDA Workshop,* United States Department of Agriculture, Agricultural Research Service, ARS-87, pp. 120–126.

48. Huckaba, R.M., and Coble, H.D., 1991, Effect of soybean thrips (Thysanoptera: Thripidae) feeding injury on penetration of acifluorfen in soybean, *J. Econ. Entomol.* **84**:300–305.

49. Hunter, W.B., and Ullman, D.E., 1989, Analysis of mouthpart movements during feeding of *Frankliniella occidentalis* (Pergande) and *F. schultzei* Trybom (Thysanoptera:Thripidae), *Intl. J. Insect Morphol. & Embryol.* **18**:161–171.

50. Hunter, W.B., Ullman, D.E., Moore, A., Electronic monitoring: Characterizing the feeding behavior of western flower thrips, *Frankliniella occidentalis* (Pergande) (Thysanoptera: Thripidae), in Ellsbury, M ., Backus, E.A., and Ullman, D.E. (eds): *Proceedings of an Informal Conference on Electronic Monitoring of Insect Feeding Behavior, Misc. Publ. ESA.* (In press)

51. Ie, T.S., 1970, Tomato spotted wilt virus, CMI/AAB Plant virus descriptions No. 39.

52. Ie, T.S., 1971, Electron microscopy of developmental stages of tomato spotted wilt virus in plant cells, *Virology* **43**:468–479.
53. Ie, T.S., 1982, A sap transmissible, defective form of tomato spotted wilt virus, *J. gen. Virol.* **59**:387–391.
54. Imms, A.D., 1957, A general textbook of entomology, 9th edition, London, 886 pp.
55. Irwin, M.E., and Ruesink, W.G., 1987, Vector intensity: A product of propensity and activity, in McLean, G.D., Garrett, R.G., and Ruesink, W.G. (eds): Plant Virus Epidemics: Monitoring, Modelling and Predicting Outbreaks, pp. 13–34.
56. Iwaki, M., Honda, Y., Hanada, K., Tochihara, H., Yanaha, T., Hokama, K., and Yokoyama, T., 1984, Silver mottle disease of watermelon caused by tomato spotted wilt virus, *Plant Dis.* **68**:1006–1008.
57. Johnson, M.W., 1986, Population trends of a newly introduced species, *Thrips palmi* (Thysanoptera:Thripidae) on commercial watermelon plantings in Hawaii, *J. Econ. Entomol.* **79**:718–720.
58. Johnson, M.W., 1989, Foliar pests of watermelon in Hawaii, *Tropical Pest Management* **35**:90–96.
59. Jones, T., 1954, The external morphology of *Chirothrips hamatus* (Trybom) (Thysanoptera), *Trans. R. Entomol . Soc.* London, **105**:163–187.
60. Kimsey, R.B., and McLean, D.L., 1988, Versatile electronic measurement system for studying probing and feeding behavior of piercing and sucking insects, *Ann. Entomol. Soc. Am.* **80**:118–129.
61. Kitajima, E.W., 1965, Electron microscopy of viracabeca (Brazilian tomato spotted wilt virus) with the host cell, *Virology* **26**:89–99.
62. Kormelink, R., Kitajima, E.W., De Haan, P., Zuidema, D., Peters, D., and Goldbach, R., 1991, The nonstructural protein (NSs) encoded by the ambisense S RNA segment of tomato spotted wilt virus is associated with fibrous structures in infected plant cells, *Virology* **181**:459–468.
63. Kunkel, H., 1987, Membrane feeding systems in aphid research, in Harris, K.F., and Maramorosch, K. (eds.), Aphids As Virus Vectors, Academic Press, New York, pp. 311–333.
64. Law, M.D., and Moyer, J.W., 1990, A tomato spotted wilt-like virus with a serologically distinct N protein, *J. gen. Virology* **71**:933–938.
65. Lewis, T., 1973, Thrips. Their biology, ecology, and economic importance, Academic Press, New York, 349 pp.
66. Lin, N.S., Hsu, Y.H. and Hsu, H.T., 1990, Immunological detection of plant viruses and a mycoplasma-like organism by direct tissue blotting on nitrocellulose membranes, *Phytopathology* **80**:824–828.
67. Linford, M.B., 1932, Transmission of the pineapple yellow-spot virus by *Thrips tabaci, Phytopathology* **22**:301–324.
68. Locke, M., 1984, The structure and development of the vacuolar system in the fat body of insects, in King, R.C., and Akai, H. (eds): Insect Ultrastructure, Volume 2 pp. 151–194.
69. Ludwig, G.V., Christensen, B.M., Yuill, T.M., and Schultz, K.T., 1989, Enzyme processing of la crosse virus glycoprotein G1: A bunyavirus-vector infection model, *Virology* **171**:108–113.
70. Maiss, E., Ivanova, L., Breyel, E., and Adam, G., 1991, Cloning and sequencing of the S RNA from a Bulgarian isolate of tomato spotted wilt virus, *J. gen. Virology* **72**:461–464.

71. Markham, P.G., Pinner, M.S., and Boulton, M., 1984, The transmission of maize streak virus by leafhoppers, a new look at host adaptation, *Bull. Soc. Entomol. Suisse* **57**:431–432.

72. Marsh, M., and Helenius, A., 1980, Adsorptive endocytosis of semliki forest virus, *J. Mol. Biol.* **142**:439–454.

73. Martoja, R., and Ballan-DuFrancais, C., 1984, The ultrastructure of the digestive and excretory organs, in King, R.C., and Akai, H. (eds): Insect Ultrastructure, Volume 2, Plenum Press, New York, London, pp. 199–268.

74. Matlin, K.S., Reggio, H., Helenius, A., and Simons, K., 1982, Pathway of vesicular stomatitis virus entry leading to infection, *J. Mol. Biol.* **156**:609–631.

75. Mau, R.F.L., Bautista, R., Cho, J.J., Ullman, D.E., Gusukuma-Minuto, L.R., and Custer, D., 1991, Factors affecting the epidemiology of TSWV in field crops: Comparative virus acquisition efficiency of vectors and suitability of alternate hosts to *Frankliniella occidentalis* (Pergande), in Hsu, H., and Lawson, R.H. (eds): *Virus-Thrips-Plant Interaction of Tomato Spotted Wilt Virus, Proceedings of a USDA Workshop*, United States Department of Agriculture, Agricultural Research Service, ARS-87, pp. 21–27.

76. McLean, D.L., and Kinsey, M.G., 1984, The precibarial valve and its role in the feeding behavior of the pea aphid, *Acyrthosiphon pisum, Bull. Entomol. Soc. Amer.* **30**:26–31.

77. Metcalf, C.L., Flint, W.P., and Metcalf, R.L., 1962, Destructive and Useful Insects, McGraw-Hill, New York, pp. 142, 211–213.

78. Mickoleit, E., 1963, Untersuchungen zur Kopfmorphologie der Thysaopteran, *Zool Jahrb. Ant.* **81**:101–150.

79. Milne, R.G., and Francki, R.I.B., 1984, Should tomato spotted wilt virus be considered as a possible member of the family Bunyaviridae?, *Intervirology* **22**:72–76.

80. Mitchell, F.L., and Smith, J.W. Jr., 1991, Epidemiology of tomato spotted wilt virus relative to thrips populations, in Hsu, H., and Lawson, R.H. (eds): *Virus-Thrips-Plant Interaction of Tomato Spotted Wilt Virus, Proceedings of a USDA Workshop*, United States Department of Agriculture, Agricultural Research Service, ARS-87, pp. 46–52.

81. Moore, E.S., 1933, The Kromnek or Kat River disease of tobacco and tomato in the East Province (South Africa), *Bull. Dept. Agr. S. Africa Sci.* No. 123.

82. Mound, L.A., 1971, The feeding apparatus of thrips, *Bull. Entomol. Res.* **60**:547–548.

83. Moritz, G., 1989, Die ontogenese der Thysanoptera (Insecta) unter besonderer berucksichtigung des fransenfluglers *Hercinothrips femoralis* (Reuter, O.M., 1891) (Thysanoptera, Thripidae, Panchaetothripinae) III. Mitteilung: Praepupa und pupa, (The ontogenesis of Thysanoptera (Insecta) with special reference to the Panchaetothripinae *Hercinothrips femoralis* (Reuter, O.M., 1891) (Thysanoptera, Thripidae, Panchaetothripinae) III. Prepupa and pupa), *Zool. Jb. Anat.* **118**:15–53.

84. Moritz, G., 1989, Die ontogenese der Thysanoptera (Insecta) unter besonderer berucksichtigung des fransenfluglers *Hercinothrips femoralis* (Reuter, O.M., 1891) (Thysanoptera, Thripidae, Panchaetothripinae) IV. Mitteilung:Imago-Kopf (The ontogenesis of Thysanoptera (Insecta) with special reference to the Panchaetothripinae *Hercinothrips femoralis* (Reuter, O.M., 1891) (Thysanoptera, Thripidae, Panchaetothripinae) IV. Imago—Head), *Zool. Jb. Anat.* **118**:273–307.

85. Moritz, G., 1989, Die ontogenese der Thysanoptera (Insecta) unter besonderer berucksichtigung des fransenfluglers *Hercinothrips femoralis* (Reuter, O.M., 1891) (Thysanoptera, Thripidae, Panchaetothripinae) V. Mitteilung: Imago Thorax, (The ontogenesis of Thysanoptera (Insecta) with special reference to the Panchaetothripinae *Hercinothrips femoralis* (Reuter, O.M., 1891) (Thysanoptera, Thripidae, Panchaetothripinae) V. Imago–Thorax), *Zool. Jb. Anat.* **118**:393–429.

86. Moritz, G., 1989, Die ontogenese der Thysanoptera (Insecta) unter besonderer berucksichtigung des fransenfluglers *Hercinothrips femoralis* (Reuter, O.M., 1891) (Thysanoptera, Thripidae, Panchaetothripinae) VI. Mitteilung:Imago— Abdomen, (The ontogenesis of Thysanoptera (Insecta) with special reference to the Panchaetothripinae *Hercinothrips femoralis* (Reuter, O.M., 1891) (Thysanoptera, Thripidae, Panchaetothripinae) VI. Imago—Abdomen), *Zool. Jb. Anat.* **119**:157–218.

87. Moyer, J.W., Law, M.D., and Urban, L.A., 1991, Characteristics of a serologically distinct TSWV-like virus from impatiens, in Hsu, H., and Lawson, R.H. (eds): *Virus-Thrips-Plant Interaction of Tomato Spotted Wilt Virus, Proceedings of a USDA Workshop*, United States Department of Agriculture, Agricultural Research Service, ARS-87, pp. 53–59.

88. Oetting, R.D., 1991, The effect of host species and different plant components on thrips feeding and development, in Hsu, H., and Lawson, R.H. (eds): *Virus-Thrips-Plant Interaction of Tomato Spotted Wilt Virus, Proceedings of a USDA Workshop*, United states Department of Agriculture, Agricultural Research Service, ARS-87, pp. 15–19.

89. Paliwal, Y.C., 1974, Some properties and thrips transmission of tomato spotted wilt virus in Canada, *Can. J. Botany* **52**:1170–1182.

90. Paliwal, Y.C., 1976, Some characteristics of the thrips *Frankliniella vector* relationship of tomato spotted wilt virus in Canada, *Can. J. Bot.* **54**:402–405.

91. Pedigo, L.P., 1989, Entomology and Pest Management, MacMillan Publishing Company, New York, 646 pp.

92. Peters, D., de Avila, A.C., Kitajima, E.W., Resende, R. de O., de Haan, P., and Goldbach, R.W., 1991, An overview of tomato spotted wilt virus, in Hsu, H., and Lawson, R.H. (eds): *Virus-Thrips-Plant Interaction of Tomato Spotted Wilt Virus, Proceedings of a USDA Workshop*, United States Department of Agriculture, Agricultural Research Service, ARS-87, pp. 1–14.

93. Pesson, P., 1951, Super-ordre des Thysanopteroides, Ordre des Thysanoptera, in Grasse, P. (ed): Traite de Zoologie, Anatomie, Systematique, Biologie, Insectes Superieurs et Hemiptereoides, Masson Cie, Paris, France, pp. 1805–1869.

94. Pittman, H.A., 1927, Spotted wilt of tomatoes. Preliminary note concerning the transmission of the 'spotted wilt' of tomatoes by an insect vector (*Thrips tabaci* Lind.), *J. Council Sci. Ind. Research (Australia)* **1**:74–77.

95. Razvyazkina, G.M., 1953, The importance of the tobacco thrips in the development of outbreaks of tip chlorosis of Makhorka (in Russian), Doklady Vsesoyuz. Akad. Sel'skokhoz, *Nauk im. V.I. Lenina* **18**:27–31, (Abstr. in *Rev. Appl. Entomol.* **A42**:146).

96. Reddy, D.V.R., and Wightman, J.A., 1988, Tomato spotted wilt virus: Thrips transmission and control, in Harris, K.F. (ed): Advances in Disease Vector Research, Volume 8 Springer-Verlag, New York, pp. 203–220.

97. Reddy, D.V.R., Sudarshana, A.S., Ratna, A.S., Reddy, A.S., Amin, P.W., Kumar,

I.K., and Murthy, A.K., 1991, The occurrence of yellow spot virus, a member of tomato spotted wilt virus group, on peanut (*Arachia hypogaea* L.) in India, in Hsu, H., and Lawson, R.H. (eds): *Virus-Thrips-Plant Interaction of Tomato Spotted Wilt Virus, Proceedings of a USDA Workshop*, United States Department of Agriculture, Agricultural Research Service, ARS-87, pp. 77–88.

98. Resende, R. de O., Kitajima, E.W., de Avila, A.C., Goldbach, R.W., and Peters, D., 1991, Defective isolates of tomato spotted wilt virus, in Hsu, H., and Lawson, R.H. (eds): *Virus-Thrips-Plant Interaction of Tomato Spotted Wilt Virus, Proceedings of a USDA Workshop*, United States Department of Agriculture, Agricultural Research Service, ARS-87, pp. 71–76.

99. Rice, D.J., German, T.L., Mau, R.F.L., and Fujimoto, F.M., 1990, Dot blot detection of tomato spotted wilt virus RNA in plant and thrips tissues by cDNA clones, *Plant Dis.* **74**:274–276.

100. Rosenheim, J.A., Welter, S.C., Johnson, M.W., Mau, R.F.L., and Gusukuma-Minuto, L.R. 1990, Direct feeding damage on cucumber by mixed-species infestations of *Thrips palmi* and *Frankliniella occidentalis* (Thysanoptera: Thripidae), *J. Econ. Entomol.* **83**:1519–1525.

101. Rothschild, M., Schlein, Y., and Ito, 5., 1986, A colour atlas of insect tissues via the flea, Wolfe Publishing Ltd., London, 184 pp.

102. Sakimura, K., 1940, Evidence for the identity of the yellow-spot virus with the spotted-wilt virus: Experiments with the vector, *Thrips tabaci, Phytopathology* **30**:281–299.

103. Sakimura, K., 1962a, *Frankliniella occidentalis* (Thysanoptera: Thripidae), a vector of the tomato spotted wilt virus, with special reference to color forms, *Ann. Entomol. Soc. Am.* **55**:387–389.

104. Sakimura, K., 1962b, The present status of thrips-borne viruses, in Maramorosch, K. (ed): Biological Transmission of Disease Agents, Academic Press, New York, pp. 33–40.

105. Sakimura, K., 1963, *Frankliniella fusca*, an additional vector for the tomato spotted wilt virus, with notes on *Thrips tabaci*, another vector, *Phytopathology* **53**:412–415.

106. Sakimura, K., 1969, A comment on the color forms of *Frankliniella schultzei* (Thysanoptera: Thripidae) in relation to transmission of the tomato spotted wilt virus, *Pacific Insects* **11**:761–762.

107. Samuel, G., Bald, J.G., and Pittman, H.A., 1930, Investigations on "spotted wilt" of tomatoes, *Commonwealth Council Sci. Ind. Research Bull.* Australia, **44**.

108. Sdoodee, R., and Teakle, D.S., 1987, Transmission of tobacco streak virus by *Thrips tabaci*: a new method of plant virus transmission, *Pl. Path.* **36**:377–380.

109. Sharga, U.S., 1933, On the internal anatomy of some Thysanoptera, *Trans. R. Entomol. Soc.* London **81**:185–204.

110. Simons, K., and Fuller, S.D., 1985. Cell surface polarity in epithelia, *Ann. Rev. Cell Biol.* **1**:243–88.

111. Skehel, J.J., Bayley, P.M., Brown, E.B., Martin, S.R., Waterfield, M.D., White, J.M., Wilson, I.A., and Wiley, D.C., 1982, Changes in the conformation of influenza virus hemagglutinin at the pH optimum of virus-mediated membrane fusion, *Proc. Natl. Acad. sci. USA*, **79**:968–972.

112. Sherwood, J.L., Sanborn, M.R., Keyser, G.C., and Myers, L.D., 1989, Use of monoclonal antibodies in detection of tomato spotted wilt virus, *Phytopathology* **79**:61–64.

113. Smith, K.M., 1932, Studies on plant virus diseases. XI. Further experiments with

ringspot virus: Its identification with tomato spotted wilt of tomato, *Ann. Appl. Biol.* **19**:305–330.

114. Steiner, M.Y., 1990, Determining population characteristics and sampling procedures for the western flower thrips (Thysanoptera: Thripidae) and the Predatory Mite *Amblyseius cucumeris* (Acari: Phytoseiidae) on greenhouse cucumber, *Environ. Entomol.* **19**:1605–1613.

115. Stoner, K.A., and Shelton, A.M., 1988a, Role of nonpreference in the resistance of cabbage varieties to the onion thrips (Thysanoptera: Thripidae), *J. Econ. Entomol.* **81**:1062–1067.

116. Stoner, K.A., and Shelton, A.M., 1988b, Influence of variety on abundance and within-plant distribution of onion thrips (Thysanoptera: Thripidae) on cabbage, *J. Econ. Entomol.* **81**:1190–1195.

117. Strassen, R. Zur. 1960, Catalogue of the known species of South African Thysanoptera, *J. Entomol. Soc. S. Afr.* **23**:321–367.

118. Ullman, D.E., Westcot, D.M., Hunter, W.B., and Mau, R.F.L., 1989, Internal anatomy and morphology of *Frankliniella occidentalis* (Pergande) (Thysanoptera: Thripidae) with special reference to interactions between thrips and tomato spotted wilt virus, *Int. J. Insect Morphol. & Embryol.* **18**:289–310.

119. Ullman, D.E., Westcot, D.M., Mau, R.F.L., Cho, J.J., and Custer, D.M., 1991, Tomato spotted wilt virus and one thrips vector: *Frankliniella occidentalis* (Pergande) Internal Morphology and Virus Location, in Hsu, H., and Lawson, R.H. (eds): *Virus-Thrips-Plant Interaction of Tomato Spotted Wilt Virus, Proceedings of a USDA Workshop*, United States Department of Agriculture, Agricultural Research Service, ARS-87, pp. 127–136.

120. Ullman, D.E., Cho, J.J., Mau, R.F.L., Westcot, D.M., and Custer, D.M., A Midgut Barrier to Tomato Spotted Wilt Virus Acquisition by Adult Western Flower Thrips, *Phytopathology*. (In press)

121. Urban, L.A., Huang, P., and Moyer, J.W., 1991, Cytoplasmic inclusions in cells infected with isolates of L and I serogroups of tomato spotted wilt virus, *Phytopathology* **81**:525–529.

122. Verkleij, F.N., and Peters, D., 1983, Characterization of a defective form of tomato spotted wilt virus, *J. gen. Virology* **64**:677–686.

123. Vernon, R.S., and Gillespie, D.R., 1990, Spectral responsiveness of *Frankliniella occidentalis* (Thysanoptera: Thripidae) determined by trap catches in greenhouses, *Environ. Entomol.* **19**:1229–1241.

124. Wang, M., and Gonsalves, D., 1990, ELISA detection of various tomato spotted wilt virus isolates using specific antisera to structural proteins of the virus, *Plant Dis.* **74**:154–158.

125. Welter, S.C., 1989, Arthropod impact on plant gas exchange, in Bernays, E.A. (ed): Insect-Plant Interactions, Volume 1, pp. 135–150.

126. Welter, S.C., Rosenheim, J.A., Johnson, M.W., Mau, R.F.L., and Gusukuma-Minuto, L.R., 1990, Effects of *Thrips palmi* and western flower thrips (Thysanoptera: Thripidae) on the yield, growth, and carbon allocation pattern in cucumbers, *J. Econ. Entomol.* **83**:2092–2101.

127. Wiesenborn, W.D., and Morse, J.G., 1985, Feeding rate of Thysanoptera estimated using C14 inulin, *J. Econ. Entomol.* **78**:151–158.

128. Wiesenborn, W.D., and Morse, J.G., 1988, The mandible and maxillary stylets of *Scirtothrips citri* (Moulton) (Thysanoptera: Thripidae), *Pan-Pacific Entomologist* **64**:39–42.

129. Yudin, L.S., Cho, J.J., and Mitchell, W.C., 1986, Host range of western flower

thrips, *Frankliniella occidentalis* (Thysanoptera: Thripidae), with special reference to Leucaena glauca, *Environ. Entomol.* **15**:1292–1295.

130. Yudin, L.S., Mitchell, W.C., and Cho, J.J., 1987, Color preference of thrips (Thysanoptera:Thripidae) with reference to aphids (Homoptera: Aphididae) and leafminers in Hawaiian lettuce farms, *J. Econ. Entomol.* **80**:51–55.

131. Yudin, L.S., Tabashnik, B.E., Cho, J.J., and Mitchell, W.C., 1988, Colonization of weeds and lettuce by thrips (Thysanoptera: Thripidae), *Environ. Entomol.* **17**:522–526.

132. Yudin, L.S., Tabashnik, B.E., Cho, J.J., and Mitchell, W.C., 1990, Disease-prediction and economic models for managing tomato spotted wilt virus disease in lettuce, *Plant Dis.* **74**:211–216.

7
Mealybug Wilt of Pineapple

Thomas L. German, Diane E. Ullman, and
U.B. Gunashinghe

Introduction

Mealybug wilt of pineapple (MBW), first described in Hawaii in the early 1900's (46), is now reported in most areas of the world where pineapple is grown (6, 7, 13/14, 22). The circumstances surrounding MBW epidemics are complex involving multi-trophic interactions between mealybugs (MB), ants, mealybug predators and parasites, pineapple plants and other plant species (58). Symptom expression is variable and apparently linked to environmental conditions as well as variations in mealybug populations. Several generations of entomologists and plant pathologists have been challenged by the study of MBW and the etiology is yet to be fully explained. A number of hypotheses involving mealybug salivary toxins, "latent transmissible factors" or viruses have been proposed to explain the cause of the disease but none of these have been substantiated. Recently, a pineapple closterovirus (PCV) was described that appears to be associated with the disease (35, 36, 38, 39) although its role in the mealybug wilt syndrome, if any, is not fully understood. In this chapter we will present the historical background concerning the biology and epidemiology of MBW, data showing the association of PCV with pineapple plants and mealybugs and describe an apparent influence of PCV on pineapple growth.

Thomas L. German, Department of Plant Pathology, University of Wisconsin, 1630 Linden Drive, Madison, Wisconsin 53706, USA.

Diane E. Ullman, Department of Entomology, University of Hawaii, 3050 Maile Way, Honolulu, Hawaii 96822, USA.

U.B. Gunashinghe, Department of Plant, Soil and Entomological Sciences, University of Idaho, Moscow, Idaho 83843, USA.

Pineapple: Characteristics and Worldwide Significance

Pineapple (*Ananas comosus* L.) is a member of the Bromeliaceae, a large plant family of the American tropics. Unlike most bromeliads, pineapple is not epiphytic on trees but is planted in the soil and grown on small farms as well as plantations. Collins (22) suggests that pineapple was originally cultivated in Brazil where wild pineapple was developed by the Tupi-Guarani Indians who named several varieties which they selected for fruit size and abundance of seeds. However, Brucher (22) is of the opinion that *Ananas sativius* cv "Cayenne", originated in the highlands of Guiana. Columbus and his crew were probably the first Europeans to taste pineapple. In 1493 they landed on the island of Guadeloupe where they found a fruit that "astonished and delighted them". Later explorers describe pineapple growing near many tropical American coasts (60). Pineapple is now grown in tropical and subtropical regions throughout the world. Presently, major pineapple growing areas include the Hawaiian Islands, South Africa, South and Central America, Southeast Asia and Australia. Hawaii once produced more than half of the world's pineapple crop but by the early 1960s its share had dropped to less than one quarter although actual production had increased. As Hawaiian land values increased and labor costs rose, world distribution of production shifted to less expensive tropical areas and Hawaii currently contributes less than 6% of the total world crop. Average world production from 1969 to 1971 was 4.9×10^6 metric tons and has remained steady at about 10^7 metric tons since 1979. Trends in production for recent years can be seen in Table 7.1. By far the greater part of the world's pineapple production is canned. Slices are the most valuable product, then juice, chunks and dice. Recently, the export of fresh pineapples to temperate countries has increased considerably and there is every indication that it will remain an important world fruit crop in the future.

There are many *Ananas* cultivars which are divided into the groups (1) Cayenne, (2) Queen, (3) Spanish, (4) Abacaxi , and (5) Maipure (Sal). The

TABLE 7.1. Major producers of pineapple[a].

	1979–1981	1987	1988	1989
World	9296	10148	9601	9791
Africa	1193	1333	1220	1296
C. America	765	723	661	727
USA (Hawaii)	597	628	598	617
S. America	823	1419	1484	1327
Asia	5771	5872	5464	5639
Philippines	1059	1303	1181	1179
Thailand	2857	1510	1771	1900

[a] Times 1000 metric tons
Source: Food and Agriculture Organization Production Yearbook, Volume 43, 1989

predominant cultivar is "Smooth Cayenne" or "Cayenne Lisse" (in French), which belongs to the Cayenne group (45). It has several favorable characteristics such as spineless leaves, high production, high fruit quality and resistance to gummoses. It has been selected to produce cylindrical fruit which minimize waste when canned as rings. The other groups are considered superior as fresh fruit (52).

Pineapple is a perennial, monocarpic herb. Each stem flowers only once and dies after fruiting, a side shoot, commonly referred to as ratoon growth, then takes over. Since the natural cycle of flowering in pineapple is rather unreliable, it is customary to induce bloom by application of hormones or products that release ethylene or acetylene. Depending on cultivar, size of propagule, date of planting, climate and soil, the time between planting and formation of the inflorescence varies between 6 and 16 months. The fruit is formed by fusion of the parthenocarpic fruitlets with the bracts and the central axis of the inflorescence. The resulting compound fruit takes 5–6 months to ripen. On top of the fruit is a crown of leaves, which continue to grow until the fruit is mature. Planting of crowns for vegetative propagation of pineapple is practiced worldwide and is the primary means by which pineapple is grown. The resistance of the crown to desiccation probably contributed to the dissemination of pineapple clones or cultivars throughout the world. Vegetative slips sometimes present below the fruit can also be used for propagation as they are really the crowns of small fruits (60). The root system of pineapple is shallow and limited. In the best media roots seldom go deeper than 50 cm and in soil rarely extend below 30 cm.

There are a number of phytophagous pests and an array of pathogens that affect pineapple. Among these are heart rot caused by *Phytophthora cinnamomi*, rot of planting material caused by *Thielaviopsis paradoxa*, root-knot nematodes (*Meloidogyne*), lesion nematodes (*Pratylenchus* and *Rotylenchus*), yellow spot caused by tomato spotted wilt virus and a variety of Symphilids, scales, mites and fruit flies (58, 60). The pineapple industry has successfully limited the incidence of many problems through the use of fungicides applied to crowns prior to planting, application of nematicides to soil and a variety of cultural practices. In contrast, MBW remains widely distributed in pineapple in most places where it is cultivated and is especially serious for the widely cultivated variety "Smooth Cayenne" (22). Major epidemics occur in Hawaii, Jamaica, Central America (8), Puerto Rico (57), Mauritius, the Ivory Coast of West Africa, Florida and in the islands of Loocho, Palau, Saipan and Bonin (10). It has also been reported from South Africa, East Africa, Malaya, Java, Bali, Borneo, the Philippine Islands, Australia, Fiji, and South America (13/14). Sastry and Singhe (61) have reported MBW incidence in India.

Control has been achieved in the past by eliminating ant populations with chlorinated hydrocarbons. In the absence of ants, mealybug populations do not establish in pineapple and MBW does not occur. Most efficacious formicides are currently unavailable to pineapple producers in many places due to negative environmental impacts; hence, MBW is again rated as one of the most severe problems limiting pineapple production in many parts of the world (22, 58).

Mealybug Wilt of Pineapple

Disease Symptoms

Initial foci of MBW epidemics usually occur at the field edge and spread inward. The disease is characterized by preliminary leaf tip dieback, reddening along the leaf length followed by a progressive change from red to pink, inward reflexing of the leaf margins, loss of rigidity and ultimately collapse of the leaves along most of their length. Roots cease to grow causing the plant to wilt (11, 15, 60). Plants affected in immature growth stages do not form fruit or produce very small fruit. Remarkably, some plants, even those that appear fully wilted, undergo apparent recovery, during which the oldest symptomatic leaves senesce, younger symptomatic leaves grow out and wilt at the tip and new growth appears normal. These plants can produce apparently healthy fruit and crowns. Crowns from such plants have often been used for planting material and can produce marketable pineapple fruit.

Association of Mealybugs and Disease

The consistent association of the disease with mealybug populations gave rise to the name mealybug wilt of pineapple (4, 6, 40). Early workers (6, 7, 68) generally referred to *Pseudococcus brevipes* (Cockerell) as the pineapple mealybug; however, Ito (41) observed two distinct types of pineapple mealybugs associated with MBW in Hawaii, which he referred to as the pink form and the gray form. Since that time, mealybugs previously classified as *P. brevipes* have been placed in the genus *Dysmicoccus* and described as two distinct species, *D. brevipes* (the pink form, *sensu* Ito) and *D. neobrevipes* (the gray form, *sensu,* Ito) (2, 13/14, 58). The gray mealybug (GMb) occurs primarily on aerial portions of the plant including the crown and fruit, while the pink mealybug (PMb), often referred to as the subterranean form, is confined largely to the lower portions of the plant, near or below ground level. Both the GMb and the PMb infest many other host plants including sisal, banana, and a variety of grasses (53). Both species are widely distributed throughout the pineapple growing areas of the world. Morphologically similar species with distinct biological traits have been described in West Africa, Madagascar, The Dominican Republic, Martinique and tropical America.

These mealybugs are parthenogenic, giving birth to live offspring rather than laying eggs. The first larval instar is mobile and called a crawler. Crawlers can walk some distance as implied by their name and are widely dispersed by wind (3, 44). Ants have often been thought to move mealybugs

(54); however, recent evidence suggests this rarely occurs and that crawlers are the predominant means of spread (44). Ants tending GMbs and PMbs protect them from predation and remove honeydews produced by the mealybugs, preventing development of sooty mold, a fungal disease that causes high mortality in mealybug populations (54). Much remains to be learned about the ant mealybug relationship; however, it is clear that the number of mealybug infested pineapple plants and the number of ants in a field are correlated (4). Indeed, controlling ants in pineapple fields is concomitant with controlling mealybug wilt of pineapple. The long tailed mealybug, *Pseudococcus longispinus* (Targioni-Tozzetti), which gets its name from the long waxy filaments extending tail-like from its posterior end, has also been associated with mealybug wilt. This species is not as common in pineapple as GMb and PMb; however, the long tailed mealybug can survive well without tending by ants (58).

Etiology: Historical Perspective

Since the early 1930s, MBW was thought to be caused by toxins in mealybug saliva that were injected into plant tissue during feeding (6, 7, 9, 12, 13/14, 18). This concept was reinforced by positive correlations between mealybug numbers and MBW incidence and by the recovery that occurs when mealybugs are removed from symptomatic plants. This "toxin" hypothesis was espoused by Carter for several decades (7, 9, 15, 17). The foundation for the toxin hypothesis was built on data showing that, apparently, plants do not become diseased when infested by mealybugs from what Carter called "negative source plants" such as sisal, yucca, grasses or seedling pineapple not previously infested with mealybugs. In contrast, if mealybugs are placed on healthy plants after feeding on plants that Carter considered positive sources, such as wilt recovered pineapple clones, MBW could be induced with a single mealybug. Thirty years after proposing the toxin hypothesis, Carter reviewed his data and expanded his evaluation to include a "latent transmissible factor", perhaps viral, that in conjunction with mealybug salivary toxins caused MBW (16).

In later years, even in the absence of any candidate, the cause was assumed to be viral (42, 43) and Sampson (60) states that "the real cause of the disease seems to be a virus, but this has not yet been proven."

Identification of a Pineapple Closterovirus (PCV)

In the face of the biological and historical data just presented we decided to explore the possibility that at least one component of MBW etiology was viral. Recently, we used a variety of techniques in an attempt to identify a candidate (35, 36, 38, 39). Because the vast majority of plant viruses have single-stranded RNA genomes (49) and produce double-strand RNA (dsRNA) during the course of their replication (50) the first step in our investigation focused on isolating these transient replicative structures from MBW infected tissues. Since uninfected plant tissues rarely contain dsRNA, its presence provides strong evidence that replication of an RNA virus has occurred (25, 26, 27). Using standard techniques (24), nucleic acids were extracted from wilt-affected pineapple plants obtained from commercial plantations on the Hawaiian islands of Oahau, Maui and Lanai as well as wilt free plants supplied from the breeding stock collection of Maui Land and Pineapple Co. When total nucleic acid preparations from these plants were treated to obtain the fraction potentially containing dsRNA and analyzed by polyacrylamide gel electrophoresis, intensely staining bands were observed from extracts of roots and leaves of symptomatic plants, but not from wilt free pineapple plants (Figure 7.1) (39). Further characterization of these bands demonstrated that the molecules extracted from MBW affected plants had properties consistent with their being the replicative form of a single stranded RNA virus and that they could be reproducibly recovered from diseased pineapple plants (38).

Several standard virus purification procedures were assessed for isolating the putative virus responsible for the synthesis of the dsRNA molecules (38). The procedure finally adopted was a modification from Zee et al. for grape-

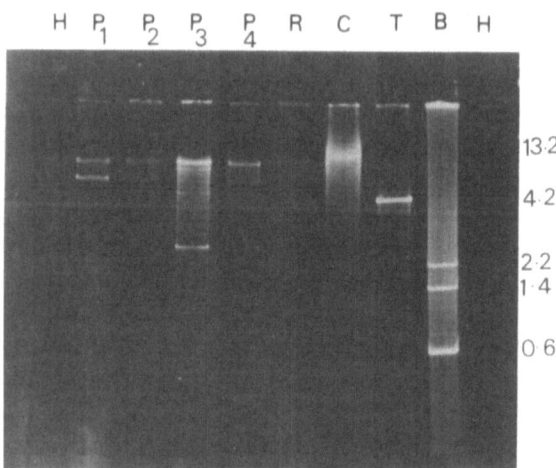

FIGURE 7.1. Polyacrylamide gel electrophoresis of double-stranded RNA (dsRNA) isolated from pineapple. Lane H: preparation from healthy pineapple. Lanes P1-P4: dsRNA isolated from four different plants. Lane R: dsRNA isolated from diseased roots. Lane C: dsRNA isolated from citrus infected with citrus tristeza virus (CTV) Lane T: dsRNA isolated from tobacco infected with tobacco mosaic virus (TMV) and lane B: dsRNA isolated from barley infected with brome mosaic virus (BMV). The numbers on the right side of the figure are the molecular weights of the dsRNA of CTV, TMV and BMV in millions of daltons. (From Gunashinghe and German, 1989 (39).)

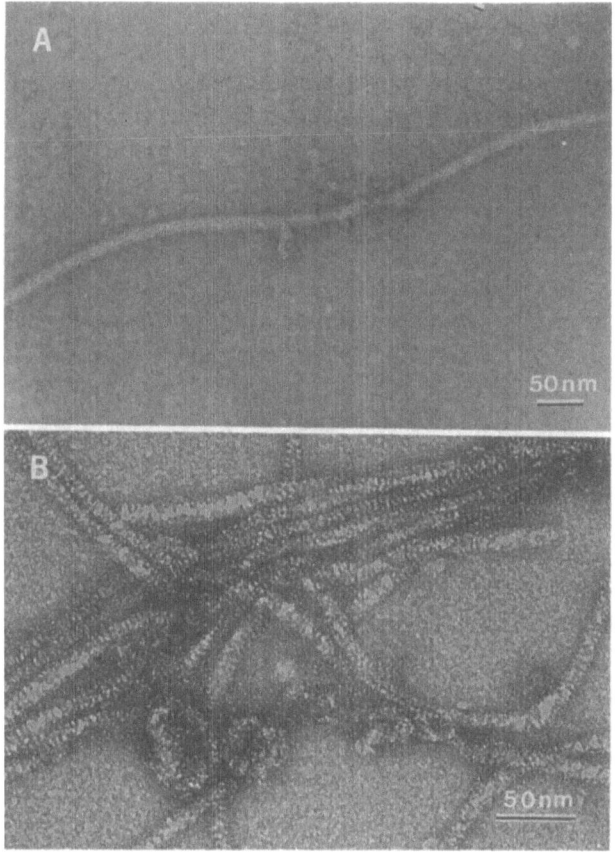

FIGURE 7.2. (A) Electron micrograph of virus particles stained with 2% phospho-tungstic acid, 200 ug/ml bacitracin in water. B, Electron micrograph of virus particles stained with saturated uranyl formate in methanol. The open structure and bending of the filaments, characteristic of closteroviruses, can be seen easily. (A, from Gunashinghe, 1989, (38) (B) from Gunashinghe and German, 1989 (39).)

vine leafroll virus (67). Following extraction the partially purified material was concentrated by centrifuging it through a 0.48M Cs_2SO_4 cushion. When the pelleted material was dissolved in buffer and viewed in the electron microscope particles such as those shown in figure 7.2A were observed. These long flexuous virus particles have a modal length of 1,200 nm and a width of 12 nm. Sodium dodecyl sulfate polyacrylamide gel electrophoretic analysis of the viral proteins indicated that the coat protein has a molecular weight of approximately 24 kD. When the particles were stained with uranyl formate in methanol and viewed at high magnification the open helical structure shown in figure 7.2B was observed. These physical properties and the pattern formed by the dsRNA bands shown in figure 7.1 are very similar to the those of the type II closterovirus group (1, 19, 32). Interestingly, grapevine virus A, also a closterovirus, is one of the few RNA plant viruses known to be transmitted by mealybugs (59).

The procedures described above were designed to detect and isolate viruses with an RNA genome. Although plant viruses with a DNA genome are relatively rare, there is a group of non-enveloped bacilliform viruses which have circular double-stranded DNA genomes (51). Members of this group infect banana, sugarcane, taro, yam, canna, kalanchoe and yucca all of which are propagated vegetatively. It is likely that natural transmission of these viruses occurs to a significant extent by clonal propagation. Furthermore, five of the viruses in this group are transmitted by mealybugs including: Cacao swollen shoot virus (CSSV) (5), *Colocasia esculenta* virus (CBV) (31), commelina yellow mottle virus (CoYMV) (47), kalanchoe top-spotting virus (KTSV) (28) and Banana streak virus (BSV) (Bouhida and Lockhart, unpublished). *Pseudococcus longispinus* and *Dysmicoccus brevipes* are vectors of CSSV and *P. longispinus* transmits CBV. CBV, KTSV, CoYMV and BSV have been transmitted by *P. citri*. Interestingly, a number of these viruses are found in association with closteroviruses (Lockhart, personal communication).

Because of the obvious potential for involvement of non-enveloped bacilliform viruses with mealybug wilt of pineapple, samples of infected material were subjected to purification procedures known to detect them. No bacilliform particles were observed in any of the material tested although, long flexuous particles were seen (Lockhart, personal communication).

Development of Diagnostic Tests for PCV

PRODUCTION OF SPECIFIC ANTISERUM

Discovery and characterization of PCV renewed scientific interest in MBW and its etiology. As a prerequisite to studying PCV distribution and association with wilt of pineapple it was necessary to develop diagnostic tests. Analysis of dsRNA and electron microscope observations of purified plant extracts enabled us to check for virus presence; however, we needed simple, non-destructive, sensitive and accurate tests for large scale testing of plants. A polyclonal antiserum was produced by injecting rabbits with preparations of partially purified pineapple virus (38, 55). The initial specificity of the antiserum was determined by the Ouchterlony double diffusion method (Figure 7.3) (65). After preabsorption of antiserum with healthy plant extracts, the antiserum reacted with either infected plant material or purified virus preparations, but not with sap from healthy plants. Because of the similarity of PCV and grapevine leafroll virus (GLV) with respect to mealybug association, we tested our antiserum against tissue infected with GLV but found it not to cross react under these conditions of sensitivity (65). Further demon-

FIGURE 7.3. Reactions of pineapple virus antiserum against leaf extracts from: well 1, mealybug wilt-diseased pineapple (cultivar Smooth Cayenne); well 2, buffer control; well 3, leaf extracts from healthy grapevine; well 4, leaf extracts from healthy pineapple (cultivar Smooth Cayenne); well 5, partially purified pineapple virus; well 6, leaf extracts from grapevine leafroll virus-infected tissue. Arrows indicate precipitin reactions to partially purified pineapple virus and leaf extract from diseased pineapple. (From Ullman et. al., 1989 (65).)

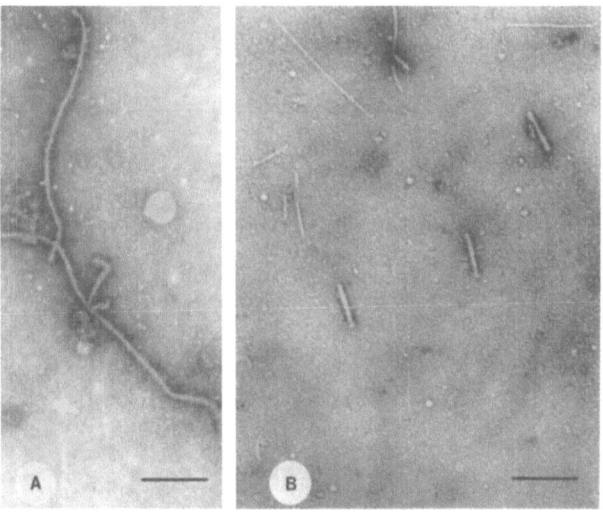

FIGURE 7.4. Serologically specific electron micrograph showing decoration of pineapple closterovirus (PCV) with antiserum prepared to partially purified preparations of PCV. Bar = 457 nm. B, In reciprocal tests, antiserum to tobacco mosaic virus (TMV) decorated TMV but not PCV. Bar = 457 nm. (From Ullman et. al, 1989 (65).)

stration that the antiserum was reacting specifically with PCV was provided using serologically specific electron microscopy (65). In this procedure PCV or tobacco mosaic virus (TMV) control particles were placed on electron microscope grids and allowed to react with their respective specific antisera (IgG). Afterwards the grids were washed free of unbound IgG and stained with phosphotungstic acid. When IgG reacted with the virus the staining procedure produced a "decoration" of electron dense immunoglobulin. Figure 7.4 shows the results of these experiments. In panel A the shadow produced around the PCV by reaction with the antibody raised against it is shown. In panel B, the specificity of the reaction is demonstrated by a control in which antiserum to TMV is shown to "decorate" the TMV but not PCV.

DEVELOPMENT OF AN ELISA ASSAY

An indirect enzyme-linked immunosorbent assay (ELISA) technique (21, 48) was adapted for use with PCV (65) because it is more sensitive and less labor intensive than the agar diffusion test, amenable to large scale testing and requires less antiserum per test sample. IgG from pineapple virus antiserum reacted to pineapple virus antigens from crude leaf extracts but not buffer controls or crude leaf extracts from virus-free pineapple plants. Optimum reactivity was obtained using crude leaf extracts from 114 mg of plant tissue/ml of buffer, 1/1000 (v:v) dilution of purified pineapple virus IgG, and 1/2000 (v:v) dilution of goat antirabbit IgG enzyme conjugate. When buffer, uninfected and infected standards from 12 different ELISA plates were com-

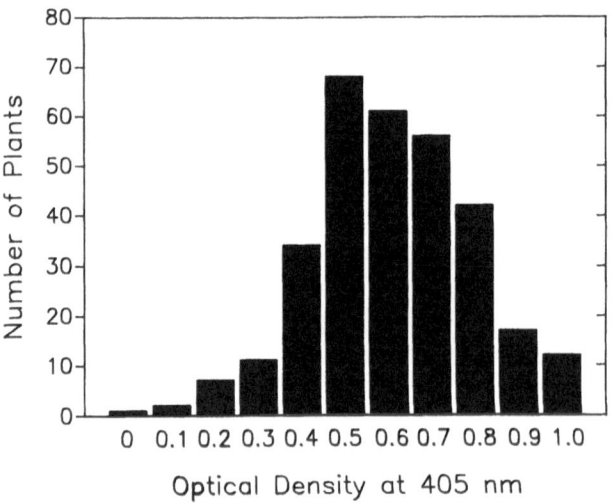

FIGURE 7.5. Variation in the reactivity (expressed as optical density at 405 nm) of crude leaf extracts from 330 pineapple plants tested with enzyme-linked immunosorbent assay. (From Ullman et. al., 1989 (65).)

pared, final absorbance was found to be very consistent. Background reactions were kept to a minimum by preabsorbing the IgG with healthy leaf extracts before and after fractionation and again just prior to use. The standard deviation around mean optical density of positive standards was ± 0.08; therefore, plant samples resulting in optical density readings of greater than 0.3 with the ELISA were considered positive. It should be noted that strong positive reactions were observed using infected plant samples kept frozen for over 6 months in contrast to the decreased reactivity found with plant tissue stored at room temperature in low light for as little as two weeks. The reliability of the ELISA assay is confirmed by the observation that virus and dsRNA can be consistently recovered from ELISA positive plants but not from ELISA negative plants. When the ELISA was used to test 330 randomly selected pineapple plants from fields in Hawaii a wide range of reactivity was observed (Figure 7.5), presumably because of differences in virus titer or the presence of mild strains which were not serologically distinguishable by the technique used.

DEVELOPMENT OF NUCLEIC ACID PROBES

DNA complementary to PCV (cDNA) was synthesized using gel purified dsRNA as the template for reverse transcription. This cDNA was then used in dot-blot assays to detect viral RNA in pineapple plants, mealybugs and other plant hosts (30, 37, 38). This procedure included the isolation of dsRNA and treatment with ribonuclease under conditions of high salt that permitted the selective removal of contaminating single-stranded (host) RNA. After enzymatic removal of the ribonuclease, the dsRNA strands were separated by boiling and the mixture was rapidly cooled to prevent reannealing. Complimentary DNA was synthesized using reverse transcriptase and the random primer method of Taylor et al. (63). For the synthesis of labeled cDNA a-α^{32}P dCTP was included in the reaction. For dot blot analysis total RNA was isolated from symptomatic or uninfected (as described above from Maui Land and Pineapple Co. breeding stock) pineapple tissue, and from various grass species by the guanidium thiocyanate procedure of Chomczynski and Sacchi (20). For isolation of RNA from mealybugs, 2-3 mature insects were crushed in a microfuge tube in the presence of buffer and extracted with phenol-chloroform prior to recovery of total nucleic acids by ethanol precipitation (29). Nucleic acids were denatured, spotted onto nitrocellulose membranes and assayed using standard procedures (56, 64).

The results presented in figure 6 show that viral RNA could be detected with the cDNA probes in diseased pineapple while virus free tissues do not react. Signal intensity varied among MBW-affected samples (Figure 7.6) presumably because of differences in virus concentrations. These results are consistent with earlier findings showing a wide range of ELISA readings from randomly selected samples (Figure 7.5). Viral RNA was also detected in mealybugs taken from diseased pineapple fruit (Df) or from diseased leaves

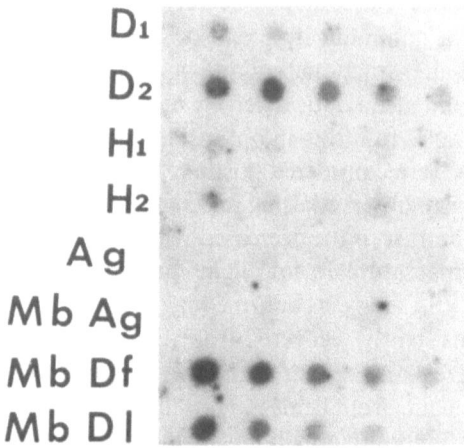

FIGURE 7.6. Results of dot-blot hybridization with cDNA probes synthesized from ribonuclease A treated dsRNA obtained from diseased pineapple plants. Dark spots are signal from total RNA extracted from: D1 and D2, leaf extracts of two different diseased pineapple plants; H1 and H2, leaf extracts from two different healthy plants; Ag, leaf extract of healthy Agave spp.; MbAg, two mealybugs taken from Agave (Ag); MbDf, two mealybugs taken from diseased pineapple fruit and MbD1, two mealybugs taken from diseased pineapple leaves. All samples were applied in a dilution series from left to right. (From Gunasinghe, 1989 (38).)

(D1) (Figure 7.6). In contrast, mealybugs (Mb Ag) from a virus free Agave plant (Ag) gave a negative hybridization signal.

Adult female mealybugs from PCV affected plants and their offspring produced in petri dishes lined with moist filter paper were assayed using the cDNA probes. The offspring of mealybugs that produced positive hybridization signals were also positive. Remarkably, similar results were obtained after several generations of rearing on squash (which is not a virus host). These data suggest that PCV persists with adult mealybugs and can be passed to offspring by an as yet unknown mechanism. The potential implications of this information are great as they suggest the mealybugs themselves represent an important virus reservoir. Furthermore, mealybug crawlers are widely dispersed by the wind (44). Infected crawlers, blown by wind to field edges may establish in the presence of ants and explain spread of the disease from field edges. These data may also explain Carter's early observations that some groups of mealybugs caused MBW, while equal numbers of mealybugs from other populations could not (18). Verification of these preliminary results is a topic of current research (Hu and Ullman, personal communication).

Andropogon insularis L. and *Paspalum urvullei* L. are grass species that occur commonly in the areas surrounding commercial pineapple plantations in Hawaii. Since these plants are mealybug hosts and could serve as wild

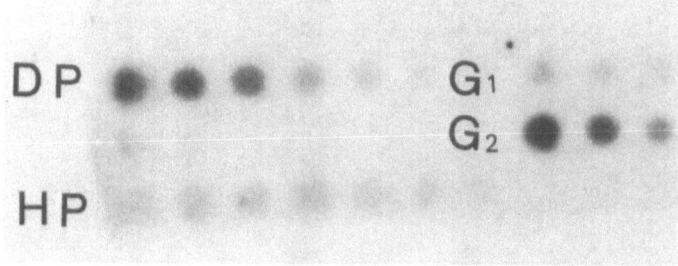

FIGURE 7.7. Results of dot-blot hybridization with cDNA probes synthesized from ribonuclease A treated dsRNA from diseased pineapple plants. Dark spots are signal from total RNA extracted from: DP, diseased pineapple; HP, virus free pineapple; G_1 and G_2, *Paspalum urvullei*. The faint signal from HP has been determined to be background. At least one of the grass samples is positive for virus.

reservoirs of PCV, they were tested for virus presence with both ELISA and cDNA probes. Results from cDNA probe assays (Figure 7.7) indicated that one of the two grass species (*P. urvullei* L.) was infected (37, 39). However, tissues from the same plants were negative using ELISA. This may have occurred due to the extremely low virus titers which were below the detection level of the serological assay or because serologically distinct virus strains are present in nature. These results raise the fascinating possibility that a virulent strain persists in various mealybug hosts and that interactions resulting in disease occur when strains mix in commercial pineapple.

PRODUCTION OF VIRUS FREE PLANTS

The results of ELISA tests (58) suggested that cultivated pineapple (cv. Smooth Cayenne) in the Hawaiian Islands is widely infected with the pineapple virus (Figure 7.5) (65). This is not surprising in view of the fact that all of the pineapple produced in Hawaii is Smooth Cayenne, and that increase of this clone has been through the vegetative propagation of crowns for over 70 years. The likely occurrence of strains and the unavailability of virus free plants make it difficult to conduct transmission experiments or complete Koch's postulates.

To address this problem, pineapple plants were freed of virus by heat treatment (66) to provide a source of experimental material. Pineapple crowns heat treated by submersion in a water bath at 50 C for 30 minutes have a 100% survival rate and are rendered 100% free of closterovirus-like particles as judged by ELISA testing and failure to recover virus particles from tissue samples through purification. These plants remained free of closterovirus-like particles for more than 18 months and produced virus free crowns and ratoon growth (66).

Plants from these treatments appeared to grow more vigorously than

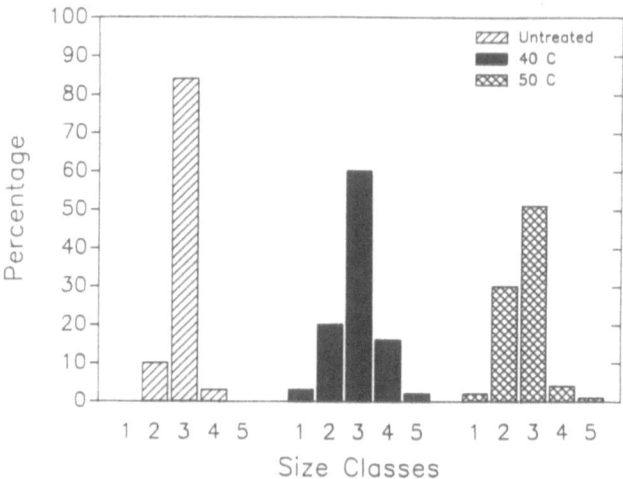

FIGURE 7.8. Evaluation of plant size in nontreated control and heat treated field plantings. Ratings are based on a visual scoring system with size classes as follows: 1 = exceptionally large, 2 = slightly smaller, but well above average, 3 = average, 4 = smaller than average with signs of active growth and 5 = stunted, no sign of active growth. (From Ullman et al., 1991 (66).)

non-treated plants and the industry became interested in the commercial utility of heat treating pineapple. Large numbers of pineapple crowns were heat treated (approximately 11,000 plants) at 40 C for 60 min. or at 50 C for 30 min. The growth rate of these plants was quantified by means of a size rating scale utilizing 5 classes, with 1 being the largest and 5 being extremely stunted with little sign of growth. Measurement of D-leaf subsamples were used to verify visual size ratings (62). Heat treatment resulted in significantly more plants in the largest size classes of 1 and 2 than observed in the non-heat treated control (Figure 7.8) Non-treated controls were very uniform in size with nearly 90% of the plants falling in the average size class of 3 and less than 10% in plant sizes 1 or 2. Heat treated plants in size classes 1 and 2 did not yield visible virus particles following virus purification, whereas heat treated and non-treated plants in size class 3 yielded virus particles upon purification (66). This data suggests that PCV reduces plant vigor; however, the influence of heat on other undescribed microorganisms or physiological processes in the plants has to be considered.

Discussion

While it might appear contradictory to find that a high percentage of symp-tomless plants are positive for PCV (Figure 7.5) (65), it seems reasonable that vegetative propagation would result in a crop that is consistently affected by a mild strain of any virus that was present at the start of the cycle. The phenomenon of cross protection, in which prior infection of a plant with a mild strain prevents the development of symptoms when it is later infected by a severe strain of the same virus, is well known (29). As planting material (crowns) are carefully selected for replanting, only asymptomatic material will be retained, resulting in the selection of crowns infected with a mild strain. Early workers suggested that the "wilt-inducing factor" may exist in symptomless plants and these plants can serve as positive source plants (13/14, 18). Perhaps what they observed was the transmission of virus taken from a plant that contained the severe strain but which was asymptomatic as a result of cross protection with the mild strain. This may explain more recent data (38, 65, this chapter) since none of the diagnostic tests described herein would distinguish between closely related strains of the virus.

How then, does one account for the occurrence of severe disease upon re-infestation with mealybugs as they move in from the edge of fields? It is well known that mild strains do not protect against all severe strains and it may be that a severe strain occurs in an alternate host (Grass?) that can be brought into the field and cause the severe wilt disease of pineapple. This has been observed in citrus plantations in California and Texas where orange trees were intentionally infected with mild strains of citrus tristeza virus (a closterovirus) in an effort to protect them from severe disease (23, 33, 34, 55). Indeed, Ito (38) suggested some 30 years ago that a mild strain of some virus may exist in field grown pineapple plants and that a severe strain of the virus may be transmitted by mealybugs, producing disease.

There is ample information linking the closterovirus-like particle found in pineapple, mealybugs and grass to mealybug wilt of pineapple. The association of both virus and mealybugs with disease is very consistent, the demonstration that mealybugs can acquire virus from infected plants, the failure to find other virus suspects in infected plants, the apparent transfer of virus to mealybug offspring and the "reverse Koch's postulates experiment" in which an increase in growth and vigor is observed when plants are freed of virus by heat treatment, all strengthen the case for involvement of the closterovirus in MBW.

Future Directions

Definitive proof for the cause of mealybug wilt awaits the completion of Koch's postulates. The facts that the virus is not mechanically transmissible by sap inoculation or grafting, that inoculation of virus with mealybugs is laborious, and that symptoms of mealybug wilt are unreliable indicators of disease and take 1–18 months to appear after inoculative feeding have been major obstacles to the completion of Koch's postulates. MBW etiology will only be revealed if complex interactions between the trophic levels and environmental interactions are considered. It seems unlikely that a simple, single pathogen, cause and effect scenario will explain MBW epidemiology. The recent development of diagnostic tools that do not rely on symptom expression and the ability to generate large amounts of virus free material with heat treatments make such experiments possible. By using these new tools and refining the techniques to produce monoclonal antibodies or cDNA probes capable of distinguishing between strains, it will be possible to separate some of the many variables associated with disease onset. The existence of severe and mild strains of PCV, alternate virus host reservoirs, involvement of other viruses, viroids or microorganisms and possible interactions between virus and mealybugs resulting in toxin production are all possibilities that can be addressed.

Expectations generated by the developments described in this chapter and the revitalized interest in MBW etiology brought about by the need for ecologically compatible control procedures, provides a strong motivation for continued research to solve this important, interesting and complex problem.

Acknowledgments. The authors gratefully acknowledge the assistance of: Mr. Steve Vicen for help with preparation of the figures. Mr. Andrew Witherell for help with the figures and reading the manuscript. Dr. B.E.L. Lockhart for looking for non-enveloped bacilliform virus particles in MBW infected tissue and providing a manuscript prior to publication. David Fisher and Marylin Dunlap for providing facilities. Dr. J. Hu for advice on virus nomenclature. Financial assistance from the following sources was used to support research described in this article: Maui Land and Pineapple Co., Ltd., 870 Haliimaile Hwy, Haliimaile, Hawaii 96768, The State of Hawaii Governor's Agriculture Coordinating Committee (contract no. 87-12), Biomedical Research Support Grant, University of Hawaii, Manoa and a Research Centers in Minority Institutions Grant RR-03061, Division of Research Resources, National Institutes of Health.

References

1. Bar-Joseph, M., Lobenstein, G., and Cohen, J., 1972, Further purification and characterization of threadlike particles associated with the tristeza disease, *Virology* **50**:821–828.
2. Beardsley, J.W., 1959, On the taxonomy of pineapple mealybugs in Hawaii, with a description of previously unnamed species (Homoptera: Pseudococcidae), *Proc. Hawaii Entomol. Soc.* **17**:29–37.
3. Beardsley, J.W., 1959, Observations on sugar cane mealybugs in Hawaii, *Proceedings of the 10th Congress on the International Society of Sugarcane Technologists,* p. 954–961.
4. Beardsley, J.W., Su, T.H., McEwen, F.L., and Gerling, D., 1982, Field investigations of the interrelationships of the big-headed ant, the grey pineapple mealybug, and pineapple mealybug wilt disease in Hawaii, *Proc. Hawaii Entomol. Soc.* **24**:51–67.
5. Brunt, A.A., and Kenten, R.H., 1971, Viruses affecting cacao, *Review of Plant Pathology* **50**:591–602.

6. Carter, W., 1932, Studies of populations of Pseudococcus brevipes (Ckl.) occuring on pineapple plants, *Ecology* **13**:296–304.

7. Carter, W., 1933, the pineapple mealybug, *Pseudococcus brevipes*, and wilt of pineapples, *Phytopathology* **23**:207–242.

8. Carter, W., 1934, Mealybug wilt and green spot in Jamacia and Central America, *Phytopathology* **24**:424–426.

9. Carter, W., 1935, Studies on the biological control of *Pseudococcus brevipes* and wilt of pineapples, *Phytopathology* **28**:1037–1041.

10. Carter, W., 1942, Geographical distribution of mealybug wilt with some other insect pests of pineapple, *J. Econ. Entomol.* **35**:10–15.

11. Carter, W., 1944, Some etiological aspects of mealybug wilt, *Phytopathology* **35**:305–315.

12. Carter, W., 1945, Some etiological aspects of mealybug wilt, *Phytopathology* **35**:305–315.

13/14. Carter, W., 1946, Insect notes from South America with special reference to *Pseudococcus brevipes* and mealybug wilt, *J. Econ. Entomol.* **42**:761–766.

15. Carter, W., 1948, The affect of mealybug feeding on pineapple crowns in finely atomized nutrient solutions, *Phytopathology* **38**:645–657.

16. Carter, W., 1963, Mealybug wilt of pineapple; a reappraisal, *Ann. N.Y. Acad. Sci.* **105**:741–764.

17. Carter, W., and Collins, J.L., 1947, Resistance of mealybug wilt of pineapple with special reference to a Cayenne-Queen hybrid, *Phytopathology* **37**:332–348.

18. Carter, W., and Ito, K., 1956, Study of the value of pineapple plants as sources of the mealybug wilt factor, *Phytopathology* **46**:601–605.

19. Chevallier, D., Engle, A., Wurtz, M., and Putz, C., 1983, The structure and characterization of a closterovirus, beet yellow virus, and a luteovirus, beet mild yellowing virus, by scanning transmission electron microscopy, optical diffraction of electron images and acrylamide gel electrophoresis, *J. Gen. Virol.* **64**:2289–2293.

20. Chomozybski, P., and Sacchi, N., 1987, Single step method of RNA isolation by acid guanidium thiocyanate-phenol-chloroform extraction, *Analytical Biochem.* **162**:156–159.

21. Clark, M.F., and Adams, A.N., 1977, Characteristics of the microplate method of enzyme-linked immunosorbant assay for the detection of plant viruses, *J. Gen. Virol.* **34**:475–483.

22. Collins, J.L., 1960, The Pineapple: Botany, Cultivation and Utilization, Leonard Hill, London, pp. 295.

23. Costa, A.S., Grant, T.T., and Moreira, S., 1954, Behavior of various citrus rootstock-scion combinations following inoculation with mild and severe strains of tristeza virus, *Fla. State Hort. Soc. Proc.* **67**:26–30.

24. Dale, J.L., Phillips, D.A., and Parry, J.N., 1986, Double-stranded RNA in banana plants with bunchy top disease, *J. Gen. Virol.* **67**:371–376.

25. Dodds, J.A., 1986, The potential for using double-stranded RNAs as diagnostic probes for plant viruses, in Jones, R.A.C., and Torrance, L. (eds): *Development and application in virus testing*, Wellsbourne, Warwick, England, pp. 71–86.

26. Dodds, J.A., and Bar-Joseph, M., 1983, Double-stranded RNA from plants infected with closteroviruses, *Phytopathology* **73**:419–423.

27. Dodds, J.A., Morris, T.J., and Jordan, R.L., 1984, Plant viral double-stranded RNAs, *Ann. Rev. Phytopathology* **22**:151–168.

28. Ferji, Z., and Lockhart, B.E.L., Transmission and serology of kalanchoe top-spotting virus, *Plant Disease.* (Submitted)
29. Fulton, R.W., 1982, in Wood, R.K.S. (ed): *Active Defense Mechanisms in Plants,* Plenum, New York, pp. 231–245.
30. Garger, S.J., and Turpen, T.H., 1986, Use of RNA probes to detect plant RNA viruses, *Methods in Enzymology* **118**:718–722.
31. Gollifer, D.E., Jackson, G.V.H., Dabek, A.J., Plumb, R.J., and May, Y.Y., 1977, The occurrence and transmission of viruses of edible aroids in the Solomon Islands and the southwest Pacific, *PNAS*:23–171.
32. Gonsalves, D., Purcifull, D.E., and Garnsey, S.M., 1978, Purification and serology of citrus tristeza virus, *Phytopathology* **68**:553–559.
33. Grant, T.J., and Costa, A.S., 1951, A mild strain of the tristeza virus of citrus, *Phytopathology* **46**:336–347.
34. Grant, T.J., and Higgins, R.P., 1957, Occurrence of mixture of tristeza virus strains in citrus, *Phytopathology* **47**:272–276
35. Gunashinghe, U.B., and German T.L., 1986, Association of virus particles with mealybug wilt of pineapple, *Phytopathology* **76**:1073 (Abstr.).
36. Gunashinghe, U.B., and German, T.L., 1987, Further characterization of a virus associated with mealybug wilt of pineapple, *Phytopathology* **77**:1776 (Abstr.).
37. Gunashinghe, U.B., and German T.L., 1988, Use of cDNA probes to characterize and detect virus in mealybug wilt affected pineapple plants, *Phytopathology* **78**: 1584 (Abstr.).
38. Gunasinghe, U.B., 1989, Characterization of a new virus isolated from pineapple, Ph.D. Thesis, University of Hawaii, Honolulu.
39. Gunasinghe, U.B., and German, T.L., 1989, Purification and partial characterization of a virus from pineapple, *Phytopathology* **79**:1337–1341.
40. Illingworth, J.F., 1931, Preliminary reports on evidence that mealybugs are an important factor in mealybug wilt, *J. Econ. Entomol.* **24**:877–889.
41. Ito, K., 1938, Studies on the life history of the pineapple mealybug, *Pseudoccous brevipes* (Ckll.), *J. Econ. Entomol.* **31**:291–298.
42. Ito, K., 1959, Terminal mottle as a symptomological aspect of mealybug wilt; evidence supporting the hypothesis of a virus etiology, PRI Research Report No. 2.
43. Ito, K., 1962, Additional immunological evidence supporting the virus nature of mealybug wilt, *J. Econ. Entomol.* **24**:877–889.
44. Jahn, G., and Beardsley, J., (1991), The role of the big-headed ant in the pineapple agroecosystem, *Proc. XVII Pacific Science Congress,* Honolulu, Hawaii.
45. Knight, R., Jr., 1980, Orgin and importance of tropical and subtropical fruit crops, in Nagy, S., and Shaw, P.E. (eds): *Tropical and Subtropical Fruits,* AVI Publishing, Westport, Connecticut, pp. 570.
46. Larsen, L.D., 1910, Diseases of pineapple, *Hawaii Sugar Plant. Assoc. Pathol. Physiol. Ser. Exp. Stn. Bull.* **10**:1–72.
47. Lockhart, B.E.L., 1988, Occurance of canna yellow mottle virus in North America, *Acta Hort.* **234**:64.
48. Lommel, S.A., McCain, A.H., and Morris, T.J., 1982, Evaluation of indirect enzyme-linked immunosorbent assay for detection of plant viruses, *Phytopathology* **72**:1018–1020.
49. Matthews, R.E.F., 1979, Classification and nomenclature of viruses, Third report of the ICTV, *Intervirology* **72**:131–296.

50. Matthews, R.E.F., 1991, *Plant Virology* 3rd ed, Academic Press., New York, pp 835.
51. Medberry, S.L., Lockhart, B.E., and Olszewsji, N.E., 1990, Properties of *Commelina* yellow mottle virus' complete DNA sequence, genomic discontinuities and transcript suggest that it is a pararetrovirus, *Nucl. Acids Res.* **18**:5505–5513.
52. Morley-Bunker, M., 1986, Subtropical and Tropical Fruits, in Jackson, D.I., (ed): *Temperate and Subtropical Fruit Production*, 294 pp.
53. Nakahara, S., 1981, List of the Hawaiian Coccoidea (Homoptera: Sternorhyncha), *Proc. Hawaii Entomol. Soc.* **23**:387–424.
54. Nixon, G.E.J., 1951, The association of ants with aphids and coccids, *Commonwealth Institute of Entomology*, London, 36 pp.
55. Olson, F.O., 1956, Mild and severe strains of tristeza virus in Texas citrus, *Phytopathology* **45**:336–341.
56. Owens, R.A., and Diener, T.O., 1984, Spot hybridization for detection of viroids and viruses, *Methods in Virology* **7**:173–187.
57. Plank, H.K., and Smith, M.R., 1940, A survey of the pineapple mealybugs in Puerto Rico and preliminary studies of its control, *J. Agr. Puerto Rico* **24**:49–76.
58. Rohrbach, K.G., Beardsley, J.W., German, T.L., Reimer, N.J. and Sanford, W.G., 1988, Mealybug wilt, mealybugs, and ants on pineapple, *Plant Disease* **72**:588–565.
59. Rosciglione, B., Catellana, M.A., Martelli, G.P., Savino, V., and Cannizzaro, G., 1983, Mealybug transmission of grapevine virus, *A. Vitis* **22**:331.
60. Samson, J.A., 1986, *Tropical Fruits*, Tropical Agriculture Series, Longman, London, pp. 250.
61. Sastry, K.S.M., and Singhe, S.J., 1974, Wilt of pineapple. A new virus disease in India, *Indian Phytopathol.* **27**:298–303.
62. Sideris, C.P., and Krauss, B.H., 1936, The classification and nomenclature of groups of pineapple leaves, sections of leaves and sections of stems based on morphological and anatomical differences, *Pineapple Quarterly* (Hawaii), **6**:135–147. ·
63. Taylor, J.M., Illmansee, R., and Summers, J., 1976, Efficient transcription of RNA into DNA by Avian Sarcoma virus polymerase, *Biochem. Biophys. Acta.* **422**:324–330.
64. Thomas, P.S., 1983, Hybridization of denatured RNA transferred or dotted to nitrocellulose paper, *Methods in Enzymology* **100**:255–265.
65. Ullman, D.E., German, T.L., Gunasinghe, U.B., and Ebesu, R.H., 1989, Serology of a closteroviruslike particle associated with mealybug wilt of pineapple, *Phytopathology* **79**:1341–1345.
66. Ullman, D.E., German, T.L., McIntosh, C.E., and Williams, D.D.F., Heat treatment produces pineapple free of a closteroviruslike particle associated with mealybug wilt of pineapple, *Plant Disease* **75**:859–861.
67. Zee, F., Gonsalves, D., Goheen, A., Kim, K.S., Pool, R., and Lee, R.F., 1987, Cytopathology of leafroll-diseased grapevines and the purification and serology of associated closteroviruslike particles, *Phytopathology* **77**:1427–1434.
68. Zimmerman, E.C., 1948, Insects of Hawaii, Volume 5, Homoptera: Sternorhyncha, University of Hawaii Press, Honolulu, 464 pp.

8
Ilarvirus Vectors

Gaylord I. Mink

Introduction

The term Ilar is derived from the description "isometric labile ringspot" first used to identify a few viruses of stone fruits that were unstable in sap and thought to be spherical (32). Lister and Saksena (56) proposed enlarging the group to include unstable viruses having three or more nucleoprotein components, a proposal approved by the International Committee on Taxonomy of Viruses (ICTV) (79). The group currently includes at least 16 distinct viruses (Table 8.1) some of which have several named strains, that can create confusion when one tries to distinguish among viruses and strains in the literature. Subgroup divisions are based on serological interrelationships (25, 90). So far no uniform criteria have been adopted for subdividing viruses and strains. At least 10 ilarviruses are found predominately, if not exclusively, in woody hosts and one (TSV) causes diseases of both woody and herbaceous plants.

Even though many ilarviruses occur naturally in woody plants, all can be transmitted by rub inoculation to a wide range of herbaceous plants. Most viruses in this group spread naturally in the field, some rapidly. Yet few vectors have been identified with certainty at this time. Perhaps one reason for this is the relatively little effort put into transmission attempts with most members of this group. Extensive attempts were made to transmit Prunus necrotic ringspot (PNRSV) and prune dwarf (PDV) viruses by a variety of insects (69, 83). For most other ilarviruses, however, transmission efforts have involved only a few insect species.

The role of thrips as natural vectors will be discussed in later sections. The possible interaction of thrips, pollen, and pollinating insects will be examined in light of recent information, some of which is, as yet, unpublished. In particular, the role of pollen in plant-to-plant spread of ilarviruses will be critically reexamined.

Gaylord I. Mink, Department of Plant Pathology, Washington State University, Irrigated Agriculture Research and Extension Center, Prosser, Washington 99350, USA.

TABLE 8.1. Distinct Ilarviruses and their suggested subgroups.[a]

Subgroup	Virus
1	Tobacco streak virus (TSV)
2	Asparagus virus 2 (AV-2)
	Citrus leaf rugose virus (CLRV)
	Citrus variegation virus (CVV)
	Elm mottle virus (EMoV)
	Tulare apple mosaic virus (TAMV)
3	Prunus necrotic ringspot virus (PNRSV)
	Apple mosaic virus (ApMV)
	Blueberry shock virus (BBSV)
	Humulus japonica virus (HJV)
4	Prune dwarf virus (PDV)
5	American plum line pattern virus (APLPV)
6	Spinach latent virus (SPLV)
7	Lilac ring mottle virus (LRMV)
8	Hydrangea mosaic virus (HydMV)
9	Parietaria mottle virus (ParMV)

[a] Extended from tables presented in references 25 and 90.

Ilarvirus Properties and Subgroups

In general, the biomolecular properties of ilarviruses are less well studied than those of many virus groups (24). This is due in part to the perception that their instability in plant extracts make ilarviruses exceptionally difficult to handle. Although the yield of some ilarviruses may be consistently low, once purified these viruses are usually quite stable. Nevertheless, key properties have been determined for many ilarviruses. All are multicomponent and separate as three or four centrifugal components having sedimentation coefficients between 75 and 125 S (25). Nucleoproteins contain a single species of protein with Mr between 24 and 30×10^{3d}. All ilarviruses have three or four positive-sense, single-stranded RNAs ranging in size between 0.3 and 1.3×10^{6d}. The largest 3 RNAs constitute the complete genome. Where 4 RNAs occur, all four RNAs are required for infection. However, the coat protein can usually be substituted for the smallest RNA. Nucleoprotein components vary from isometric (20–25 nm diameter) to ovate to bacilliform particles up to 70 nm in length. Not all isolates have bacilliform particles (33).

Vectors of Subgroup 1

Subgroup 1 consists of a single virus, TSV, which exists as many strains. Some strains can be differentiated on the basis of symptomatology and serology (33). Synonyms include asparagus stunt virus (7) and black raspberry latent virus (49) both of which are serologically similar to but not identical with the type strain. While several insect species have been tested as potential vectors (14, 50, 72) only thrips have been shown to be involved in transmission.

Thrips as Vectors

An association between the natural occurrence of TSV in weed populations and spread of the virus into adjacent crop plants has been reported in many countries over a period of seven decades (5, 13, 15, 17, 23, 43, 44, 47, 50, 51, 85, 88, 91). Despite this awareness that TSV-induced diseases spread in the field, the first report of successful vector transmission did not appear until 1976. Costa and Lima Neto (16) reported infection of soybean and tobacco plants in Brazil using thrips (*Frankliniella sp*) taken from field-grown, naturally-infected plants of *Ambrosia polystachya*. They obtained transmission in four out of ten tests infecting a total of seven soybean and three tobacco plants. However, subsequent studies in Brazil failed to substantiate thrips transmission (55).

In 1982, Kaiser et al. (50) reported experimental transmission of TSV in the U.S. using thrips (species not identified) collected from flowers of naturally-infected white sweet clover (*Melilotus alba* Medik) growing in the Central Ferry area of Washington state. When field-collected thrips were transferred in groups of 25–50 to screened cages containing different test species, an undisclosed number of *Chenopodium quinoa* Willd. plants became infected but in only one out of six experiments. In a second series of tests four out of ten *C. quinoa* plants were infected when thrips were collected from healthy bean plants, given 6-10 days acquisition access feeding on TSV-infected *M. alba* and then fed for 4–6 days on test plants. Thrips used in this test were found to be a mixture of *F. occidentalis* Perg. and *Thrips tabaci* Lind. In midsummer a third test used a natural colony of *T. tabaci* collected from TSV-infected *M. alba* plants grown outdoors near Pullman, WA. The virus was subsequently transmitted to *M. alba* (2 out of 8 plants) and *C. quinoa* (4 out of 8 plants). In all experiments the appropriate control plants remained virus-free (50).

Like the Brazilian results before it, the Washington report describing successful experimental transmissions of TSV by thrips seemed convincing when published, even though transmission results were somewhat erratic and only small numbers of plants were infected in successful tests. However, subsequent attempts to repeat and extend these results have been singularly unsuccessful (Kaiser, personal communication), even though various refinements have been made in caging and handling techniques. In those recent experiments, thrips were given acquisition feeding directly on infected leaf tissues. Yet no transmission was obtained. Such inconsistency in results of TSV transmission tests is not without precedent. In an earlier study, Converse and Lister (14) reported the infection of 5 out of 12 healthy black raspberry plants by hand pollination of flowers with pollen contaminated with the black raspberry latent strain of TSV. However, more extensive tests conducted later failed to confirm infection of plants by hand application of pollen (12). The reason for the inconsistencies in these early reports remained a puzzle until the recent work of Sdoodee and Teakle (77) which demonstrated not only that thrips play a role in TSV transmission but also that the mechanism for this transmission differs from that of most other insect vectors.

The Role of Pollen in Seed Transmission of TSV

Like most ilarviruses, TSV is transmitted through seed. Variable levels of seed transmission have been demonstrated using seed harvested from virus-infected plants (Table 8.2). In these tests, virus in the seedlings was presumably derived mainly from the mother plant. As is the case with most seed-transmitted viruses, the amount of transmission depends upon the species and the cultivar used.

Relatively little is known about the effectiveness of pollen from TSV-infected plants in producing virus-contaminated seed on healthy plants. First of all, the exact relationship between TSV and pollen has not yet been determined. However, the fact that virus-infected seedlings can be obtained from seed produced on healthy plants whose flowers were pollinated with pollen from infected plants can be taken as evidence that some pollen-associated virus is borne internally, presumably associated with sperm cells. Hand pollination of black raspberry plants with pollen from infected plants gave rise to about 10 percent infected seedlings (14). Similar levels of infected seedlings were found in crosses of strawberry (48) and tomato (78) that involved flowers on healthy, greenhouse-grown, female plants and pollen taken from TSV-infected plants. These results clearly demonstrate that TSV

TABLE 8.2. Transmission of tobacco streak virus through seed from infected plants.

Species	% Transmission	Reference
Cicer arietinum	1.5–22*	51
Chenopodium amaranticolor	48	80
C. quinoa	74	80
	86	50
Glycine max	3–31*	37
	91	50
Gomphrena globosa	1.4	50
Melilotus alba	<3	50
Nicotiana clevelandii	10	50
Phaseolus vulgaris	1.4	87
	<1–15*	51
Vigna angularis	29	51
V. unguiculata	0.7	50
	100	80

* dependent upon variety.

can be transmitted from pollen through the embryo to the resulting seedling, i.e., vertical transmission as defined by Mandahar (58).

It is curious that in the strawberry test (48), the incidence of infected seedlings was similar (11% vs. 9%) whether the infected pollen was used to pollinate female flowers on infected or healthy plants. Roughly twice as many (24%) infected seedlings occurred when healthy pollen was applied to infected female flowers. In the tomato study (78), no virus-infected seedlings were obtained from infected female plants regardless of the source of pollen and despite the fact that viral antigen was readily detectable in seed parts by enzyme-linked immunosorbent assay (ELISA). The significance of these divergent results is not yet clear.

Results from the hand pollination studies suggest that pollen carried on pollinating insects might contribute to a significant incidence of TSV in seed produced on healthy plants located in fields near infected plants. In our own studies involving field spread of TSV in Washington State asparagus fields, we have recently transmitted TSV to *C. quinoa* by rub inoculation with asparagus pollen collected from honeybees foraging in an asparagus field where TSV-infected plants were known to occur (Stark and Mink, unpublished data). This raises the possibility of bees as potential vectors in seed transmission of TSV. However, we have yet to detect TSV in asparagus seedlings grown from commercial seed lots harvested in areas where TSV occurs naturally.

Plant-to-Plant Spread

Except for the single early report of Converse and Lister (14) which subsequently could not be verified (12), there is no evidence from the hand pollination studies conducted thus far that would support an assumption that plant-to-plant transmission of TSV actually occurs during the pollination process, i.e. through back inoculation as defined by Mandahar (58, 59). Sdoode and Teakle (77), on the other hand, demonstrated that when they mixed any of three feeding instars of *T. tabaci* with TSV-contaminated tomato pollen and then transferred the mixture to leaves of *C. amaranticolor* Coste and Reyn seedlings or when they placed thrips directly on pollen-dusted leaves of *C. amaranticolor* they obtained transmission of virus to 40–100 percent of the plants. Plants that were treated with virus-contaminated pollen only or thrips only remained healthy. Thrips that fed on TSV-infected *C. amaranticolor* leaves for 1–5 days failed to transmit virus to healthy plants even though microscopic observations verified that the insects had fed on both the infected and healthy leaves. Likewise, if thrips were fed first on virus-contaminated pollen and then brushed lightly to remove adhering pollen and pollen debris, they failed to transmit virus when fed on *C. amaranticolor* leaves. These results strongly suggest that pollen from TSV-infected plants can be involved in plant-to-plant spread by a mechanism that is basically unrelated to pollination. Consequently, the term back-inoculation as proposed by Mandahar (58, 59) seems inappropriate to define the infection process involved in field spread of TSV.

A Possible Mechanism for Thrips-Mediated Spread

Results of light and electron microscopy (76) indicate the following: 1) TSV can exist as either individual particles or crystalline aggregates in cells that form the tomato anther wall, 2) these cells undergo degeneration beginning shortly after microsporogenesis and ultimately form part of the cellular debris found in the anther cavity at pollen maturity, 3) TSV aggregates and individual TSV particles become part of the debris present when thrips feed on mature pollen grains. From this it seems logical to assume that as thrips feed on pollen, they become coated not only with pollen grains but also with debris that can include virus particles and aggregates. While the data are mainly circumstantial at the moment, it seems logical to further assume that some of the contaminating virus particles might be introduced into tissues, especially flower parts, either through thrips feeding wounds or mechanical abrasion that might occur during feeding. In our own studies, we have shown that C. quinoa plants can be readily infected with TSV by lightly rubbing pollen-dusted leaves with a cotton swab. As yet, similar inoculation tests have not been reported using flowers.

While parts of this hypothesis remain to be verified by direct experimentation, it provides the framework for a transmission mechanism that appears distinct from the mechanisms proposed for thrips transmission of tomato spotted wilt (2) and tobacco ringspot viruses (61). Basically, thrips-mediated transmission of TSV appears at this time to be little more than mechanical transmission via feeding wounds. Because adult thrips are weak flyers, their ability to provide field spread over distances beyond just the adjacent plants would be dictated largely by their movement with prevailing air currents. Another possibility for longer range spread might be as erratic passengers on mobile flower-foraging insects such as bees. A passive association of virus-contaminated pollen, flower-inhabiting thrips and mobile pollinating insects would give the appearance that plant-to-plant spread of TSV was directly related to pollination.

An Epidemiological Example

Significant events in the epidemiology of a TSV-induced disease of an annual crop are illustrated in a recent report from Australia (44). These authors encountered a tobacco field with a gradient of TSV-infected plants that extended at least 12 rows into the field from a roadway containing numerous TSV-infected weeds, the predominant weed species being *Ageratum houstonianum* Mill. Flower heads of *A. houstonianum* not only contained large populations of the thrips *Microcephalothrips abdominalis* Crawford, but also produced copious amounts of a finely dispersed, easily wind blown, virus-contaminated pollen. Transmission experiments demonstrated that *M. abdominalis* taken from mature flower heads of virus-infected *A. houstonianum* plants or thrips mixed with contaminated pollen before transfer to leaves of various test plants readily infected 'Xanthi' tobacco, cucumber, *Nicotiana clevelandii* Gray, and *C. amaranticolor*. The studies indicated that wind-assisted pollen and thrips from infected weeds could provide the gradient of inoculum and wounding, respectively, needed to infect the crop plants.

Vectors of Subgroup 2

Five distinct viruses have been included in subgroup 11 (Table 8.1) because of their serological interrelationships (90). These are asparagus virus 2 (AV-2), citrus leaf rugose virus (CLRV), citrus variegation virus (CVV), elm mottle virus (EMoV) and Tulare apple mosaic virus (TAMV). Citrus crinkly leaf virus (93), sometimes included in this list (24, 25), is now considered a mild strain of CVV (34). There are no published reports confirming vector transmission of any of these viruses.

Thus far, the only virus in subgroup 2 reported to be seed-transmitted is AV-2 (89). In that report, seed were harvested from an infected plant. Subsequently, virus has been detected in pollen (22) from infected plants and in seedlings grown from seed harvested from non-infected plants (Mink, unpublished data). Circumstantial evidence is accumulating in Washington state that AV-2 spreads in some areas of the state by what appears to be an aerial vector. Although the vector has not yet been determined, plant-to-plant spread seems to occur mainly, if not exclusively, during the asparagus flowering season (Mink, unpublished data).

The lack of vector information for CLRV, CVV, EMoV, and TAMV may reflect the fact that these viruses occur only rarely (CLRV, CVV, and TAMV) and are of little or no economic importance. Of these four, only EMoV has been reported to be detected in pollen (75).

Vectors of Subgroups 3 and 4

Subgroup 3 includes Prunus necrotic ringspot virus (PNRSV), apple mosaic virus (ApMV) (90), and possibly two newly described viruses, blueberry shock virus (BBSV) (57) and *Humulus japonicus* virus (HJV) (1). It is the most taxonomically complex of the ilarvirus subgroups. The literature of subgroup 3 members includes a bewildering array of names (65). Most of the names applied to PNRSV and ApMV derivatives were assigned before techniques were available to study the viruses themselves. Consequently, many of the so-called viruses are now considered distinctive strains or biotypes of either PNRSV or ApMV (Table 8.3). To further complicate matters, while PNRSV and ApMV are still regarded as distinct, they are rather closely related

TABLE 8.3. Names applied to viruses in subgroup 3.

Prunus necrotic ringspot virus
   Almond calico virus
   Cherry necrotic ringspot virus
   Cherry recurrent ringspot virus
   Cherry rugose mosaic virus
   Danish line pattern virus
   Hop virus C
   Necrotic ringspot virus
   Prunus ringspot virus
   Ringspot virus
   Rose mosaic (some isolates)
   Stecklenberger virus
   Stone fruit ringspot virus
   Tatter leaf virus
Apple mosaic virus
   Apple variegation virus
   European plum line pattern virus
   Hop virus A
   Rose mosaic (some isolates)
Blueberry shock virus
Humulus japonicus virus

---

serologically. Their distinct names were preserved mainly because they were initially isolated from distinct hosts (31, 32). However, a few plants such as rose and hop are hosts to both viruses (65). It is ironic that had PNRSV and ApMV been isolated initially from either rose or hop, they undoubtedly would have been considered serotypes of a single virus. Because of this, the name Prunus necrotic ringspot has been used to identify the subgroup (65) until the discovery of BBSV and HIV. There is a major difference between PNRSV and ApMV, however, in that natural hosts of PNRSV appear limited to hop, rose, and most Prunus species, while ApMV is found in a variety of hosts that include almond, apple, birch, hop, hazel nut, horse chestnut, plum, rose and Rubus (65). Despite the wider natural host range for ApMV, natural spread of this virus is rare if it occurs at all. Essentially all information about natural spread has accumulated with PNRSV.

Prune dwarf virus (PDV), the only member of subgroup 4, spreads naturally by methods similar to PNRSV (32). In fact, the two viruses often occur together, sometimes in the same plant (32). Because they can cause similar symptoms in several hosts, the early literature on natural spread and vector results often confused the two viruses. Attempts will be made here to use the currently accepted names to reduce confusion.

Arthropod Vector Tests

Unlike most other ilarviruses, a wide range of arthropods have been tested as possible vectors of PNRSV and PDV (36, 69, 83). In tests over three years,

George and Davidson caged virus-infected and healthy Montmorency cherry trees with honeybees, 25 species of Cicadellidae, 3 species of Miridae, 3 species of Fulgaridae together with numerous Thripidae collected from cherry blossoms, buckwheat bloom and bloom of several weed species (36). The only transmission occurred in one cage which included honeybees and thrips taken from cherry blossoms. These tests will be discussed in more detail below.

In an eight-year study, Swensen and Milbrath (83) attempted transmission of PDV from Prunus to Prunus (mainly Lovell peach seedlings and some Montmorency trees), Prunus to cucurbits (squash cv. Buttercup or cucumber), cucurbit to Prunus, or cucurbit to cucurbit with 32 aphid species, 18 leafhopper species, and 12 other species of mites and insects including two thrips species, F. occidentalis and Taenothrips frici (Uzel). A single infected peach plant was detected in tests with the aphid Amphorophora rubitoxica (Knowl.). However, no further transmission was obtained with nearly 10,000 A. rubitoxica and several PNRSV strains. Five infected Montmorency cherry trees were found in several tests with the leafhopper Draeculacephala crassicornis (Van D.). While these plants were reported to have been healthy prior to the leafhopper experiment, the treatments in which they occurred raised suspicions about the validity of the results. Attempts to repeat these results were also unsuccessful.

Among the mites tested by Swenson and Milbrath (83) was the Eriophyid mite Vasates fockeui (Wat and Trt). In these tests, no transmission was obtained when 114 Prunus and cucurbit plants were exposed to over 9,600 mites taken from infected plants. Although these tests were reported in 1974, some recent reviews of ilarvirus vectors report V. fockeui as a confirmed vector (25, 33, 45, 65) citing a 1968 article by Proeseler (70). However, this article describes transmission experiments with a "latent Prunus virus" that had been transmitted earlier to Chenopodium foetidum from a plum tree located near Aschersleben in eastern Germany. Symptomatic C. foetidum leaves were found to contain rigid elongated particles about 750 nm in length (71). These two reports give no indication that any ilarvirus was actually transmitted using V. fockeui. Furthermore, this article also indicates that the "latent" Prunus virus was only found around two cities in eastern Germany. Consequently, there seems to be no justification to perpetuate this citation as evidence for V. fockeui transmission of PNRSV.

Nematode Test

Transmission of PNRSV by the nematode Longidorus macrosoma has been reported once (26). This report has not been verified.

Recently, Yuan et al. (96) reported that detectably elevated color reactions were obtained by ELISA using extracts of the ring nematode Criconemella xenoplax (Raski) Lua and Raski that had been hand picked from the root zone of PNRSV-infected peach trees in South Carolina. However, these nematodes failed to transmit the virus to cucumber or peach seedlings. Furthermore, nematodes that fed on PNRSV-infected cucumber root explants failed to transmit virus to roots of *Prunus besseyi* in agar culture. While the elevated ELISA readings may be interpreted to suggest that ring nematodes may aquire PNRSV antigen during feeding, there is, as yet, no evidence using techniques to support this hypothesis. The authors indicate that if *C. xenoplax* does transmit PNRSV, the event must be rare.

Pollen Transmission

POLLEN-VIRUS ASSOCIATION

Buffered extracts of pollen from PNRSV- and PDV-infected trees were found to be infectious when rub inoculated to various herbaceous hosts (21, 95). Later, Cole et al. (11) demonstrated that PNRSV-specific antigens were easily removed from intact cherry and almond pollen by washing. These antigens appeared to include both intact virus and subviral antigens. Results suggested that most PNRSV antigens occurred on the pollen surface. Subsequent reports (10, 46) and unpublished results by two different individuals in our lab support the original contention (11) that most PNRSV occurs on the outside of the pollen grain whereas substantial amounts of PDV are borne internally. Using a different cherry cultivar, Kelley and Cameron (53) presented results suggesting that some PNRSV and PDV occurred both externally and internally. To my knowledge, similar studies using pollen from virus-infected sour cherry trees have not yet been done.

Immunofluorescence studies indicated that antigens of PNRSV, PDV, and ApMV were located on the outer exine of plum pollen from infected trees with some antigens of PNRSV and PDV found within the plum pollen (62).

Various abnormalities have been reported in pollen from PNRSV-infected peach (60) and sour cherry (4, 66). The type and severity of these effects varied according-to virus isolate and cultivar examined. Whether these abnormalities adversely influence pollination has been a point of speculation but, so far, there is no specific evidence to support this speculation.

ROLE OF POLLEN IN SEED TRANSMISSION

Both PNRSV and PDV have been transmitted to seedlings grown from seed taken from infected trees of several Prunus species. The list includes *P. americana, P. amygdalus, P. avium, P. cerasus, P. domestica, P. mahaleb, P. pennsylvanica, P. persica, and P. salicina* (94) as well as *P. armeniaca* (3, 73) and *P. spinosa* (74). In these cases, it is assumed that most infections result from virus provided by the female parent.

Both viruses have also been detected in seedlings grown from seed produced on healthy plants of several Prunus species after hand pollination with pollen from infected plants (35, 36, 40, 42, 94). Infections in these cases would appear to result from virus delivered to the embryo during fertilization. This is taken as evidence that virus is borne within pollen grains in close association with the pollen sperm cells. No microscopic evidence is available as yet to substantiate this.

Hand pollination clearly demonstrates that both PNRSV and PDV can be transmitted to seedlings through pollen. No such information is available for ApMV.

PLANT-TO-PLANT TRANSMISSION

Although field spread of diseases caused by PNRSV and PDV was recognized during the 1930s (52, 85), the possible role of pollen in field spread was not examined until the late 1950s and 1960s. Beginning in 1957, both viruses were transmitted to herbaceous plants, studied by conventional virological techniques, and their unique differences defined (27, 28, 29, 30, 31). Concurrently, the demonstration that pollen was a source of virus (21) led to studies on the role of pollen in virus dissemination. Several extensive studies were conducted during this period and these provide most of the circumstantial evidence that is still used today to extrapolate roles for pollen in tree-to-tree spread.

In the absence of much basic information about pollen and its exact role in plant-to-plant transmission, the term back inoculation was introduced by Mandahar (59). This term implies that virus borne on or in the pollen invades the female flower parts sometime during the act of fertilization and then ultimately invades the flower bearing plant. However, attempts to provide experimental evidence for this or other alternatives have produced erratic, often equivocal results. On the other hand, the recent recognition that thrips can play a mediating role in pollen transmission of TSV (77) suggests that results of some earlier experiments with PNRSV and PDV should be reevaluated.

In a series of field and screenhouse experiments, Milbrath and co-workers provided some of the first experimental evidence for plant-to-plant spread of a Prunus virus using Buttercup squash (18, 19, 63). Although these three reports describe work done with "stone fruit ringspot virus," the description of the biological properties (18) of their Oregon RS-31 strain demonstrate clearly that the virus was what is now considered to be Prune dwarf virus (39).

In their greenhouse experiment, Milbrath et al. (63) artificially pollinated 204 flowers on 97 greenhouse-grown squash plants whose flowers had been tied shut to prevent any chance of early pollination. Flowers were opened briefly while the stigmas were air-dusted with pollen collected a short time before from greenhouse-grown plants (presumably the male flowers were not tied). The artificially pollinated flowers were retied until they dropped. A total of 9 of the 97 plants (8.3%) became infected. From seed produced on all plants, the authors calculated that approximately 2,000 fertilized ovules had been involved in the infection of nine plants while some 18,000–19,000 potential infection sites (= fertilized ovules) on the remaining 88 plants had not resulted in infection. Their conclusion was that this type of infection (presumably infection through fertilization) is uncommon. They prophetically reasoned that since a cherry fruit develops from a single ovule, it would be difficult to demonstrate this type of transmission from cherry tree to cherry tree (18).

Whether the few infections reported in the above experiment resulted from chance fertilizations with virus-infected sperm cells or from the actions of possibly unobserved thrips trapped in a few flowers (at least one flower of each of the nine infected plants) is impossible to deduce in retrospect. In a subsequent report, however, these same authors were unable to transmit stonefruit ringspot virus (= PDV) from squash to squash using two species of thrips (83). They did not indicate, however, whether these insects were given their acquisition feedings on leaves or on flowers prior to test feeding.

In a series of caged tree experiments that extended over three seasons, George and Davidson (36) attempted to associate insects with spread of PNRSV and PDV among sour cherry trees. A total of 5 screened compartments were used, each initially containing four healthy trees and two infected trees; one each with PNRSV and PDV. A range of possible arthropod vectors were introduced into three compartments. These are of no concern here as no transmission occurred over the three seasons. However, the results from two compartments in which honeybees were placed along with a variety of thrips are highly significant.

In 1957, honeybees (*Apis mellifera* L.) were placed in compartments 1 and 3 during the cherry bloom period (May 3 to May 10). During this period Thripidae (species not specified) from cherry blossoms were introduced into compartment 1 while four aphid species were placed in compartment 3. In August, 2,000 Thripidae from buckwheat bloom were also placed in compartment 1. PNRSV spread to two of four healthy trees in compartment 1 during 1957; a third infected tree was detected in 1958. No spread occurred in compartment 3 in 1957 when both bees and aphids were present or in 1958 when bees were supplemented in August with "innumerable" Thripidae from buckwheat bloom (36). These results seem to be the first report where a role for thrips from cherry flowers might be suspected in tree-to-tree spread. This possibility was discussed but ultimately dismissed by the authors. Their reasons were 1) that one infected tree was detected in 1958 when bees were caged without thrips, and 2) they failed to obtain transmission in unpublished studies when "large numbers of Thripidae" were caged on cucumber plants. It should be pointed out that these workers and others reported that some stone fruit trees do not express symptoms until one, sometimes two years after infection (41, 42). Consequently, the one tree in compartment 1 which expressed symptoms in 1958 could easily have been infected in 1957, a possibility not acknowledged by the authors. If transmission did occur in 1957, that would result-in infection of three of four trees exposed to bees and thrips during the bloom period. Since thrips from cherry flowers were not included with the bees in 1958 tests, the conditions in compartment 1 during bloom resembled those in compartment 3 where no transmission occurred in either of two seasons with exposure to bees and pollen alone.

TABLE 8.4. Summary of data from hand pollination experiments involving Prunus necrotic ringspot and prune dwarf viruses and various Prunus species.

Species	CV	Experiment location[a]	Virus	Number of trees Pollinated	Number of trees Infected	Infected seedlings produced	Reference
P. cerasus	English morello	GH	NRSV + PDV	42	0	+	40
P. cerasus	English morello	F	NRSV + PDV	?[b]	0	+	41
P. cerasus	Montmorency	GH	NRSV + PDV	3	1 (NRS)	+	36
P. cerasus		SC	NRSV + PDV	3	1 (PDV)	+	36
P. cerasus		OC	NRSV; PDV	2	1 (NRS)	+	36
P. cerasus		F	NRSV	3	2 (NRS)	+	36
P. cerasus		F	PDV	3	1 (PDV)	+	36
P. cerasus	Montmorency	GH	NRSV + PDV	48	2 (PDV)	+	36
P. avium	Yellow Glass	F	PDV	1	1	NR	38
P. fructicosa	—	F	PDV	6	1	NR	38
P. persicae	Ha-12	?[b]	NRSV	1	1	NR	54
P. persicae	Golden Queen	N	PDV	10	10	NR	81

[a] Abbreviations: GH = greenhouse, F = field, SC = screened compartment, OC = outside cage, N = nursery, NR = not reported.
[b] Information not presented.

The unpublished tests reported to involve "large numbers of Thripidae" apparently were conducted with cucumber plants (36) without bees or additional pollen sources being included. There is no indication that the authors attempted to repeat the test conditions of compartment 1 that gave between 50–75% transmission. They were unable to provide an explanation for why two years of tests with bees and pollen alone during bloom resulted in no transmission. As a consequence of the authors' dismissal of their own results, this work has been cited often but only as proof that tree-to-tree transmission occurs as a result of pollen carried by bees.

The failure of PNRSV and PDV to spread into debloomed trees (9, 20, 36) is often cited as strong supportive evidence that pollen moved by pollinating insects such as bees is the primary, if not the only, means of tree-to-tree spread (58, 59, 64). This strong correlation between flowering and the period for natural spread of PNRSV and PDV obviously does not exclude a mediating role for the flower-inhabiting thrips.

Hand pollination experiments that should provide unequivocal data on the role of pollen in tree-to-tree spread have so far failed to do so. While there seems to be little doubt that trees of several Prunus species have become infected and have expressed distinctive symptoms within 4-8 weeks after hand pollination (Table 8.4), the numbers of plants infected are remarkably low considering the numbers of flowers pollinated on each tree. The exception to this is peach where all 11 pollinated test plants were reported to have been infected (54, 81).

In our own unpublished studies conducted over a period of six years, we have applied pollen from infected sweet cherry trees to flowers on 84 caged, virus free 'Bing' trees either by hand or by use of honeybees. We have repeatedly demonstrated transmission of PNRSV or PDV to seed on these caged, healthy trees but we have consistently failed to infect any of the 84 trees used.

THE ROLE OF BEES IN POLLEN TRANSMISSION

Although PNRSV and PDV have been detected in association with pollen that was collected and stored by honeybees (10, 46, 64), such pollen does not germinate (64). The stored pollen is eventually consumed by bee larvae as a protein source (67). Bees appear to coat pollen they collect almost immediately with a substance, probably an enzyme complex, which degrades virus at normal hive temperatures (10). Nevertheless, our unpublished data demonstrates that bees emerging from hives that were previously exposed to flowering Prunus trees usually carry enough viable virus-contaminated pollen on their bodies to pollinate occasional fruit on caged trees. Some of these fruit develop virus-infected seeds which produce virus-infected seedlings. However, these experimental tests have failed to demonstrate that tree infections can result from bees carrying virus-contaminated pollen. In this respect, our results are similar to the bee pollination results reported by George and Davidson (36) which occurred in the absence of thrips during bloom (compartment 3).

AN HYPOTHESIS

The results discussed above suggest the following as a working hypothesis. Tree-to-tree (plant-to-plant) spread of PNRSV or PDV seems to require the interaction of four components; 1) infectious virus particles associated with mature pollen but probably not borne internally; 2) thrips present in flowers of the infection source which become contaminated with pollen grains and virus during feeding; 3) movement of virus and pollen-contaminated thrips from infected to healthy trees either by direct flight or in association with pollinating insects; and 4) virus transmission to flower parts as a result of thrip feeding wounds. Except for the possible role of pollinating insects in moving thrips from plant to plant, this hypothesis is similar to that already demonstrated for experimental transmission of TSV. This hypothesis can be tested by direct experimentation.

NOTE IN ADDED PROOF

A 1991 article by Greber, R.S., Klose, M.J., Milne, J.R., and Teakle, D.S. entitled, "Transmission of Prunus necrotic ringspot virus using plum pollen and thrips" reports PNRSV transmission rates from plum pollen to cucumber seedlings of 50% using *thrips tabaci* and 66% using a mixture of five thrips species collected from *Ageratum houstonianum* flowers (Annals of Applied Biology, **118**:589–593). More recently, PNRSV and PDV were each transmitted from sweet cherry pollen to cucumber seedlings using *Frankliniella occidentalis*. Transmission of PDV was four times greater than that of PNRSV (Greber, Teakle, and Mink, manuscript in press).

Vectors of Other Ilarviruses

Spinach latent virus is seed-transmitted in spinach and a few experimental hosts (6), and pollen-transmitted in spinach (82). Lilac ring mottle virus (92) and hydrangea mosaic virus (84) are also seed-transmitted, but there are no reports of pollen transmission. There is as yet no report of seed, pollen, or arthropod transmission of American plum line pattern virus (68) or the newly described parietaria mottle virus (8).

Concluding Comments

Pollen transmission is a recognized, albeit not unique, characteristic of ilarviruses. However, it has yet to be demonstrated for more than one-half of the 16 distinct viruses in this group. This probably reflects, to some extent, the minor economic importance of the diseases these viruses cause.

While pollen can function as a vector for some ilarviruses by delivering them to embryos produced in fruit on healthy trees, pollen seems to play more of an indirect role in plant-to-plant spread.

There is no evidence that any arthropod acts as an ilarvirus vector in the conventional sense; i.e., that the vector acquires virus by feeding upon virus-infected cells and which it subsequently introduces into healthy cells through feeding. There is, however, increasing evidence that thrips are an essential component of plant-to-plant spread. Their contamination while feeding on pollen from infected plants and their presumed mechanical inoculation while feeding on other tissues of healthy plants provides a plausible explanation of experimental results and some field observations. However, numerous details about the interactions of thrips, viruses, and plant tissues and pollinating insects are still missing. More information is needed to formulate strategies for controlling field spread of economically important ilarviruses such as TSV, PNRSV, PDV, and possibly AV-2.

References

1. Adams, A.N., Clark, M.F., and Barbara, D.J., 1989, Host range, purification, and some properties of a new ilarvirus from *Humulus japonicus, Ann. Appl. Biol.* **114**:497–508.

2. Amin, P.W., Reddy, D.V.R., Ghanekar, A.M., and Reddy, M.S., 1981, Transmission of tomato spotted wilt virus, the causal agent of bud necrosis of peanut by *Scirtothrips dorsalis and Frankliniella schultzei, Plant Dis.* **65**:663–665.

3. Barba, M., Pasquini, G., and Quacquarelli, A., 1986, Role of seeds in the epidemiology of two almond viruses, *Acta Horticulturae,* **193**:127–130.27

4. Basak, W., 1966, Influence of prunus necrotic ringspot virus on pollen of sour cherry, *Polska Akad. Nauk. Bull. Ser. Sci. Biol.* **14**:797–800.

5. Berkeley, G.H., and Phillips, J.H.H., 1943, Tobacco streak, Can. *J. Res.* **21**:181–190.

6. Bos, L., Huttinga, H., and Matt, D.Z., 1980, Spinach latent virus, a new ilarvirus seed-borne in *Spinacia oleracea, Neth. J. Pl. Path.* **86**:79–98.

7. Brunt, A.A., and Paludan, N., 1970, The serological relationship between "asparagus stunt" and tobacco streak virus, *Phytopathol. Z.* **69**:277–282.

8. Caciagli, P., Boccardo, G., and Lovisolo, O., 1989, Parietaria mottle virus, a possible new ilarvirus from *Parietaria officinalis (Urticaceae), Plant Pathol.* **38**:577–584.

9. Cameron, R.R., Milbrath, J.A., and Tate, L.A., 1973, Pollen transmission of prunus necrotic ringspot in prune and sour cherry orchards, *Plant Dis. Rep.* **57**:241–243.

10. Cole, A., and Mink, G.I., 1984, An agent associated with bee-stored pollen that degrades intact virus, *Phytopathology* **74**:1320–1324.

11. Cole, A., Mink, G.I., and Regev, S., 1982, Location of Prunus necrotic ringspot virus on pollen grains from infected almond and cherry trees, *Phytopathology* **72**:1542–1545.

12. Converse, R.H., 1979, Transmission of tobacco streak virus in *Rubus, Acta Horticulturae* **95**:53–61.

13. Converse, R.H., 1984, Natural spread of tobacco streak virus in red raspberry, *Phytopathology* **74**:1137 (Abstr.).

14. Converse, R.H., and Lister, R.J., 1969, The occurrence and some properties of black raspberry latent virus, *Phytopathology* **59**:325–333.

15. Costa, A.S., and Carvalho, A.M.B., 1961, Studies on Brazilian tobacco streak, *Phytopath. Z.* **42**:113–138.

16. Costa, A.S., and Lima Neto, V.daC., 1976, Transmissao do virus de necrose branca do fumo por *Frankliniella* sp., *Congresso de Sociedade Brasileira de Fitopatologia* No. 9, 1 p.

17. Cupertino, F.P., Grogan, R.G., Peterson, L.J., and Kimble, K.A., 1984, Tobacco streak virus infection of tomato and some natural weed hosts in California, *Plant Dis.* **68**:331–333.

18. Das, C.R., and Milbrath, J.A., 1961, Plant-to-plant transfer of stone fruit ringspot virus in squash by pollination, *Phytopathology* **51**:489–490.

19. Das, C.R., Milbrath, J.A, and Swenson, K.G., 1961, Seed and pollen transmission of Prunus ringspot virus in Buttercup squash, *Phytopathology* **51**:64 (Abstr.).

20. Davidson, R.T., 1976, Field spread of prunus necrotic ringspot in sour cherry orchards in Ontario, *Plant Dis. Rep.* **60**:1080–1082.

21. Ehlers, L.G., and Moore, J.D., 1957, Mechanical transmission of certain stone fruit viruses from Prunus pollen, *Phytopathology* **47**:519–520.

22. Evans, T.A., and Stephens, C.T., 1988, Association of asparagus virus 11 with pollen from infected asparagus, (*Asparagus officinalis*), *Plant Dis.* **72**:195–198.

23. Finlay, J.R., 1974, Tobacco streak virus in tobacco, *Australian Plant Pathology Society Newsletter* **3**:71.

24. Francki, R.I.B., 1985, The viruses and their taxonomy, in Francki, R.I.B. (ed): The Plant Viruses, Volume I. Polyhedral Viruses with Tripartite Genomes, Plenum Press, New York.

25. Francki, R.I.B., Milne, R.G., and Hatta, T., 1985, An atlas of plant viruses, Volume. II. CBC Press, Boca Raton, Florida.

26. Fritzsche, R., and Kegler, H., 1968, Nematoden als Vektoren von Viruskrankheiten der Obstgewächse, *Dtsch. Akad. Landwirtsch. Wiss. Berlin Tagungsber* **97**:289–295.

27. Fulton, R.W., 1957, Comparative host ranges of certain mechanically transmitted viruses of Prunus, *Phytopathology* **47**:215–220.

28. Fulton, R.W., 1957, Properties of certain mechanically transmitted viruses of *Prunus, Phytopathology* **47**:683–687.

29. Fulton, R.W., 1957, Mechanical transmission of Prunus viruses to cherry, *Phytopathology* **47**:12 (Abstr.).

30. Fulton, R.W., 1958, Identity of and relationships among certain sour cherry viruses mechanically transmitted to Prunus species, *Virology* **6**:499–511.

31. Fulton, R.W., 1968, Serology of viruses causing cherry necrotic ringspot, plum line pattern, rose mosaic and apple mosaic, *Phytopathology* **58**:635–638.

32. Fulton, R.W., 1968, Relationships among ringspot viruses of Prunus, *Dtsch. Akad. Landwestschaf wiss Berl. Tagungsber.* **97**:123–138.

33. Fulton, R.W., 1981, Ilarviruses, in Kurstak, E. (ed): Handbook of Plant Virus Infections and Comparative Diagnosis, pp. 387–413.

34. Garnsey, S.M., Baksh, N., Davino, M., and Agostini, J.P., 1984, A mild isolate of citrus variegation virus found in Florida citrus, *Proc. Ninth IOCV Conference* pp. 188–195.

35. George, J.A., 1962, A technique for detecting virus infected Montmorency cherry seeds, *Can. J. Plant Sci.* **42**:198–203.

36. George, J.A., and Davidson, T.R., 1963, Pollen transmission of necrotic ringspot and sour cherry yellows viruses from tree to tree, *Can. J. Pl. Sci.* **43**:276–288.

37. Ghanekar, A.M., and Schwenk, F.W., 1974, Seed transmission and distribution of tobacco streak virus in six cultivars of soybeans, *Phytopathology* **64**:112–114.

38. Gilmer, R. M., 1965, Additional evidence of tree-to-tree transmission of sour cherry yellows virus by pollen, *Phytopathology* **55**:482–483.

39. Gilmer, R.M., Nyland, G., and Moore, J.D., 1976, Prune dwarf, in *Virus Diseases and Noninfectious Disorders of Stone Fruits in North America*, U.S. Dept. Agric. Handbook **437**:179–190.

40. Gilmer, R.M., and Way, R.D., 1960, Pollen transmission of necrotic ringspot and prune dwarf viruses in sour cherry, *Phytopathology* **50**:624–625.

41. Gilmer, R.M., and Way, R.D., 1961, Pollen transmission of necrotic ringspot and prune dwarf viruses in cherry, *Tidsskr. Planteavl.* **65**:111–117.

42. Gilmer, R.M., and Way, R.D., 1963, Evidence for tree-to-tree transmission of sour cherry yellows virus by pollen, *Plant Dis. Rep.* **47**:1051–1052.

43. Greber, R.S., 1971, Some characteristics of tobacco streak virus isolates from Queensland, *Queensland Journal of Agriculture and Animal Sciences* **28**:105–114.

44. Greber, R.S., Klose, M.J., Teakle, D.S., and Milne, R.J., 1991, A high incidence of tobacco streak virus in tobacco and its transmission using *Microcephalothrips abdominalis* and pollen from *Ageratum houstonianum, Plant Disease.* (In press)

45. Hamilton, R.I., 1985, Virus transmission, in Francki, R.I.B. (ed): The Viruses, I. The Plant Viruses, Polyhedral Virions with Tripartite Genomes, Plenum Press, New York.

46. Hamilton, R.I., Nichols, C., and Valentina, B., 1984, Survey for Prunus necrotic ringspot and other viruses contaminating the exine of pollen collected by bees, *Can. J. Plant Pathol.* **6**:196–199.

47. Johnson, J., 1936, Tobacco streak, a virus disease, *Phytopathology* **26**:285–292.

48. Johnson, H.A., Converse, R.H., Amorao, A., Espejo, J.I., and Frazier, N.W., 1984, Seed transmission of tobacco streak in strawberry, *Plant Dis.* **68**:390–392.

49. Jones, A.T., and Mayo, M.A., 1975, Further properties of black raspberry latent virus and evidence for its relationship to tobacco streak virus, *Ann. Appl. Biol.* **79**:297–306.

50. Kaiser, W.J., Wyatt, S.D., and Pesho, G.R., 1982, Natural hosts and vectors of tobacco streak virus in eastern Washington, *Phytopathology* **72**:1508–1512.

51. Kaiser, W.J., Wyatt, S.D., and Klein, R.E., 1991, Epidemiology and seed transmission of two tobacco streak virus pathotypes associated with seed increases of legume germ plasm in eastern Washington, *Plant Dis.* **75**. (In press).

52. Keitt, G.W., and Clayton, C.N., 1943, A destructive virus disease of sour cherry, *Phytopathology* **33**:449–468.

53. Kelly, R.D. and Cameron, H.R., 1986, Location of prune dwarf virus and prunus necrotic ringspot virus in sweet cherry pollen and fruit, *Phytopathology* **76**:317–322.

54. Kishi, K., Takanashi, K., and Abiko, K., 1973, Studies on virus diseases of stone fruits, 111. Pollen transmission of peach necrotic ringspot virus and prune dwarf virus on peach trees and results of several related experiments, *Bul. Hort. Res. Sta. Hiratsuka* **12**:185–196.

55. Lima Neto, V.daC., Colturato, L.C., Nasser, L.C., Gumaraes, O.A., and Thomas J.C., 1979, Epifitia de guerma dos brotos da soja em culturas da regias centro sul do parana, *Revesta do Setor de Ciencias Agrarias* **1**:9–17.

56. Lister, R.M., and Saksena, K.N., 1976, Some properties of Tulare apple mosaic and ILAR viruses suggesting grouping with tobacco streak virus, *Virology.* **70**:440–450.

57. MacDonald, S.G., Martin, R.R., and Bristow, P.R., 1991, Characterization of an ilarvirus associated with a necrotic shock reaction in blueberry, *Phytopathology.* (In press)

58. Mandahar, C.L., 1985, Vertical and horizontal spread of plant viruses through seed and pollen—an epidemiological view, in Gupta, V.M., Singh, B.P, Verma, H.N., and Srivastava, K.M. (eds): Perspectives in Plant Virology, Print House, Lucknow, 23 pp.

59. Mandahar, C.L., 1990, Virus transmission, in Mandahar, C.L. (ed.), Plant Viruses, Volume II. Pathology, CRC Press, Boca Raton, Florida, pp. 205–253.

60. Marenaud, C., and Desvignes, J.C., 1965, Effets de divers virus sur le pollen de *Prunus persica* GF305, *C. R. Acad Agric.* **51**:782–790.

61. Messieha, M., 1969, Transmission of tobacco ringspot virus by thrips, *Phytopathology* **59**:943–945.

62. Meyer, S., Casper, R., and Bünemann, G., 1986, Die Lokalisierung von Ringfleckenviren im Pflaumenpollen durch Immunfluoreszenz, *Gartenbauwissenschaft* **51**:125–130.

63. Milbrath, J.A., and Swenson, K.G., 1959, Field spread of ringspot virus of stone fruits from squash to squash, *Plant Dis. Rep.* **43**:705–709.

64. Mink, G.I., 1983, The possible role of honeybees in long distance spread of Prunus necrotic ringspot virus from California into Washington sweet cherry: orchards, Plumb, R.T., and Thresh, J.M. (eds): Plant Virus Epidemiology, Blackwell Scientific Publications, Oxford, pp. 85–91.

65. Mink, G.I., 1991, Prunus necrotic ringspot virus, *Plant Diseases of International Importance*, Mukhopadhyay, A.N., Chauke, H.S., Kumar, J., and Singh, U.S. (eds): Prentice Hall, New Jersey. (In press)

66. Nyeki, J. and Vertesy, J., 1974, Effects of different ringspot viruses on the physiological and morphological properties of Montmorency sour cherry pollen, *Acta Phytopathologica Academiae Scientiarum Hungaricae* **9**:23–29.

67. Oertal, E., 1967, Nectar and pollen plants, *Beekeeping in the United States, US Dept. Agric. Handbook* U.S. Government Printing Office, Washington, D.C., 335 pp.

68. Paulson, A.Q. and Fulton, R.W., 1969, Purification, serological relationships and some characteristics of plum line pattern virus, *Ann. Appl. Biol.* **63**:233–240.

69. Phillips, J.H.H., 1951, An annotated list of Hemiptera inhabiting sour cherry orchards in the Niagara Peninsula, Ontario, *Can. Entomol.* **83**:194–205.

70. Proeseler, G., 1968, Ubertragungsversuche mit dem latenten Prunus-virus und der Gallmilbe Vasates fockeni Nal, *Phytopath. Z.* **63**:1–9.

71. Proeseler, G. and Kegler, H., 1966, Übertragung eines latenten Virus von Pframe durch Gallmilben (*Eriophyidae*), *Monatsber. Dtsch. Akad. Wiss.* **8**:472–476.

72. Salazar, L.F., Abad, J.H., and Hooker, W.J., 1982, Host range and properties of a strain of tobacco streak virus from potatoes, *Phytopathology* **72**:1550–1554.

73. Schimanski, H.-H., and Fuchs, E., 1984, Seed transmission of Prunus necrotic ringspot virus in apricot (*Prunus armeniaca* L.), *Zbl. Mikrobiol.* **139**:649–651.

74. Schimanski, H.H., Fuchs, E. and Kegler, H., 1984, Seed transmission of prunus necrotic ringspot virus in blackthorn, *Zentralblatt für Mikrobiologie* **139**:213–216.

75. Schmelzer, K., 1969, Das Ulmenschekungs-Virus, *Phytopath. Z.* **64**:39–67.

76. Sdoodee, R., 1989, Biological and biophysical properties of tobacco streak virus, Ph.D. Thesis, University of Queensland, Queensland, Australia.

77. Sdoodee, R., and Teakle, D.S., 1987, Transmission of tobacco streak virus by *Thrips tabaci*: a new method of plant virus transmission, *Plant Pathol.* **36**:377–380.

78. Sdoodee, R., and Teakle, D.S., 1988, Seed and pollen transmission of tobacco streak virus in tomato (*Lycopersicon esculentum* cv. Grosse Lisse), *Aust. J. Agric. Res.* **39**:469–474.

79. Shepherd, R.G., Francki, R.I.B., Hirth, L., Hollings, M., Inouye, T., Macleod, R., Purcifull, D.E., Sinha, R.C., Tremaine, J.H., Valenta, V., and Wetter, C., 1975, New groups of plant viruses approved by the International Committee on Taxonomy of Viruses, *Intervirology* **6**:181–184.

80. Shukla, D.D., and Gough, K.H., 1983, Tobacco streak, broad bean wilt, cucumber mosaic, and alfalfa mosaic viruses associated with ringspot *Ajuga reptans* in Australia, *Plant Dis.* **67**:221–224.

81. Smith, P.R., and Stubbs, L.L., 1976, Transmission of prune dwarf virus by peach pollen and latent infection in peach trees, *Aust J. Agric. Res.* **27**:839–843.

82. Stefanac, Z., 1978, Investigations of viruses and virus diseases of spinach in Croatia, *Acta. Bot. Croat.* **37**:39–46.

83. Swenson, K.G., and Milbrath, J.A., 1964, Insect and mite transmission tests with Prunus necrotic ringspot virus, *Phytopathology* **54**:399–404.

84. Thomas, B.J., Barton, R.J., and Tuszynski, A., 1983, Hydrangea mosaic virus, a new ilarvirus from *Hydrangea macrophylla* (*Saxifragaceae*), *Ann. Appl. Biol.* **103**:261–270.

85. Thomas, H.E., and Rawlins, T.E., 1939, Some mosaic diseases of prunus species, *Hilgardia* **12**:623–644.

86. Thomas, W.D., Jr., 1949, Pinto bean diseases in Colorado, *Proc. Western Colo. Hortic. Soc.* **6**:137–149.

87. Thomas, W.D., Jr., and Graham, R.W., 1951, Seed transmission of red node in pinto beans, *Phytopathology* **41**:959–962.

88. Thomas, H.R., and Zaumeyer, W.J., 1950, Red node, a virus disease of beans, *Phytopathology.* **40**:832–846.

89. Uyeda, I., and Mink, G.I., 1981, Properties of asparagus virus 11, a new member of the Ilarvirus group, *Phytopathology* **71**:1264–1269.

90. Uyeda, I., and Mink, G.I., 1983, Relationships among some Ilarviruses: proposed revision of subgroup A, *Phytopathology.* **73**:47–50.

91. Valleau, W.D., 1940, Sweet clover, a probable host of tobacco streak virus, *Phytopathology* **30**:438–440.

92. Van Der Meer, F.A., Huttinga, H., and Matt, D.Z., 1976, Lilac ring mottle virus: isolation from lilac, some properties and relation to lilac ringspot disease, *Neth. J. Pl. Path.* **82**:67–80.

93. Wallace, J.M., 1969, Psorosis A., blind pocket, concave gum, crinkly leaf, and infectious variegation, in *Indexing Procedures for 15 Virus Diseases of Citrus Trees*, Agriculture Handbook No. 333, USDA-ARS, Washington, D.C., pp. 5–15.

94. Williams, H.E., Jones, R.W., Traylor, J.A., and Wagnon, H.K., 1970, Passage of necrotic ringspot virus through almond seed, *Plant Dis. Rep.* **54**:822–824.

95. Williams, H.E., Traylor, J.A., and Wagnon, H.K., 1963, The infectious nature of pollen from certain virus-infected stone fruit trees, *Phytopathology* **53**:1144 (Abstr.).

96. Yuan, W.-Q, Barnett, O.W., Westcott III, S.W., and Scott, S.W., 1990, Tests for transmission of Prunus necrotic ringspot and two nepoviruses by *Criconemella xenoplax, Journal of Nematology* **22**:489–495.

9
Immunoelectron Microscopy of Plant Viruses and Mycoplasmas

Robert G. Milne

Introduction

Serology and electron microscopy of plant viruses have been associated for more than 50 years (4, 147, 170) and their interaction is still astonishingly fruitful. The antigen-antibody bond provides very high specificity while the electron microscope furnishes both high resolution and fine discrimination between objects of different morphology or electron density. In combination, the two approaches can generate very direct and hence relatively unequivocal evidence; if you can see the antigen (or antigen-associated structure), and see the antibody (or faithfully label it with gold), and witness their interaction at the nanometre level, then you have a ringside seat. This chapter will attempt to review briefly the contribution of transmission electron microscopy (TEM) to this spectacle.

There are two main kinds of question that immunoelectron microscopy (IEM) can attempt to answer. The first is, "To what degree are certain antigens related?" Replies to this question are invaluable in taxonomy and diagnostics. Secondly, "Where (in what plants or vectors, in which tissues and cells, in what intracellular or intercellular location) does the antigen in question occur, and in what quantities?" Replies to this type of question are invaluable in studying the cell biology of the pathogens, hosts, and vectors involved.

Where the antigenic particle or its morphological context not only survive extraction and manipulation in vitro, but are small enough to be viewed entire by TEM, immunosorbent electron microscopy (ISEM) and accompanying decoration techniques such as gold label antibody decoration (GLAD) can be applied. The advantage here is that resolution can be relatively high, labeling intensity and fidelity can be very good, and specimen preparation is

Robert G. Milne, Istituto di Fitovirologia Applicata, CNR, Strada delle Cacce 73, I-10135 Torino, Italy.
© 1992 by Springer-Verlag New York, Inc. *Advances in Disease Vector Research,* Volume 9.

usually simple and rapid. Alternatively or in addition, the antigen can be embedded and examined in thin sections, either by pre- or post-embedding labeling methods applied to whole cells or tissues.

Each of these subjects will be discussed in relation to plant viruses and plant-infecting mycoplasmalike organisms (MLOs), bearing in mind of course that many plant viruses and MLOs multiply in their insect vectors, and that others may have strict vector associations, even where there is no evidence of multiplication. A promising new area of research and application that is outside the author's experience, and will not be discussed, is immunolabeling of cryosections (see for example 155). A good review on detecting plant viruses in their vectors is that of Plumb (135), and Paliwal (129) has reviewed the subject of the present article, five years ago. Viruses will be discussed first.

Viruses

In Vitro Methods Involving Extracted Particles

Classical Antibody-Virus Mixtures

In the classical method, antiserum and virus preparation were incubated together, and the resulting complexes were sampled on grids (see for example 3, 86, 136). A theoretical aim was to examine the nature of the antibody-antigen bridge and to see why virus particles formed precipitates only in the presence of correct proportions of antibody; a practical aim was to detect particles of a given serological type. For all experiments, a compromize had to be made between three sets of not necessarily compatible conditions: those best for virus-antibody interaction (e.g. presence of sodium chloride or other ions, and of buffering capacity), best for optimal attachment to the grid (little studied at this time), and best for viewing in the EM (absence of crystallizing salts, presence of materials giving electron contrast). The methods were developed initially using purified TMV, but later progress was made examining preparations from human and animal virus infections.

Leaf-Dip Serology

Leaf-dip serology was developed in the above context for examination of crude virus-infected plant sap. Ball and Brakke (12, 13) and Ball (10, 11) based their method on the leaf-dip procedure of Brandes (28, 29). The procedure has been extended by Langenberg (87, 89, 93) and Lin (101). The freshly sliced edge of a virus-infected leaf was drawn through a drop of diluted antiserum placed on a grid; the resulting mixture of sap, serum and buffer was dried down on the grid, which was then negatively stained. This simple and rapid method gave valuable results and could also be used to quantitate serum dilution endpoints, but with today's insight we can see weaknesses in a number of steps, that are worth discussing.

1. Mixing and incubating together the infected plant extract and the antibody in its diluting buffer led to simultaneous clumping and decoration of the virus particles, whereas these two processes are best separated, and indeed clumping is best avoided.

2. Because the preparation necessarily included buffer during the drying phase, the totally volatile ammonium acetate was used, although the authors acknowledged that phosphate buffer was a much more effective medium for serological reactions.

3. For fear that the reactants could wash off the grid, they were dried down, prior to negative staining (or originally, metal shadowing). We now know that rinsing the grid before negative staining is largely beneficial in that (a) relatively little material of interest washes off, (b) the preparation becomes cleaner and is much more easily interpreted, (c) in case of doubt or necessity, an immunosorbent step can be used to bind the antigen to the grid, and (d), drying in the absence of negative stain, and *then* applying the stain generally causes distortion or breakage of virus particles, and is always to be avoided.

4. It was the fashion then to use phosphotungstate (PTA) as negative stain. The disadvantages of this stain, compared especially with uranyl acetate, and perhaps ammonium molybdate, are now well recognized (51, 71, 113, 116, 143).

Partly in recognition of some of these problems, the newer methods described below have been developed. The guiding principle is the separation of the different phases (such as binding to the grid, decoration, gold labeling, and negative staining) so that each may be optimized and used appropriately, without compromize to the others.

Immunosorbent Electron Microscopy

The principle of coating an electron microscope grid with antibody in order to trap serologically related virus particles was introduced by Derrick (41) under the name "serologically specific electron microscopy" or SSEM. The idea proved to be a breakthrough in the sensitive detection of virus particles in extracts of plants and of vectors, and the technique was soon improved in a number of ways (20, 30, 39, 64, 67, 85, 97, 108, 117, 120, 124, 126, 127, 142–144). All these papers concerned detection or serological relationships of plant viruses in plants, but there were also reports on detection of plant viruses in aphids (9, 60, 128, 142, 171), in nematode vectors (141) and in planthoppers (52). I shall attempt to summarize the technical aspects of this work as follows.

1. Precoating EM grids with antibody, and then incubating them with virus preparations can increase numbers of particles seen on the grid by several-fold up to over ten thousand-fold.

2. At the same time, antigenically unrelated material (e.g. host tissues or

proteins from plants or vectors) is rejected by the grid, and can be rinsed away leaving a very clean background.

3. Two kinds of control grid are possible: grids without pretreatment, and grids coated with pre-immune serum; numbers of particles adsorbed to pre-immune coated grids are depressed, compared with those on uncoated grids.

4. ISEM can be used simply to trap increased numbers of particles, or can be used to estimate serological relationships between different virus isolates or strains.

5. ISEM can be used to increase particle numbers of elongated viruses for the purpose of measuring modal lengths; however, quite strong selection for shorter particles or particle fragments occurs (130, 132).

6. Optimal antiserum dilutions are in the region of 1/1000; with undiluted sera or those diluted less than about 1/500, numbers of particles trapped are depressed below the optimum. This is almost certainly because at high concentrations, serum proteins other than immunoglobulins compete successfully for anchorage on the grid surface. With purified IgGs, an optimal concentration might be 1–10 μg/ml, or alternatively, the same concentration that is found optimal for coating in ELISA.

7. Serum (antibody) coating times can conveniently be 5 min at 25°C, though sometimes up to 60 min has been preferred.

8. Virus incubation times can be as short as 15 min at 25°C, but for high sensitivity, longer times such as 3h at 25°C or overnight at 4°C are usually better (see 133, 138, 142). Very long incubation times of the order of days may increase particle counts or may reduce them, depending on rather unforeseeable factors such as possible detachment or degradation of virus particles; particular attention should be paid to inhibiting bacterial growth and preventing the action of proteolytic enzymes undoubtedly present in the preparation.

9. Various buffers, particularly phosphate and Tris have been favored, both for diluting the antiserum and suspending the virus preparation: almost any buffer over a range of molarities from 0.02 to 0.3 M, and pH values from 6.5 to 9.5, can be used, though for best results, each system needs to be optimized by experiment (39).

10. Adsorption of virus particles to grids is normally a result of competition for attachment sites between the virus and "impurities" (97). An antibody layer eliminates this competition and so especially favors the virus when it is to be trapped from dirty preparations. Conversely, trapping of purified virus preparations may not be much improved by antibody coating, though there is some evidence that virus particles bound to antibody which is in turn bound to the grid surface adhere more tenaciously (during rinsing steps) than virus adsorbed directly to the grid.

11. With some viruses, especially more unstable ones, application of ISEM may not appreciably increase the numbers of particles adsorbed to grids from unpurified preparations. A major cause of this failure is the presence

in the virus preparation (or crude sap) of free coat protein subunits or oligomers, that diffuse rapidly to the antibody, and block its capacity to bind intact virus.

All the above work was done with plant viruses in their role as protein antigens, but there are a few papers indicating that double-stranded RNA can be trapped by ISEM (40, 42, 44, 53). This procedure should prove useful in diagnostics, especially in conjunction with a uranyl acetate-phosphate positive staining technique that gives contrast to the RNA without resort to metal shadowing (44). However, there do not so far appear to have been very many applications.

The next improvement in ISEM technology arrived with the use of protein A (PA) to precoat grids before adding the antibody layer. This protein, produced by *Staphylococcus aureus*, binds up to two IgG molecules by the Fc portion, leaving the serologically active Fab arms exposed (62). PA binds well to human and rabbit IgGs, less well to goat, rat and mouse IgGs, and not to chicken IgG, whereas the analogous protein G (derived from human group G *Streptococcus*) binds better than PA to goat IgG (21, 22, 137, 159).

In theory, precoating with PA followed by antibody coating should present a concentrated and correctly oriented layer of IgG to the antigenic particles being incubated with the grid, and should therefore make ISEM more efficient. Shukla and Gough (149) were the first to demonstrate the system, and later work has led toward optimization of conditions for its use (1, 27, 33, 47, 63, 76, 78, 95, 96, 109, 112, 114–116, 118, 125, 138, 139, 146, 164, 167).

This work, in turn, may be summarized as follows.

1. Normally 10 μg/ml of PA is used, diluted in 0.1 M phosphate buffer, pH 7 (though within limits the type, molarity and pH of the buffer do not appear to be critical); coating time can be 5–20 min at 25°C. The grid is then rinsed with buffer and drained but not allowed to dry, and is then coated with the antibody (again, for 5–20 min at 25°C).
2. As already noted, inhibition is encountered with antiserum-coated grids if the serum is diluted less than about 1/500. With PA-precoated grids, this inhibitory effect is much reduced, so that antiserum dilutions of, say, 1/50 can be used. Consequently, antisera of low titer can be effective in trapping virus particles, whereas without PA they would not have been.
3. With PA, the plateau of near-optimal serum dilutions is broad, extending over 5–6 two-fold dilution steps. Without PA, the plateau extends over 1–3 steps, giving somewhat less flexibility,
4. When ISEM with or without PA are compared for sensitivity, each at its optimal dilution, and using a reasonably high titer antiserum, it is found that with PA precoating, up to five times more virus particles can be trapped. With sugarcane mosaic potyvirus, a 25-fold increase has been reported (63). However, any advantage offered by PA becomes small if virus concentration rather than antibody availability is limiting (a common case). For example, it was found with the small isometric particles of

maize chlorotic mottle virus that concentrations lower than 12.5 μg/ml were detected equally well or better by ISEM without rather than with PA precoating (95). Pares and Whitecross (133) noted that PA is a large molecule taking up a lot of room, so it may not in fact allow significantly more IgG to attach to the grid (the mean orientation of IgG molecules directly coated to grids is unknown); they concluded that precoating grids with PA did not appear to offer any advantage.

5. Trapping particles by ISEM using different antisera, and then counting them is a laborious way to estimate serological differences between viruses, especially as variability in counts is usually high. However, the method has been used successfully (33, 50, 54, 66, 96, 109, 117, 133, 139, 143, 144, 150, 167).

   Fribourg et al. (54) noted that ISEM could be used as a broad or narrow spectrum test for relationships (in this case, of tobamoviruses). Narrow-spectrum conditions were coating of the grids with antiserum diluted 1/1000, and incubation of the grids with the virus extracts for 15 min, whereas for a broad-spectrum test, grids were precoated with 100 μg/ml of PA, followed by antiserum diluted 1/50. The virus extract was incubated with these grids for 17 h at room temperature.

6. It is clear that the main reason to use PA precoating is to concentrate IgGs on the grid, starting from a crude antiserum that may not have a very high titer. If the purified IgG fraction (or better, affinity-purified specific IgG) is available, coating this over PA should offer little advantage. It is interesting, therefore, that Paliwal (128) reported that the most sensitive trapping of barley yellow dwarf luteovirus from extracts of single vector aphids was obtained by coating purified IgG over PA. This may have been due to improved orientation of the IgG.

The conditions necessary for successful ISEM tests, with or without PA, and the results obtained have been extensively reviewed (65, 78, 84, 95, 114–116, 118, 129, 133, 138, 139).

Decoration

Decoration procedures were introduced to plant virology by Luisoni et al. (107) and Milne and Luisoni (119), though a similar approach had earlier and unknown to them been used with bacteriophages (18, 162, 174). Later (120), ISEM and decoration were combined into one procedure in which virus particles were initially trapped on the grid and rinsed, to remove unwanted salts, sugars, and host materials (visible debris, and also invisible but potentially destructive enzymes) and were then further incubated with the same or a different antibody to give a visible antibody halo (decoration).

The difference between this procedure and classical antibody-antigen mixture methods (including leaf-dip serology) is that the trapping and decorating phases are separated; this means that each can be carried out under (their differing) optimal conditions. It also means that particle breakage in the presence of antiserum, noted by Ball and Brakke (12) is largely avoided, because once virus particles become adsorbed to the grid (in the presence of highly dilute antiserum, which does not cause breakage) they become stabilized against the more disruptive forces of higher serum concentrations that may be used for decoration.

Lin (101) attempted to evaluate levels of decoration after employing leaf-dip procedures, and obtained poor micrographs that were not easy to interpret. She stated "Decoration of isometric virus particles in leaf-dip preparations has not been found reliable...". Similarly, Langenberg (89), after leaf-dip experiences, noted "Decoration is more difficult to observe with unpurified icosahedral viruses, particularly when virus concentration in crude sap is low". In contrast, using ISEM and decoration procedures in sequence, it has been possible to obtain very clean decoration results starting with crude sap, and working with small icosahedral particles occurring in very low concentration (26, 123).

The basic steps for decoration are the following.

1. Virus particles are adsorbed to the grid so that they are well separated, and neither too many nor too few (about 5 particles on the screen at 40,000 X magnification, is convenient). The particles are adsorbed to the support film using the best system available for this phase of the procedure—such as choice of buffer and protectants, optional use of ISEM, and possible fixation with glutaraldehyde in case the virus particles or surface antigens are labile (93, 168). Where appropriate, internal controls (other virus particles of distinguishable morphology and differing antigenicity) should be added at this point.
2. Grids are rinsed thoroughly and drained. Here the buffer may be changed to that most suitable for diluting the antibody, which may be the same as or different from that used in the virus adsorption phase.
3. Grids are incubated under standard conditions (usually 15 min at room temperature) with the diluted antibody. This is a good opportunity to make a dilution series, as first done by Ball and Brakke (12), using one grid for each dilution, and so titrate the antibody.
4. The grids are rinsed thoroughly with water, then negatively stained, in uranyl acetate for choice. They are alternatively rinsed again with buffer and incubated with a second antibody ("double decoration"; 80, 81), or incubated with gold particles attached either to a second antibody or to protein A (see Gold Labeling).

Decoration of plant viruses has been widely reviewed (9, 65, 78, 84, 113–116, 133, 138), and reference to some of the very many examples of its use can be found in these discussions.

Gold Labeling

Decoration, if appropriately carried out, is a very useful method, but is limited in the sense that individual IgG molecules are barely visible in routine negative stain preparations, especially when attached to and in the "stain shadow" of the larger varieties of virus. In addition, some viruses (rhabdoviruses. bunyaviruses) have spiked envelopes that can be confused with antibody haloes. In this context, ability to label IgG molecules with gold is a great extension of the technique. One should not forget, however, that gold labeling brings two extra links into the chain (the second antibody, or the protein A, and the gold itself). This contributes some uncertainty and necessitates tighter controls.

The literature on gold labeling of in vitro particulate preparations is already enormous, and there are many recent reviews (32, 70, 83, 116, 165, 166), so I will merely try to illustrate some of the principles and give a few examples.

Gold-label antibody decoration (GLAD) was introduced to plant virology by Pares and Whitecross (130, 131), using rabbit antivirus antibody followed by protein A-gold (PAG). The principle was good but the results not exceptional. Also in 1982, Giunchedi and Langenberg (61) used an interesting system in which the carbohydrate associated with the barley stripe mosaic virus particle was exploited. The virus particles were trapped by ISEM on grids coated with antiserum against the coat protein, and were then treated with a purified lectin with affinity for the virus-associated carbohydrate. There followed treatment with gold-conjugated anti-lectin antibody, and this resulted in convincing and specific labeling of the virus particles. More general applications were developed by Lin (101), although unfortunately still within the leaf-dip context, and especially by Louro and Lesemann (106).

The general protocol established by these last authors was as follows (all operations done at room temperature). It can be noted that where protein A-gold (PAG) is employed, a second antibody labeled with gold (goat anti-rabbit, or goat anti-mouse) can be substituted with equally good results. For labeling mouse primary antibodies, protein G can be used instead of protein A. In particular cases (e.g. 46) a monoclonal antibody has been directly conjugated with gold. Although variations in technique have emerged, such as choice of buffer or concentration of blocking agent used, the basic ideas have not changed. As for any procedure, each step must be individually optimized for the system under consideration, and for the realization of different experimental goals.

1. The virus is adsorbed to carbon-fronted plastic support films on nickel grids (in this case, without ISEM, but see below), and rinsed with phosphate buffered saline containing 0.05% Tween 20 (PBST).
2. The grids are blocked (nonspecific protein adsorption sites saturated) by incubation for 15 min with 0.1–1.0% aqueous bovine serum albumin (BSA), and rinsed again with PBST.
3. The grids are incubated for 15 min with drops of antiserum diluted in 0.1 M phosphate buffer, pH 7 (PB), and rinsed with PBST.
4. The grids are incubated with protein A-gold (PAG) in PBS for 60 min, then rinsed successively in PBST and water, before negative staining in uranyl

acetate (1% to give good contrast to the labeled structures; 0.1% to see the label better).

It was noted (106) that inclusion of BSA in the PAG preparation reduced nonspecific background labeling, but also reduced the intensity of the specific label and inactivated the PAG after 2 days of storage. BSA was therefore omitted from the PAG preparation. If, however, a gold-labeled goat anti-mouse (anti-rabbit) antibody is used, BSA or another blocking agent such as diluted normal goat serum can be incorporated.

If ISEM is used to trap the virus prior to gold labeling, the normal procedure outlined above will give some background due to recognition of the coating antibody by the PAG (or, for example, the goat anti-rabbit). One solution is to use the coating anti-body as dilute as possible (say, $1-10$ μg/ml IgG, or whole anti-serum diluted 1/5000); the more elegant answer of Louro and Lesemann (106) to this problem was to initially coat the grid with F(ab')$_2$ fragments prepared from the virus antibody. These cannot be recognized by the PAG.

The following papers (2, 45, 48, 73, 77, 98, 175), all concerned with GLAD of virus particles labeled with monoclonal or polyclonal antibodies, will give an idea of recent methodology. Different binding sites on the virus particles (beet necrotic yellow vein virus, a grapevine leafroll-associated closterovirus, plum pox virus, potato virus V, tobacco mosaic virus), the presence of virus mixtures (grapevine leafroll-associated clostero-viruses), and the several external and internal components of a complex virion (tomato spotted wilt virus) are all elegantly illustrated.

IN SITU METHODS INVOLVING THIN SECTIONING

Antibodies in or on thin-sectioned material are in themselves invisible by electron microscopy, so they need labeling, and these days gold is the label of choice. There is an enormous literature on gold labeling of thin sections, access to which can only be given by citing the following selected reviews and papers (14, 19, 21, 23, 25, 68, 69, 70, 72, 79, 156, 157, 165, 166, 169).

Pre-Embedding Techniques

With the pre-embedding approach, the tissue of interest, perhaps after optional glutaraldehyde prefixation, is exposed to labeled antibodies, and is then post-fixed and embedded in a classical resin such as Epon or Araldite. Thin sections of the reacted material are then examined. Surprisingly, there are apparently only a few reports of pre-embedding labeling being used with plant viruses (57, 58, 148).

Certainly, pre-embedding suffers from a number of disadvantages. First, only the antigens exposed on the surface of cells, or within cells that have been cut open, can be labeled, since membranes, and certainly cell walls, constitute a barrier to antibody penetration. Second, each experiment and each control must be done on a separate piece of tissue, which must then be separately embedded and sectioned; this makes experiments laborious and comparisons less precise. (With post-embedding, different experiments or controls can be done with different serial or quasi-serial sections off the same block).

However, three clear advantages also exist. One is that antibodies can be directly reacted with the antigen in a native state, without the problems introduced (in post embedding) by fixation, dehydration, and inclusion in resin. (In discussing mycoplasmas, I draw attention to an example where a monoclonal antibody, active in pre-embedding conditions, gives zero response with post-embedding). A second advantage is that labeling occurs in three dimensions and is usually intense; in post-embedding systems, only antigen exposed in the surface of the section can be labeled, although other unlabeled antigen will be visible within the thickness of the section. Finally. with pre-embedding it is possible to retain all or most of the excellent morphological preservation of detail attained with classical embedding methods (even though we bear in mind that some of this detail is artifact). With post embedding, there is always a tug-of-war between good structure and good antigenic response, and many gold-labeled post embedding micrographs are of awful quality, judged in terms of retention of fine structure. A case for pre-embedding therefore exists.

Shalla and Amici (148) investigated the distribution of tobacco mosaic virus coat protein antigen inside systemically infected tomato leaf cells. They fixed pieces of leaf in glutaraldehyde, froze them to $-20°C$ in water, and cut cryotome slices which were then brought to room temperature and incubated overnight in ferritin-conjugated antibody. After rinsing, the slices were fixed in osmium tetroxide, embedded in Araldite, and sectioned. In some experiments, cell fractions derived from sucrose density gradients were also examined.

The results were remarkably clear, with intense and accurate labeling, and excellent fine structure. Such beautiful and informative micrographs have not been rivalled since.

Gildow and Rochow (59) and Gildow (57) wished to solve a totally different problem: the fate of ingested barley yellow dwarf virus (BYDV) particles, the mode of their transcellular transport into the haemocoel of the aphid vectors (*Rhopalosiphum padi* for the RPV isolate, and *Sitobion avenae* for the MAV isolate), and the route by which the viruses arrive in the salivary glands and are finally transmitted in the saliva. An interesting question was: at which point is the RPV/MAV specificity exerted? The problem was complicated because in thin sections the small isometric particles of BYDV could be confused with those of *Rhopalosiphum padi* virus (RhPV), a virus native to this aphid, and possibly other aphid viruses also. IgGs purified from antisera to the RPV isotate of BYDV, the serologically unrelated MAV isolate, and to RhPV were employed.

Initially (59), *S. avenae* aphids were given acquisition feeds of 5 days on sources of BYDV-MAV or BYDV-RPV. The salivary glands were then dissected and exposed to MAV or PAV antiserum, followed by ferritin-labeled goat anti-rabbit IgG. The ferritin-labeled virus particles were clearly distinguished, and it was seen that ingested virus (both MAV and RPV) could be absorbed into the haemocoel and penetrate the basal lamina of the salivary glands, but that only the MAV isolate was taken up by the plasmalemma of accessory gland cells, which appeared to exercise a selective function. Thus only the appropriate BYDV isolate was in a position to be transmitted.

Gildow (57) used ferritin-labeled antibody in a different way, to observe how the virus was absorbed from the gut. R. padi were fed on RPV infected barley plants, anaesthetized with carbon dioxide, and injected with homologous or heterologous IgG. After recovery and further feeding, the aphids were again anaesthetized and injected with ferritin-labeled goat anti-rabbit IgG. The aphids were then fixed, embedded and sectioned. RPV and MAV isolates were clearly distinguished by the label, and the site of virus uptake by endocytosis was identified as cells of the hindgut. R. padi was able to acquire both RPV and MAV into the haemocoel. This and related work is reviewed by Gildow (58).

These three examples must serve to illustrate the possibilities of the pre-embedding approach.

Post-Embedding Techniques

Despite a lot of work, these methods, at least as regards plant viruses, are still at a primitive stage, and suffer from a number of fundamental technical problems. These are (a) that good ultrastructural preservation has rarely been compatible with good labeling; (b) that approaches to the conservation of active antigens in the section surface are still largely empirical; (c) that some antigen-antibody combinations give good results, and others (particularly with some monoclonal antibodies) do not; and (d) that only the antigen directly exposed in the section surface can be labeled. Clear guidelines as to which methods are best are hard to draw up, but a few comments can be made, based on the following papers (5, 8, 15–17, 21, 24, 49, 55, 56, 88–90, 92, 94, 102–105, 122, 134, 140, 145, 158, 160, 161, 163).

1. Tissue is generally fixed in glutaraldehyde, with or without formaldehyde; use of osmium tetroxide is usually reported to abolish antigenicity, even when etching the sections in metaperiodate precedes labeling. However, adequate labeling after osmium fixation has been reported (104, 158, 163). Some antigens have been labeled after formaldehyde but not glutaraldehyde fixation (56, and see also 100).

2. The resin of choice is usually a methacrylate type such as Lowicryl K4M, Lowicryl HM20, LR White, LR Gold or glycol methacrylate. These types of resin were abandoned in the early 1960's due to their destructive effects on tissues, but they have returned to haunt us because very often they allow better labeling. Epon and Araldite have their devotees, and undoubtedly, when labeling is successful, the ultrastructure retained is supe-

rior (5, 24, 55, 56, 90, 91, 104, 158, 163). Etching with ethanolic NaOH (90), 10% hydrogen peroxide (5, 55) or saturated metaperiodate (163) can improve the labeling of epoxy sections.

3. Dehydration is achieved in methanol, ethanol or acetone, with no one of these giving clearly superior results. Berryman and Rodewald (25) suggest that acetone dehydration is superior for conserving membrane phosopholipids, and also that fixation in uranyl acetate before dehydration helps to conserve membranes and their antigenicity. Some voices suggest that uranyl acetate fixation may damage antigenicity, but in our (unpublished) experience it does not.

4. Sections are usually collected on plastic films such as Formvar, mounted on nickel grids. We find (Milne and colleagues, unpublished) that the Formvar can give quite high nonspecific backgrounds (and in any case, contributes *some* background). Moreover the grid itself is a potential source of contamination. An excellent solution to these problems is to process the sections as a group floating on drops of liquid, and confined by a 1.5 mm internal diameter plastic loop (a "goldfish", Figure 9.1). The sections are only placed on the Formvar film (on copper grids) just before examination in the EM. Processing is no more laborious than when using grids. The goldfish idea is not ours (though the name is); we do not know who thought of it.

5. Sections are initially incubated for, say, 60 min with 1% BSA or ovalbumin, or 5–10% normal goat serum, to block non-specific protein-binding sites.

FIGURE 9.1. A group of thin sections (invisible) floating on a drop of buffer, and confined within the 1.5 mm diameter "eye" of a plastic "goldfish". The goldfish plus sections is transferred with forceps from one solution to another; when processing is completed, a Formvar-filmed grid is lowered onto the sections, which adhere to the grid and are ready for examination.

6. Around 10–100 µg/ml IgG is usually an optimal concentration for the primary antibody. This is generally diluted in the buffer of choice containing 0.1–1.0% BSA, 0.1% Tween 20, and 0.05% sodium azide. Buffers used have varied widely (phosphate, phosphate-citrate, cacodylate, Tris-HCl).
7. Both protein A-gold and goat anti-rabbit-IgG-gold (GARG) (or anti-mouse-gold, as the case may be) are frequently used as labels. There is some feeling that GARG gives more intense label and lower background, but this must depend greatly on the quality of the respective labels. The smallest manageable size of gold—5 nm for routine transmission EM— gives the best resolution and the highest number of label units per unit of antigen.

As to results, there was a pioneering phase in the early 1980's when uncertainties were such that virus particles were used to label the gold (102, 160).

Now more result-oriented rather than technique-oriented papers are appearing, concerning viruses with complex structures, such as tomato spotted wilt (163) or rhabdoviruses (103), or those inducing complex viral inclusions (5, 15–17, 49, 55, 56, 90–92, 94, 104, 122, 140). Some dynamic problems such as the mode of virus entry (145), the development of local lesions (24), and the mechanism of passage of virus into the seed (134) have been approached.

Finally, a phase has begun in which virus-coded nonstructural proteins are being probed in infected cells. Notable here is work on the localization of definite or probable transport proteins in the cell walls and plasmodesmata. Some examples are those of alfalfa mosaic (158), cauliflower mosaic (105), parsnip yellow fleck (49) and tobacco mosaic (8, 161). This and other related work has been excellently reviewed by Atabekov and Taliansky (7).

Mycoplasmas

MLOs (74, 110, 172) are generally larger than plant viruses, and very variable in size ("strings of pearls" are about 50 nm in diameter, filaments about 30 nm, and mature bodies from 200 to 1000 nm in diameter). Unlike virus particles, MLO bodies are morphologically both very variable and rather featureless, especially when extracted in vitro, and positively or negatively stained on EM grids. They may become flattened, extruded or fragmented, and then morphologically resemble membrane-bound host debris that may be the remains of chloroplasts, mitochondria or endoplasmic reticulum (173).

Thus recognition on EM grids of MLOs derived from in vitro preparations is very problematical, unless the bodies are also labeled, for example by using immunogold. There are apparently rather few reports of this being done (99, 115, 121). The situation has not been made easier by the fact that until fairly recently good polyclonal and monoclonal antibodies against MLOs have not been available.

The above remarks may form a background to a discussion of in vitro work. In thin sections, MLOs are much more easily recognized as such, and distinguished from host organelles. However, this distinction may be difficult if the MLOs are few and the tissue necrotic, and in any case one MLO looks much like another. Thus for thin section applications too, the need for gold immuno-labeling of MLOs is clear. As with in vitro work, reports in this area appear to be few so far (23, 99, 100, 115).

Immunological and other work on detection and identification of MLOs has been extensively reviewed (37, 75, 82, 99, 111, 115, 151, 152).

In Vitro Preparations

The first report of trapping by ISEM of an MLO was by Derrick and Brlansky (43). This report in fact concerned not an MLO but what later became known as corn stunt spiroplasma (*S. kunkelii*). The relevant differences between MLOs and spiroplasmas, in this case, are that for spiroplasmas, good antibody preparations were easier to obtain, and the morphology of the organism is very distinct, and unmistakable in the EM.

However, this successful demonstration, and the later development of MLO antisera, led to several reports that MLOs could be trapped and identified by ISEM. In all these reports, reliability of identification depended on the close application of controls (extracts of healthy material, coating of grids with pre-immune serum etc) since, as noted above, the individual MLO bodies could not be positively identified.

Sinha and Benhamou (153), using antiserum to aster yellows (AY) MLO, were able to trap bodies by ISEM, from partially purified extracts of AY-infected aster plants, that appeared to be the AY MLOs. Similar bodies could be trapped from preparations of other infected plants, including *Trifolium* carrying clover phyllody, and celery, *Catharanthus* (periwinkle), oat and wheat. On the other hand, MLOs from peach X and clover yellow edge were not trapped by this antiserum, and an antiserum to *Spiroplasma citri* failed to trap the AY MLOs. The results corresponded with those of ELISAs on the same material.

As we have noted in discussing ISEM of plant viruses, coating grids with preimmune sera, even diluted 1/1000, has the effect of inhibiting attachment of virus particles, so that fewer are found than on uncoated grids. Similarly, antiserum dilutions of the order of 1/100 are inhibitory (or at least, far from optimal) when used for trapping even the homologous virus. It is therefore curious that Sinha and Benhamou (153) reported that AY MLOs were non-specifically adsorbed on grids coated with preimmune serum at dilutions of 1/50 or less. As a consequence of this finding, an antiserum dilution of 1/100 was used for coating, and was apparently successful. It would seem that either this level of antiserum combined with a precoating of PA, or an antiserum dilution of around 1/1000, should have given better results. Sinha and Chiykowski (154) used similar methods to detect peach X MLO in partially purified extracts of celery leaves.

Sinha (152) reported unpublished data that AY MLO can be detected by

ISEM in individual vector leafhoppers, *Macrosteles fascifrons*, and that peach X MLO can be similarly detected in the leafhopper vector *Paraphlepsius irroratus*. Some hoppers exposed to infected plants did not transmit and were ISEM-negative; however, other hoppers that were not transmitters were ISEM-positive, indicating that they could acquire but could not transmit the agent. (Such behaviour appears to be not uncommon with propagative hopper vectors of plant viruses; see for example 31, 52).

Caudwell et al. (34–36) somewhat controversially claimed detection by ISEM of an MLO said to be responsible for the Flavescence Dorée (FD) disease of grapevine, in extracts prepared from the vector leafhoppers, *Euscelidius variegatus*. This work was controversial because of the general difficulties outlined above, and because the purported MLO was present at very low levels; however, a particular problem was that other workers were unable to repeat the results obtained by Caudwell's school. More recent work (99, 100) suggests that this group was on the right track all the time.

A difficulty of immunological studies with MLOs is that monoclonal antibodies are not easily obtained, and polyclonal antibody preparations initially contain significant levels of antibodies directed against the host. The interesting solution of Caudwell et al. (34–36, 100) was to raise one antiserum using FD-enriched material from infective hoppers, and another using similar material from infected *Vicia faba* plants. The insect-based antiserum was used to detect the MLO in plants, and vice versa, so that the effects of host-directed antibodies could be sidestepped.

Lherminier et al. (99) and Milne (115) report preliminary results with gold-labeling of MLOs trapped by ISEM. This method will undoubtedly become routine, reliable and convincing, especially when the possibilities of using PA precoating and optimal antibody dilutions have been explored. Milne (115), in preliminary experiments, found that most consistent gold labeling on grids was obtained, if a 10 μl aliquot of MLO preparation (primula yellows or tomato big bud) was added to 100 μl of distilled water, causing osmotic bursting. Many membrane fragments could then be labeled with the appropriate antibodies. However, this was a partial solution to the problem, as the characteristic (even if imprecise) morphology of the MLOs was lost. Lherminier et al. (99) show a convincing micrograph of intact MLOs trapped by ISEM and decorated with the same antiserum, followed by protein A-gold.

THIN SECTIONS

Post Embedding

The only work published to my knowledge on post-embedding gold-labeling of thin sectioned MLOs is that of Benhamou (cited in 23), Lherminier et al. (100), and Milne and colleagues (99, 115).

Benhamou (23), in a review of gold labeling using lectins, described the use of a gold-labeled *Ricinus communis* lectin to localize galactose residues on glutaraldehyde-osmium fixed Epon sections of MLOs associated with white clover decline. Gold particles were mainly distributed over the internal structure of the MLOs. Use of other specific lectins showed that besides galactose, glucose and to a lesser extent mannose were the main sugar residues associated with these MLOs. No doubt the lectin-gold approach will be a useful addition to serological characterization of this group of organisms.

Lherminier et al. (100) demonstrated FD MLOs in the salivary glands of the vector *E. variegatus*. The antiserum was prepared by injecting MLO-enriched plant extracts into a rabbit, and after absorption with healthy plant proteins, the IgG was purified. This was used at concentrations of 50–500 (optimal, 75–150) µg/ml. Various embedding schedules were evaluated, using Epon 812, Lowicryl K4M at 4°C or at −30°C, or LR White embedded at room temperature and cured at 55°C, or embedded and cured at −15°C. Frozen thin sections were also used. Different types of fixation (paraformaldehyde (FA), glutaraldehyde, (GA), and osmium tetroxide) were tried.

Epon gave poor results (weak or no labeling and high backgrounds) even after etching sections in sodium ethoxide and hydrogen peroxide, or sodium metaperiodate. K4M at −30°C gave better ultrastructural preservation than at 4°C, and convincing labeling. In LR White there was no labeling after room temperature embedding and curing at 55°C, but the best compromize between labeling and preservation of structures was obtained with LR White at −15°C. Similar levels of labeling, but worse ultrastructural preservation, was obtained with frozen sections. Osmium tetroxide and levels of GA above 0.1% reduced or abolished the label, and the best fixative was found to be a mixture of 4% FA and 0.1% GA (see 56 for a somewhat similar result). The only labeling obtained was over the MLO bounding membrane.

Milne and colleagues (99, 115) and Milne (unpublished results), have used LR Gold embedding at −20°C, and the MLOs of primula yellows (PY), tomato big bud (TBB), and bermudagrass white leaf (BGWL), plus their respective polyclonal antisera, and one anti-PY monoclonal, all provided by Dr. M.F. Clark (38). The MLOs were examined in infected leaf tissues of *Catharanthus roseus* (PY, TBB) and bermudagrass (*Cynodon dactylon*; BGWL). The results confirm that highly specific labeling over the periphery of the MLOs can be obtained (Figure 9.2), using a mixture of 4% FA and 0.1% GA, but not if levels of 1% or more of GA are used. Reactions were highly specific, in the sense that PY, TBB and BGWL MLOs reacted only with their own antibodies, and labeling with the heterologous antibodies was close to zero. The ultrastructure of the host was not optimally preserved, but was adequate for the purpose; fortunately, the MLOs themselves seem to be more robust than the host tissues. Although the antisera were produced from plant extracts and tested on plant tissues, background levels were very low provided the sera had been adsorbed with healthy plant extracts. Optimal levels of labeling were obtained at 50 µg/ml of IgG. The PY monoclonal, tested at levels from 50 to 200 µg/ml IgG, gave no label, although it reacted strongly in a pre-embedding system (see below).

FIGURE 9.2. Unstained transverse section of a phloem sieve tube of periwinkle (*Catharanthus roseus*) infected with primula yellows MLO, incubated with homologous polyclonal antibody, then labeled with goat anti-rabbit IgG conjugated with 5 nm gold. Note the gold particles over the periphery of the MLOs (bar, 100 nm).

In conclusion, it is possible to say that MLOs in the vector or in the plant can now be immunogold-labeled (or lectin-gold labeled) by post-embedding methods, although there is still some problem with ultrastructural preservation of host tissues, due to restriction on the use of glutaraldehyde, and although the one monoclonal antibody tested failed to react, in this system.

Pre-embedding

Where the antigen to be detected is labile or becomes damaged by fixatives, but is or can be exposed extracellularly (as is the case with MLOs), a pre-embedding approach may be useful. Milne and colleagues (99, 115; Milne and Lenzi, unpublished results) have used the following system with PY and TBB MLOs, their respective antisera, and the monoclonal against PY.

Leaf veins of infected *C. roseus* plants were isolated and, at 0°C, sliced transversely with a razorblade into discs about 0.2 mm thick. These discs were

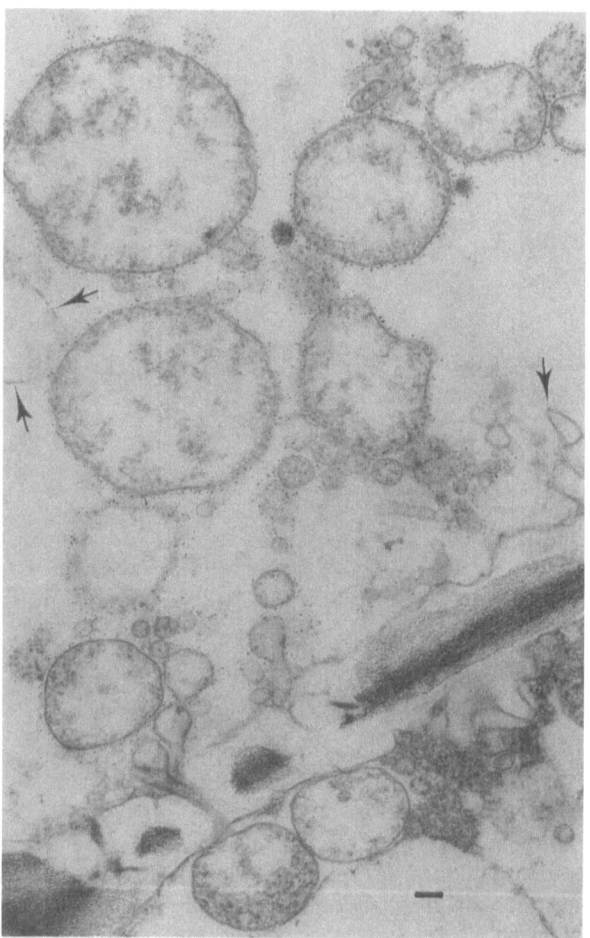

FIGURE 9.3. Pre-embedding preparation of *Catharanthus roseus* leaf vein infected with primula yellows MLO, incubated with homologous monoclonal antibody followed by goat anti-mouse IgG conjugated with 5 nm gold. The preparation was later fixed in glutaraldehyde and osmium tetroxide, embedded in Epon, sectioned longitudinally, and stained in lead citrate. Note that the exposed MLOs (upper part of the Figure) are labeled, but not host membranes (arrows) nor the two MLOs behind the sieve plate at the base of the Figure (bar, 100 nm).

incubated, in the cold room with gentle shaking, for 1–3 h with dilutions of the MLO antibody; they were then rinsed in buffer and incubated for a further 60 min in a commercial preparation of gold-labeled antibody (goat anti-rabbit or goat anti-mouse, as appropriate). Following a further rinse, the discs were processed in the classical way, by fixation in glutaraldehyde and osmium tetroxide, and embedding in Epon.

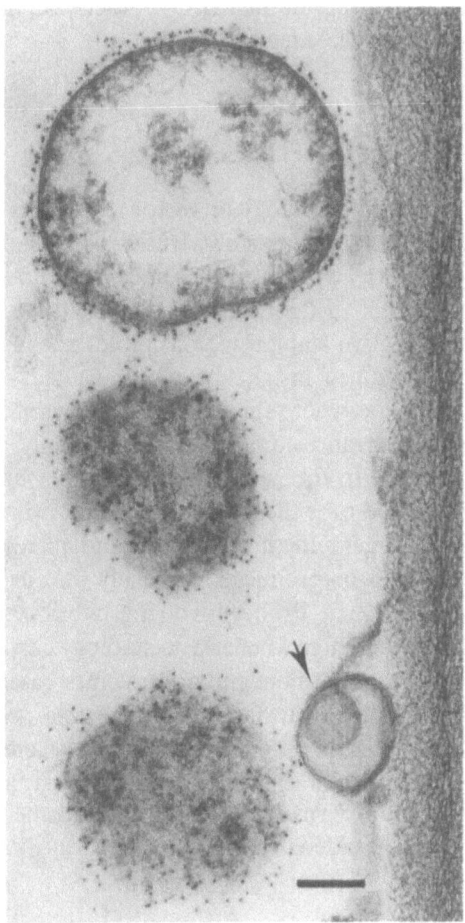

FIGURE 9.4. Preparation similar to Figure 9.3, but tomato big bud MLO-infected tissue treated with homologous polyclonal antibody followed by goat anti-rabbit IgG conjugated to 5 nm gold. Note the three MLO profiles strongly labeled, but absence of label on host membranes (arrow); (bar, 100 nm).

The results (Figures 9.3 and 9.4) showed that labeling on the outer surface of the MLOs was intense and specific (no cross-reaction between PY and TBB), but, as expected, only those MLOs exposed by the initial slicing were labeled. Ultrastructural preservation of the MLOs was good, and the mono-clonal antibody, found to be inactive using a post-embedding protocol (see previous section) gave excellent labeling. One technical difficulty was that occasionally the MLOs at or near the cut surface were washed away, and only empty phloem cell walls then remained.

It is interesting to see that labeling of the periphery of the MLO, found by post-embedding methods (see above) was confirmed; in addition, it is clear that the MLO antigen provoking the major response in immunized animals is actually exposed on the surface of the MLO—otherwise, the pre-embedding approach could not have succeeded.

Conclusion

The advance of a subject like virology or vector research is limited partly by the scarcity of geniuses (there are no slow viruses—only slow virologists—as I am personally well aware). However, the other great limitation (let us not talk about funding) is that of the technology. Every time there has been a technical advance—negative staining, ISEM, gold-labeling—floods of new results have appeared and new insights have been gained.

It is foolish to speculate where the next leaps forward in technique will appear, but sometimes advances are made incrementally, and here we can often see what is coming. In the next few years, deeper understanding and finer control will be gained over the, at present, rather primitive manoeuvers involved in labeling antigens in thin sections, and increasing numbers of interesting antibodies will be produced, especially (in our context) to non-capsid virus-encoded proteins. Probes involving hybridization with nucleic acids will be developed to a stage where the technology can be used routinely. Cryotechniques will further develop, and microscopes based on novel physical principles will become somewhat more accessible. All this is tending toward the exercise of molecular electron microscopy, whereas at the moment we are still largely struggling at the old fashioned fine-structure level. Many problems concerning plant viruses, plant MLOs, and their vectors, will no doubt be solved—or be rendered more confusing (and interesting) by this approach.

Acknowledgments. I thank Riccardo Lenzi, Vera Masenga and Eliana Ramasso, who participated in the work described from our laboratory.

References

1. Accotto, G.P., 1982, Immunosorbent electron microscopy for detection of fanleaf virus in grapevine, *Phytopath. Medit.* **21**:75–78.
2. Adam, G., Lesemann, D.E., and Vetten, H.J., 1991, Monoclonal antibodies against tomato spotted wilt virus: characterization and application, *Ann. Appl. Biol.* **118**:87–104.
3. Almeida, J.D., and Waterson, A.P., 1969, The morphology of virus-antibody interaction, *Advances in Virus Research* **15**:307–338.
4. Anderson, F.A., and Stanley, W.M., 1941, A study by means of the electron microscope of the reaction between tobacco mosaic virus and its antiserum, *J. Biol. Chem.* **139**:339–344.
5. Appiano, A., D'Agostino, G., Bassi, M., Barbieri, N., Viale, G., and Dell'Orto, P.,

1986, Origin and function of tomato bushy stunt virus-induced inclusion bodies, *J. Ultrastruct. Molec. Struct. Res.* **97**:31–38.

6. Appiano, A., Dell'Orto, P., Viale, G., and Braidotti, P., 1988, Immunogold cytochemistry of a plant potyvirus in sections of host tissue: effect of different fixation and embedding methods on labelling with polyclonal and monoclonal antibodies, *Inst. Phys. Conf. Ser. No. 93*, Volume 3, pp. 295–296.

7. Atabekov, J.G., and Taliansky, M.E., 1990, Expression of a plant virus-coded transport function by different viral genomes, *Advances in Virus Research* **38**:201–248.

8. Atkins, D., Hull, R., Wells, B., Roberts, K., Moore, P., and Beachy, R.N., 1991, The tobacco mosaic virus 30K movement protein in transgenic tobacco plants is localized to plasmodesmata, *J. Gen. Virol.* **72**:209–211.

9. Baker, K.K., Ramsdell, D.C., and Gillett, J.M., 1985, Electron microscopy: current applications to plant virology, *Plant Disease* **69**:85–90.

10. Ball, E.M., 1971, Leaf-dip serology, *Methods in Virology* **5**:445–450.

11. Ball, E.M., 1974, Serological tests for the identification of plant viruses, *Bull. American Phytopath. Soc.* St. Paul. Minnesota, 31 pp.

12. Ball, E.M., and Brakke, M.K., 1968, Leaf-dip serology for electron microscopic identification of plant viruses, *Virology* **36**:152–155.

13. Ball, E.M., and Brakke, M.K., 1969, Analysis of antigen-antibody reactions of two plant viruses by density-gradient centrifugation and electron microsocopy, *Virology* **39**:746–758.

14. Baschong, W., and Wrigley, N.G., 1990, Small colloidal gold conjugated to Fab fragments or to immunoglobulin G as high-resolution labels for electron microscopy: a technical overview, *J. Electron Microscopy Technique* **14**:313–323.

15. Baunoch, D.A., Das, P., and Hari, V., 1988, Intracellular localization of TEV capsid and inclusion proteins by immunogold labeling, *J. Ultrastruct. Molec. Struct. Res.* **99**:203–212.

16. Baunoch, D.A., Das, P., and Hari, V., 1990, Potato virus Y helper component protein is associated with amorphous inclusions, *J. Gen. Virol.* **71**:2479–2482.

17. Baunoch, D.A., Das, P., Browning, M.E., and Hari, V., 1991, A temporal study of the expression of the capsid, cytoplasmic inclusion and nuclear inclusion proteins of tobacco etch potyvirus in infected plants, *J. Gen. Virol.* **72**:487–492.

18. Beckendorf, S.K., 1973, Structure of the distal half of the bacteriophage T4 tail fiber, *J. Mol. Biol.* **73**:37–53.

19. Beesley, J.E. (ed), 1989, Colloidal Gold: A New Perspective for Cytochemical Marking, Oxford University Press/Royal Microscopical Society, Oxford.

20. Beier, H., and Shepherd, R.J., 1978, Serologically specific electron microscopy in the quantitative measurement of two isometric viruses, *Phytopathology* **68**:533–538.

21. Bendayan, M., 1987, Introduction of protein G-gold complex for high resolution immunocytochemistry, *J. Electron Microscope Tech.* **6**:7–13.

22. Bendayan, M., and Garzan, S., 1988, Protein G-gold complex: comparative evaluation with protein A-gold for high-resolution immunocytochemistry, *J. Histochem. Cytochem.* **36**:597–607.

23. Benhamou, N., 1989, Preparation and application of lectin-gold complexes, in Hayat, M.A. (ed): Colloidal Gold. Principles, Methods, and Applications, Volume 1, Academic Press, San Diego, pp. 95–143.

24. Benhamou, N., Mazan, D., Esquerre-Tugaye, M.T., and Masselin, A., 1990,

Immunogold localization of hydroxyproline-rich glycoproteins in necrotic tissue of *Nicotiana tabacum* L., cv. xanthi-nc infected by tobacco mosaic virus, *Physiol. Mol. Plant Pathol.* **36**:129–145.

25. Berryman, M.A., and Rodewald, R.D., 1990, An enhanced method for post-embedding immunocytochemical staining which preserves cell membranes, *J. Histochem. Cytochem.* **38**:159–170.

26. Boccardo, G., Milne, R.G., Luisoni, E., Lisa, V., and Accotto, G.P., 1985, Three seedborne cryptic viruses containing double-stranded RNA isolated from white clover, *Virology* **147**:29–40.

27. Bovey, R., Brugger, J.J., and Gugerli, P., 1982, Detection of fanleaf virus in grapevine tissue extracts by enzyme-linked immunosorbent assay (ELISA) and immune electron microscopy (IEM), in McGinnis, A.J. (ed): *Proc. 7th Meeting Intern. Council for the Study of Viruses and Viruslike Diseases of the Grapevine*, Niagara Falls, Canada, 1980, pp. 259–275.

28. Brandes, J., 1957, Eine elektronenmikroskopische Schnell-methode zum nachweis faden- und stäbchenförmiger Viren, insbesondere in Kartoffeldunkelkeimen, *Nachrbl. deut. Pflanzenschutzdienst (Braunschweig)* **9**:151–152.

29. Brandes, J., and Wetter, C., 1959, Classification of elongated plant viruses on the basis of particle morphology, *Virology* **8**:99–115.

30. Brlansky, R.H., and Derrick, K.S., 1979, Detection of seedborne plant viruses using serologically specific electron microscopy, *Phytopathology* **69**:96–100.

31. Caciagli, P., Roggero, P., and Luisoni, E., 1985, Detection of maize rough dwarf virus by enzyme-linked immunosorbent assay in plant hosts and in the plant-hopper vector, *Ann. Appl. Biol.* **107**:462–471.

32. Carrascosa, J.L., 1988, Immunoelectron microscopical studies on viruses, *Electron Microsc. Rev.* **1**:1–16.

33. Casper, R., Meyer, S., Lesemann, D.E., Reddy, D.V.R., Rajeshwari, R., Misari, S.M., and Subbarayudu, S.S., 1983, Detection of a luteovirus in groundnut rosette diseased groundnuts (*Arachis hypogaea*) by enzyme-linked immunosorbent assay and immunoelectron microscopy, *Phytopathol. Z.* **108**:12–17.

34. Caudwell, A., Meignoz, R., Kuszala, C., Larrue, J., Fleury, A., and Boudon, E., 1982, Purification sérologique et observation ultramicroscopique de l'agent pathogène (MLO) de la flavescence dorée de la vigne dans les extraits liquides de fèves malades, *C.R. Soc. Biol.* **176**:723–729.

35. Caudwell, A., Meignoz, R., Kuszala, C., Schneider, C., Larrue, J., Fleury, A., and Boudon, E., 1982, Purification immunologique et observation ultramicroscopique en milieu liquide de l'agent pathogène (MLO) d'une jaunisse végétale, la flavescence dorée de la vigne, *C.R. Acad. Agric. Fr.* **68**:407–415.

36. Caudwell, A., Meignoz, R., Kuszala, C., Schneider, C., Larrue, J., Fleury, A., and Boudon, E., 1983, Serological purification and visualization in the electron microscope of the grapevine Flavescence Dorée pathogen (MLO) in diseased plants and infectious vector extracts, *Yale J. Biol. Med.* **56**:936–937.

37. Chen, T.A., Lei, J.D., and Lin, C.P., 1989, in Whitcomb, R.F., and Tully, J.G. (eds): Detection and identification of plant and insect mollicutes, The Mycoplasmas Volume V, Spiroplasmas, Acholeplasmas, and Mycoplasmas of Plants and Arthropods, Academic Press, San Diego, pp. 393–424.

38. Clark, M.F., Morton, A., and Buss, S.L., 1989, Preparation of mycoplasma immunogens from plants and a comparison of polyclonal and monoclonal antibodies made against primula yellows MLO-associated antigens, *Ann. Appl. Biol.* **114**:111–124.

39. Cohen, J., Loebenstein, G., and Milne, R.G., 1982, Effect of pH and other conditions on immunosorbent electron microscopy of several plant viruses, *J. Virol. Methods* **4**:323–330.

40. Del Vecchio, V.G., Dixon, C., and Lemke, P.A., 1979, Immune electron microscopy of virus-like particles of *Agaricus bisporus*, *Exp. Mycol.* **2**:138–144.

41. Derrick, K.S., 1973, Quantitative assay for plant viruses using serologically specific electron microscopy, *Virology* **56**:652–653.

42. Derrick, K.S., 1978, Double-stranded RNA is present in extracts of tobacco plants infected with tobacco mosaic virus, *Science* **199**:538–539.

43. Derrick, K.S., and Brlansky, R.H., 1976, Assay for viruses and mycoplasmas using serologically specific electron microscopy, *Phytopathology* **66**:815–820.

44. Derrick, K.S., French, R.C., Clark, C.A., and Gabriel, C.J., 1984, Detection of double-stranded RNA by serologically specific electron microscopy, *J. Virol. Methods* **9**:293–299.

45. Dore, I., Weiss, E., Altschuh, D., and van Regenmortel, M.H.V., 1988, Visualization by electron microscopy of the location of tobacco mosaic virus epitopes reacting with monoclonal antibodies in enzyme immunoassay, *Virology* **162**:279–289.

46. Dore, I., Ruhlmann, C., Oudet, P., Cahoon, M., Caspar, D.L.D., and van Regenmortel, M.H.V., 1990, Polarity of binding of monoclonal antibodies to tobacco mosaic virus rods and stacked disks, *Virology* **176**:25–29.

47. Edwards, M.L., Kelley, S.E., Arnold, M.K., and Cooper, J.I., 1989, Properties of a hordeivirus from *Anthoxanthum odoratum*, *Plant Pathology* **38**:209–218.

48. Farmer, M.J., Milne, E.W., Roberts, I.M., and Harrison, B.D., 1989, Monoclonal antibodies to potato V potyvirus (PVV), *Scottish Crop Research Institute Report for 1988*, pp. 172–174.

49. Fasseas, C., Roberts, I.M., and Murant, A.F., 1989, Immunogold localization of parsnip yellow fleck virus particle antigen in thin sections of plant cells, *J. Gen. Virol.* **70**:2741–2749.

50. Forde, S.M.D., 1989, Strain differentiation of barley yellow dwarf virus isolates using specific monoclonal antibodies in immunosorbent electron microscopy, *J. Virol. Methods* **23**:313–320.

51. Francki, R.I.B., Milne, R.G., and Hatta, T., 1985, Atlas of Plant Viruses, Volume I, CRC Press, Boca Raton, Florida.

52. Francki, R.I.B., Ryan, C.C., Hatta, T., Rohozinski, J., and Grivell, C.J., 1986, Serological detection of Fiji disease virus antigens in the planthopper *Perkinsiella saccharicida* and its inefficient ability to transmit the virus, *Plant Pathology* **35**:324–328.

53. French, R.C., Price, M.A., and Derrick, K.S., 1982, Circular double-stranded RNA in potato spindle tuber viroid-infected plants, *Nature (London)* **295**:259–260.

54. Fribourg, C.E., Koenig, R., and Lesemann, D.E., 1987, A new tobamovirus from *Passiflora edulis* in Peru, *Phytopathology* **77**:486–491.

55. Garnier, M., Candresse, T., and Bové, J.M., 1986, Immunocytochemical localization of TYMV-coded structural proteins by the protein A-gold technique, *Virology* **151**:100–109.

56. Giband, M., Stoeckel, M.E., and Lebeurier, G., 1984, Use of the immuno-gold technique for in situ localization of cauliflower mosaic virus (CaMV) particles and the major protein of the inclusion bodies, *J. Virol. Methods* **9**:277–281.

57. Gildow, F.E., 1985, Transcellular transport of barley yellow dwarf virus into

the haemocoel of the aphid vector *Rhopalosiphum padi, Phytopathology* **75**:292–297.

58. Gildow, F.E., 1987, Virus-membrane interactions involved in circulative transmission of luteoviruses by aphids, in Harris, K.F. (ed): Current Topics in Vector Research, Volume 4, Springer-Verlag, New York, pp. 94–120.

59. Gildow, F.E., and Rochow, W.F., 1980, Role of accessory salivary glands in aphid transmission of barley yellow dwarf virus, *Virology* **104**:97–108.

60. Gillett, A.M., Morinoto, K.M., Ramsdell, D.C., Baker, K.K., Chaney, W.G., and Esselman, W.J., 1982, A comparison between the relative abilities of ELISA, RIA and ISEM to detect blueberry shoestring virus in its aphid vector, *Acta Hort.* **129**:25–29.

61. Giunchedi, L., and Langenberg, W.G., 1982, Efficacy of colloidal gold-labeled antibody as measured in a barley stripe mosaic virus-lectin-antilectin system, *Phytopathology* **72**:645–647.

62. Goding, J.W., 1978, Use of staphylococcal protein A as an immunological reagent, *J. Immunol. Methods* **20**:241–253.

63. Gough, K.H., and Shukla, D.D., 1980, Further studies on the use of protein A in immune electron microscopy for detecting virus particles, *J. Gen. Virol.* **51**:415–419.

64. Hamilton, R.I., and Nichols, C., 1978, Serological methods for detection of pea seedborne mosaic virus in leaves and seeds of *Pisum sativum, Phytopathology* **68**:539–543.

65. Hampton, R., Ball, E., and De Boer, S. (eds), 1990, Serological Methods for Detection and Identification of Viral and Bacterial Plant Pathogens: A Laboratory Manual, APS Press, St. Paul, Minnesota.

66. Harrison, B.D., Muniyappa, V., Swanson, M.M., Roberts, I.M., and Robinson, D.J., 1991, Recognition and differentiation of seven whitefly-transmitted geminiviruses from India, and their relationships to African cassava mosaic and Thailand mung bean yellow mosaic viruses, *Ann. Appl. Biol.* **118**:299–308.

67. Harville, B.G., and Derrick, K.S., 1978, Identification and prevalence of white clover viruses in Louisiana, *Plant Dis. Reptr.* **62**:290–292.

68. Hayat, M.A. (ed), 1989, Colloidal Gold. Principles, Methods, and Applications, Volume 1, Academic Press, San Diego.

69. Hayat, M.A. (ed), 1990, Colloidal Gold. Principles, Methods, and Applications, Volume 2, Academic Press, San Diego.

70. Hayat, M.A. (ed), 1991, Colloidal Gold. Principles, Methods, and Applications, Volume 3, Academic Press, San Diego.

71. Hayat, M.A., and Miller, S.E., 1990, Negative Staining, McGraw Hill, New York.

72. Herrera, G.A., and Lott, R.A., 1991, Colloidal gold labeling for diagnostic pathology, in Hayat, M.A. (ed): Colloidal Gold. Principles, Methods, and Applications, Volume 3, Academic Press, San Diego, pp. 322–345.

73. Himmler, G., Brix, U., Steinkellner, H., Laimer, M., Mattanovich, D., and Katzinger, H.W.D., 1988, Early screening for anti-plum pox virus monoclonal antibodies with different epitope specificities by means of gold-labelled immunosorbent electron microscopy, *J. Virol. Methods* **22**:351–358.

74. Hiruki, C., (ed) 1988, Tree Mycoplasmas and Mycoplasma Diseases, University of Alberta Press, Edmonton.

75. Hiruki, C., 1988, Rapid and specific detection methods for plant mycoplasmas,

in Maramorosch, K., and Raychaudhuri, S.P. (eds): Mycoplasma Diseases of Crops, Springer-Verlag, New York, pp, 77–101.

76. Hsu, H.T., and Lawson, R.H., 1984, Comparative detection of carnation etched ring virus (CERV) using mouse monoclonal antibodies and chicken and rabbit antiserum, *Phytopathology* **74**:861.

77. Hu, J.S., Gonsalves, D., Boscia, D., and Namba, S., 1990, Use of monoclonal antibodies to characterize grapevine leafroll-associated closteroviruses, *Phytopathology* **80**:920–925.

78. Katz, D., and Kohn, A., 1984, Immunosorbent electron microscopy for detection of viruses, *Advances in Virus Research* **29**:169–194.

79. Kellenberger, E., and Hayat, M.A., 1991, Some basic concepts for the choice of methods, in Hayat, M.A. (ed): Colloidal Gold. Principles, Methods, and Applications, Volume 3, Academic Press, San Diego, pp. 1–30.

80. Kerlan, C., Mille, B., and Dunez, J., 1981, Immunosorbent electron microscopy for detecting apple chlorotic leaf spot and plum pox viruses, *Phytopathology* **71**:400–404.

81. Kerlan, C., Mille, B., and Dunez, J., 1983, Immunoelectron microscopy for virus detection and virus strain identification, *Acta Hort.* **130**:173–178.

82. Kirkpatrick, B.C., 1989, Strategies for characterizing plant pathogenic mycoplasma-like organisms and their effects on plants, in Kosuge, T., and Nester, E.W. (eds): Plant-Microbe Interactions. Molecular and Genetic Perspectives, Volume 3, McGraw Hill, New York, pp. 241–293.

83. Kjeldsberg, E., 1989, Immunogold labeling of viruses in suspension, in Hayat, M.A. (ed): Colloidal Gold. Principles, Methods, and Applications, Volume 1, Academic Press, San Diego, pp. 433–449.

84. Koenig, R., 1988, Serology and Immunochemistry, in Milne, R.G. (ed): The Filamentous Plant Viruses, Plenum Press, New York, pp. 111–158.

85. Kojima, M., Chou, T., and Shikata, E., 1978, Rapid diagnosis of potato leafroll virus by immune electron microscopy, *Ann. Phytopathol. Soc. Jpn.* **44**:585–590.

86. Lafferty, K.J., and Oertelis, S., 1963, The interaction between virus and antibody III. Examination of virus-antibody complexes with the electron microscope. *Virology* **21**:91–99.

87. Langenberg, W.G., 1974, Leaf-dip serology for the determination of strain relationships of elongated plant viruses, *Phytopathology* **64**:128–131.

88. Langenberg, W.G., 1985, Immunoelectron microscopy of wheat spindle streak and soil-borne wheat mosaic virus doubly infected wheat, *J. Ultrastructure Res.* **92**:72–79.

89. Langenberg, W.G., 1986, Deterioration of several rod-shaped wheat viruses following antibody decoration, *Phytopathology* **76**:339–341.

90. Langenberg, W.G., 1986, Virus protein association with cylindrical inclusions of two viruses that infect wheat, *J. Gen. Virol.* **67**:1161–1168.

91. Langenberg, W.G., 1986, Soil-borne wheat mosaic virus protein interactions with wheat streak mosaic virus cylindrical inclusions, *J. Ult. Mol. Struct. Res.* **94**:161–169.

92. Langenberg, W.G., 1988, Barley stripe mosaic virus but not brome mosaic virus binds to wheat streak mosaic virus cylindrical inclusions in vivo, *Phytopathology* **78**:589–594.

93. Langenberg, W.G., 1989, Rapid antigenic modification of wheat streak mosaic

virus in vitro is prevented in glutaraldehyde-fixed tissue, *J. Gen. Virol.* **70**:969–973.

94. Langenberg, W.G., 1991, Cylindrical inclusion bodies of wheat streak mosaic virus and three other potyviruses only self-assemble in mixed infections, *J. Gen. Virol.* **72**:493–497.

95. Lesemann, D.E., 1983, Advances in virus identification using immunosorbent electron microscopy, *Acta Hort.* **127**:159–173.

96. Lesemann, D.E., and Paul, H.L., 1980, Conditions for the use of protein A in combination with the Derrick method of immuno electron microscopy, *Acta Hort.* **110**:119–127.

97. Lesemann, D.E., Bozarth, R.F., and Koenig, R., 1980, The trapping of tymovirus particles on electron microscope grids by adsorption and serological binding, *J. Gen. Virol.* **48**:257–264.

98. Lesemann, D.E., Koenig, R., Torrance, L., Buxton, G., Boonekamp, P.M., Peters, D., and Schots, A., 1990, Electron microscopical demonstration of different binding sites for monoclonal antibodies on particles of beet necrotic yellow vein virus, *J. Gen. Virol.* **71**:731–733.

99. Lherminier, J., Boudon-Padieu, E., Meignoz, R., Caudwell, A., and Milne, R.G., 1991, Immunological detection and localization of mycoplasma-like organisms (MLOs) in plants and insects by light and electron microscopy, in Mendgen, K., and Lesemann, D.E. (eds): Electron Microscopy of Plant Pathogens, Springer-Verlag, Berlin, pp. 177–184.

100. Lherminier, J., Preniser, G., Boudon-Padieu, E., and Caudwell, A., 1990, Immunolabeling of grapevine flavescence dorée MLO in salivary glands of *Euscelidius variegatus*: a light and electron microscopy study, *J. Histochem. Cytochem.* **38**:79–85.

101. Lin, N.S., 1984, Gold-IgG complexes improve the detection and identification of viruses in leaf dip preparations, *J. Virol. Methods* **8**:181–190.

102. Lin, N.S., and Langenberg, W.G., 1983, Immunohistochemical localization of barley stripe mosaic virions in infected wheat cells, *J. Ultrastructure Res.* **84**:16.

103. Lin, N.S., Hsu, Y.H., and Chiu, R.J., 1987, Identification of viral structural proteins in the nucleoplasm of potato yellow dwarf virus-infected cells, *J. Gen. Virol.* **68**:2723–2728.

104. Lin, N.S., Wang, N., and Hsu, Y.H., 1988, Sequential appearance of capsid protein and cylindrical inclusion protein in root-tip cells following infection with passion-fruit woodiness virus, *J. Ult. Mol. Struct. Res.* **100**:201–211.

105. Linstead, P.J., Hills, G.J., Plaskitt, K.A., Wilson, I.G., Harker, C.L., and Maule, A.J., 1988, The subcellular location of the gene 1 product of cauliflower mosaic virus is consistent with a function associated with virus spread, *J. Gen. Virol.* **69**:1809–1818.

106. Louro, D., and Lesemann, D.E., 1984, Use of protein A-gold complex for specific labelling of antibodies bound to plant viruses. I. Viral antigens in suspensions, *J. Virol. Methods* **9**:107–122.

107. Luisoni, E., Milne, R.G., and Boccardo, G., 1975, The maize rough dwarf virion II. Serological analysis, *Virology* **68**:86–96.

108. Luisoni, E., Milne, R.G., and Roggero, P., 1982, Diagnosis of rice ragged stunt virus by enzyme-linked immunosorbent assay and immunosorbent electron microscopy, *Plant Disease* **66**:929–932.

109. Makkouk, K.M., Koenig, R., and Lesemann, D.E., 1981, Characterization of a tombusvirus isolated from eggplant, *Phytopathology* **71**:572–577.

110. Maramorosch, K., and Raychaudhuri, S.P. (eds), 1988, Mycoplasma Diseases of Crops, Springer-Verlag, Berlin.

111. Markham, P.G., 1988, Detection of mycoplasmas and spiroplasmas in insects, in Hiruki, C. (ed): Tree Mycoplasmas and Mycoplasma Diseases, University of Alberta Press, Edmonton, pp. 157–177.

112. Milne, R.G., 1980, Some observations and experiments on immunosorbent electron microscopy of plant viruses, *Acta Hort* **110**:129–135.

113. Milne, R.G., 1984, Electron microscopy for the identification of plant viruses in in vitro preparations, *Methods in Virology* **7**:87–120.

114. Milne, R.G., 1986, New developments in electron microscope serology and their possible applications, in Jones, R.A.C., and Torrance, L. (eds): Developments and Applications of Virus Testing, Association of Applied Biologists, Warwick, pp. 179–191.

115. Milne, R.G., 1992, Immunoelectron microscopy for virus identification, in Mendgen, K., and Lesemann, D.E. (eds): Electron Microscopy of Plant Pathogens, Springer Verlag, Berlin, pp. 87–120.

116. Milne, R.G., 1992, Solid-phase immune electron microscopy of virus preparations, in Hyatt, A.D., and Eaton, B.T. (eds): Immune Electron Microscopy for Virus Diagnosis, CRC Press, Boca Raton, ch. 2.

117. Milne, R.G., and Lesemann, D.E., 1978, An immunoelectron microscopic investigation of oat sterile dwarf and related viruses, *Virology* **90**:299–304.

118. Milne, R.G., and Lesemann, D.E., 1984, Immunosorbent electron microscopy in plant virus studies, *Methods in Virology* **8**:85–101.

119. Milne, R.G., and Luisoni, E., 1975, Rapid high-resolution immune electron microscopy of plant viruses, *Virology* **68**:270–274.

120. Milne, R.G., and Luisoni, E., 1977, Rapid immune electron microscopy of virus preparations, *Methods in Virology* **6**:265–281.

121. Mowry, T.M., Klomparens-Baker, K., and Whalon, M.E., 1985, Effects of glutaraldehyde fixation on antigenicity and surface labelling properties of *Spiroplasma citri*, *Phytopathology* **75**:1350 (Abstr.).

122. Murphy, J.F., Järlfors, U., and Shaw, J.G., 1991, Development of cylindrical inclusions in potyvirus-infected protoplasts, *Phytopathology* **81**:371–374.

123. Natsuaki, T., Natsuaki, K.T., Okuda, S., Teranaka, M., Milne, R.G., Boccardo, G., and Luisoni, E., 1986, Relationships between the cryptic and temperate viruses of alfalfa, beet and white clover, *Intervirology* **25**:69–75.

124. Nicolaieff, A., and van Regenmortel, M.H.V., 1980, Specificity of trapping of plant viruses on antibody-coated electron microscope grids, *Ann. Virol. (Inst. Pasteur)* **131E**:95–110.

125. Nicolaieff, A., Katz, D., and van Regenmortel, M.H.V., 1982, Comparison of two methods of virus detection by immunosorbent electron microscopy (ISEM) using protein A, *J. Virol. Methods* **4**:155–166.

126. Noel, M.C., Kerlan, C., Garnier, M., and Dunez, J., 1978, Possible use of immune electron microscopy (IEM) for the detection of plum pox virus in fruit trees, *Ann. Phytopathol.* **10**:381–386.

127. Paliwal, Y.C., 1977, Rapid diagnosis of barley yellow dwarf virus in plants using serologically specific electron microscopy, *Phytopathol. Z.* **89**:25–36.

128. Paliwal, Y.C., 1982, Detection of barley yellow dwarf virus in aphids by serologically specific electron microscopy, *Can. J. Bot.* **60**:179–185.

129. Paliwal, Y.C., 1987, Immunoelectron microscopy of plant viruses and mycoplasmas, in Harris, K.F. (ed): Current Topics in Vector Research, Volume 3, Springer-Verlag, New York, pp. 217–249.

130. Pares, R.D., and Whitecross, M.I., 1982, The detection and identification of plant viruses by immuno-electron microscopy (IEM), *Micron* **13**:305–306.

131. Pares, R.D., and Whitecross, M.I., 1982, Gold-labelled antibody decoration (GLAD) in the diagnosis of plant viruses by immuno-electron microscopy, *J. Immunol. Methods* **51**:23–28.

132. Pares, R.D., and Whitecross, M.I., 1983, A critical examination of the utilization of serum-coated grids to increase particle numbers for length determination of rod-shaped plant viruses, *J. Virol. Methods* **7**:241–250.

133. Pares, R.D., and Whitecross, M.I., 1985, An evaluation of some factors important for maximizing sensitivity of plant virus detection by immuno-electron microscopy, *J. Virol. Methods* **11**:339–346.

134. Pesic, Z., Hiruki, C., and Chen, M.H., 1988, Detection of viral antigen by immunogold cytochemistry in ovules, pollen, and anthers of alfalfa infected with alfalfa mosaic virus, *Phytopathology* **78**:1027–1032.

135. Plumb, R.T., 1990, Detecting plant viruses in their vectors, in Harris, K.F. (ed):, Advances in Disease Vector Research, Volume 6, Springer-Verlag, New York, pp. 191–208.

136. Randles, Harrison, B.D., and Roberts, I.M., 1976, *Nicotiana velutina* mosaic virus: purification, properties, and affinities with other rod-shaped viruses, *Ann. Appl. Biol.* **84**:193–204.

137. Richman, D.D., Cleveland, P.H., Oxman, M.N., and Johnson, K.M., 1982, The binding of staphylococcal protein A by the sera of different animal species, *J. Immunology* **128**:2300–2305.

138. Roberts, I.M., 1986, Immunoelectron microscopy of extracts of virus-infected plants, in Harris, J.R., and Horne, R.W., (eds): Electron Microscopy of Proteins, Volume 5. Viral Structure, Academic Press, London, pp. 293–357.

139. Roberts, I.M., 1986, Practical aspects of handling, preparing and staining samples containing plant virus particles for electron microscopy, in Jones, R.A.C., and Torrance, L. (eds): Developments and Applications in Virus Testing, Association of Applied Biologists, Warwick, pp. 213–243.

140. Roberts, I.M., 1989, Indian cassava mosaic virus: ultrastructure of infected cells, *J. Gen. Virol.* **70**:2729–2739.

141. Roberts, I.M., and Brown, D.J.F., 1980, Detection of six nepoviruses in their nematode vectors by immunosorbent electron microscopy, *Ann. Appl. Biol.* **96**:187–192.

142. Roberts, I.M., and Harrison, B.D., 1979, Detection of potato leafroll and potato mop-top viruses by immunosorbent electron microscopy, *Ann. Appl. Biol.* **93**:289–297.

143. Roberts, I.M., Robinson, D.J., and Harrison, B.D., 1984, Serological relationships and genome homologies among geminiviruses, *J. Gene. Virol.* **65**:1723–1730.

144. Roberts, I.M., Tamada, T., and Harrison, B.D., 1980, Relationship of potato leafroll virus to luteoviruses: evidence from electron microscope serological tests, *J. Gen Virol.* **47**:209–213.

145. Roenhorst, J.W., van Lent, J.W.M., and Verduin, B.J.M., 1988, Binding of cowpea chlorotic mottle virus to cowpea chloroplasts and relation of binding to virus entry and infection, *Virology* **164**:91–98.

146. Russo, M., Martelli, G.P., and Savino, V., 1982, Immunosorbent electron microscopy for detecting sap-transmissible viruses of grapevine, in McGinnis, A.J. (ed): *Proc 7th Meeting Intern. Council for the Study of Viruses and Virus-like Diseases of the Grapevine*, Niagara Falls, Canada, 1980, pp. 251–257.

147. Schramm, G., and Friedrich-Freksa, N., 1941, Präcipitinreaktion des Tabak-mosaikvirus mit Kaninchen und Schweinantiserum, *Z. Physiol. Chem.* **270**:233–246.

148. Shalla, T.A., and Amici, A., 1967, The distribution of viral antigen in cells infected with tobacco mosaic virus as revealed by electron microscopy, *Virology* **31**:78–91.

149. Shukla, D.D., and Gough, K.H., 1979, The use of protein A, from *Staphylococcus aureus*, in immune electron microscopy for detecting plant virus particles, *J. Gen. Virol.*, **45**:533–536.

150. Shukla, D.D., and Gough, K.H., 1984, Serological relationships among four Australian strains of sugarcane mosaic virus as determined by immune electron microscopy, *Plant Disease* **68**:204–206.

151. Sinha, R.C., 1984, Transmission mechanisms of mycoplasmalike organisms by leafhopper vectors, in Harris, K.F. (ed): Current Topics in Vector Research, Volume 2, Praeger Scientific, New York, pp. 93–109.

152. Sinha, R.C., 1988, Serological detection of mycoplasmalike organisms from plants affected with yellows diseases, in Hiruki, C. (ed): Tree Mycoplasmas and Mycoplasma Diseases, University of Alberta Press, Edmonton, pp. 143–156.

153. Sinha, R.C., and Benhamou, N., 1983, Detection of mycoplasmalike organism antigens from aster yellows-diseased plants by two serological procedures, *Phytopathology* **73**:1199–1202.

154. Sinha, R.C., and Chiykowski, L.N., 1984, Purification and serological detection of mycoplasmalike organisms from plants affected by peach eastern X-disease, *Can. J. Plant Pathol.* **6**:200–205.

155. Steinbrecht, R.A., and Zierold, K., 1987, (eds): Cryotechniques in Biological Electron Microscopy, Berlin, Springer-Verlag.

156. Stierhof, Y.D., Schwarz, H., Dürrenberger, M., villiger, W., and Kellenberger, E., 1991, Yield of immunolabel compared to resin sections and thawed cryosections, in Hayat, M.A. (ed), Colloidal Gold. Principles, Methods, and Applications, Volume 3, Academic Press, San Diego, pp. 87–115.

157. Stirling, J.W., 1990, Immuno- and affinity probes for electron microscopy: a review of labeling and preparation techniques, *J. Histochem. Cytochem.* **38**:145–157.

158. Stussi-Garaud, C., Garaud, J.C., Berna, A., and Godefroy-Colburn, T., 1987, In situ localization of an alfalfa mosaic virus non-structural protein in plant cell walls: correlation with virus transport, *J. Gen. Virol.* **68**:1779–1784.

159. Timmons, T.S., and Dunbar, B.S., 1990, Protein blotting and immunodetection, *Methods in Enzymology* **182**:679–688.

160. Tomenius, K., Clapham, D., and Oxelfelt, P., 1983, Localization by immunogold cytochemistry of viral antigen in sections of plant cells infected with red clover mottle virus, *J. Gen. Virol.* **64**:2669–2678.

161. Tomenius, K., Clapham, D., and Meshi, T., 1987, Localization by immunogold

cytochemistry of the virus-encoded 30K protein in plasmodesmata of leaves infected with tobacco mosaic virus, *Virology* **160**:363–371.

162. Tosi, M., and Anderson, D.L., 1973, Antigenic properties of bacteriophage ⌀29 structural proteins, *J. Virol.* **12**:1548–1559.

163. Urban, L.A., Huang, P.Y., and Moyer, J.W., 1991, Cytoplasmic inclusions in cells infected with isolates of L and I serogroups of tomato spotted wilt virus, *Phytopathology* **81**:525–529.

164. van Balen, E., 1982, The effect of pretreatments of carbon-coated Formvar films on the trapping of potato leafroll virus particles using immunosorbent electron microscopy, *Neth. J. Plant Pathol.* **88**:33–37.

165. van Lent, J.W.M., and Verduin, B.J.M., 1985, Specific gold labelling of antibodies bound to plant viruses in mixed suspensions, *Neth. J. Plant Pathol.* **91**:205–213.

166. van Lent, J.W.M., and Verduin, B.J.M., 1991, Immunolabeling of viral antigens in infected cells, in Mendgen, K., and Lesemann, D.E. (eds): Electron Microscopy of Plant Pathogens, Springer-Verlag, Berlin, pp. 119–131.

167. van Regenmortel, M.H.V., Nicolaieff, A., and Burckard, J., 1980, Detection of a wide spectrum of virus strains by indirect ELISA and serological trapping electron microscopy (STREM), *Acta Hort.* **110**:107–115.

168. Vetten, H.J., Lesemann, D.E., and Dalchow, J., 1987, Electron microscopical and serological detection of virus-like particles associated with lettuce big vein disease, *J. Phytopathol.* **120**:53–59.

169. Vigil, E.L., and Hawes, C. (eds), 1989, Cytochemical and Immunological Approaches to Plant Cell Biology, Academic Press, London.

170. von Ardenne, M., Friedrich-Freksa, H., and Schramm, G., 1941, Elektronenmikroskopische untersuchung der Präcipitinreaktion von Tabakmosaikvirus mit Kaninchenantiserum, *Arch. gesamte Virusforsch.* **2**:80–86.

171. Waterhouse, P.M., and Murant, A.F., 1981, Purification of carrot red leaf virus and evidence from four serological tests for its relationship to luteoviruses, *Ann. Appl. Biol.* **97**:191–204.

172. Whitcomb, R.F., and Tully, J.G. (eds), 1989, The Mycoplasmas V. Spiroplasmas, Acholeplasmas, and Mycoplasmas of Plants and Arthropods, Academic Press, San Diego.

173. Wolansky, B., and Maramorosch, K., 1970, Negatively stained mycoplasmas: fact or artifact? *Virology* **42**:319–327.

174. Yanagida, M., and Ahmad-Zadeh, C., 1970, Determination of gene product position in bacteriophage T4 by specific antibody association, *J. Mol. Biol.* **51**:411–421.

175. Zimmermann, D., Sommermeyer, G., Walter, B., and van Regenmortel, M.H.V., 1990, Production and characterization of monoclonal antibodies specific to closterovirus-like particles associated with grapevine leafroll disease, *J. Phytopathol.* **130**:277–280.

10
Aphid-borne Rhabdoviruses—
Relationships with Their Vectors

Edward S. Sylvester and Jean Richardson

Introduction

The vector-dependent, aphid-borne, rhabdoviruses for purposes of this review are considered as a subset of the larger group of rhabdoviruses that
infect plants. Qualitatively, the uniqueness of the subset lies in the host range
of the viruses, i.e., dicots and aphids, and includes viruses that mature by
budding from the inner nuclear membrane as well as those that bud off
cytoplasmic membrane systems.

Reports indicate that at least 11 plant rhabdoviruses are transmitted by
aphids. These are broccoli necrotic yellows virus (BNYV), carrot latent virus
(CLV), coriander feathery red vein virus (CFRVV), lettuce necrotic yellows
virus (LNYV), lucerne enation virus (LEV), parsley rhabdovirus (PRV), raspberry vein chlorosis virus (RVCV), sonchus yellow net virus (SYNV), sowthistle yellow vein virus (SYVV), strawberry crinkle virus (SCV), and strawberry
latent C virus (SLCV). The diseases caused by the viruses (from which the
names of the viruses are derived), as well as their reported vectors' are given
(in chronological order of description) in Table 10.1. Information on one
additional virus, nasturtium veination virus (NVV) (Figure 10.1), has not been
reported previously.

The Viruses

Rhabdoviruses have been reviewed extensively (7, 17, 18, 35, 44). A recent
useful review limited to the plant rhabdoviruses, many of which are transmitted by arthropods, is that of Francki et al. (18). The majority of the vectors

Edward S. Sylvester, Department of Entomological Sciences, University of California, Berkeley,
California 94720, USA.
Jean Richardson, Department of Entomological Sciences, University of California, Berkeley,
California 94720, USA.

TABLE 10.1 Aphid-borne rhabdoviruses.

Disease	Date described	Virus acronym	Vector	References
Strawberry crinkle	1932	SCV	*Chaetosiphon fragaefolii*	71, 73
			C. jacobi	20
			Macrosiphum euphorbiae[a]	66
			Myzus ornatus[a, b]	62
Strawberry latent C[c]	1951	SLCV	*C. fragaefolii*	13, 72
			C. minor (?)	13
			C. thomasi (?)	36
Raspberry vein chlorosis	1952	RVCV	*Aphis idaei*	9
Lettuce necrotic yellows	1963	LNYV	*Hyperomyzus lactucae*	54
			H. carduellinus	48
Sowthistle yellow vein	1963	SYVV	*H. lactucae*	14
			M. euphorbiae[b]	2
Broccoli necrotic yellows	1968	BNYV	*Brevicoryne brassicae*	26, 68
Lucerne enation	1972	LEV	*Aphis craccivora*	1, 33
Parsley rhabdovirus[d]	1974	PRV	*Cavariella aegopodii*	35, 67
Sonchus yellow net	1974	SYNV	*Aphis coreopsis*	11
Carrot latent	1978	CLV	*Semiaphis heraclei*	41
Coriander feathery red-vein[e]	1979	CFRVV	*Hyadaphis foeniculi*	37, 38
			Macrosiphum euphorbiae[a]	
			Myzus persicae[b]	37, 38
Nasturtium veination[f]	1991	NVV	*Aphis fabae*	
			Myzus persicae	

[a] There is no evidence of acquisition by fed vectors. The insects were infected by injection. In the case of CFRVV, the aphids were injected with an extract from a diseased *Nicotiana benthamiana*, transmitted virus to coriander seedlings, and rhabdoviruslike particles were found by TEM in a negatively stained preparation from a macerated head of an infected aphid (unpublished).
[b] An inefficient vector, rarely transmits.
[c] Strawberry latent C virus is considered to be synonymous with Type 2 virus of Demaree and Marcus (13) who reported its transmission by field collections of *C. fragaefolii*, *C. minor* and *C. thomasi*. Rorie (50) failed to obtain transmission with *C. minor*, and the identification of *C. thomasi* as being the "unknown species" of Demaree and Marcus (13) is doubtful (see: Blackman, et al. (5)). Frazier failed to transmit isolates of the virus with *C. jacobi* (36).
[d] Originally called Parsley latent virus, but this name previously has been used for a small polyhedral virus (see Francki, et al. (18)). The listing of *Cavariella aegopodii* as a vector is based upon a personal communication (see Martelli and Russo (35)).
[e] Perhaps first described as celery yellow spot in 1932 (38).
[f] Not previously described. In 1986, we found rhabdoviruslike particles by TEM examination of PTA negatively-stained leaf dip preparations, and in thin-sections of leaf tissue (Fig. 10.1) from older specimens of cultivated nasturtium, *Tropaeolum major*. Symptoms consisted of an intensification of the veinlet network by a narrow veinlet-banding. Transmission was obtained by both a black aphid (*Aphis fabae* complex) commonly found on nasturtium and *Myzus persicae*. Under greenhouse conditions, however, infections were primarily latent (symptomless). No juice inoculation was obtained, and due to the lack of a suitable symptomatic host plant, work was discontinued.

are leafhoppers or planthoppers and aphids, but one is transmitted by a heteropteran, and another by a mite.

The aphid-borne rhabdoviruses, like the other plant-infecting rhabdoviruses, have large bacilliform (100–430 × 45–100 nm) or bullet-shaped virions with a central axial channel. The inner helical nucleocapsid complex consists of a nucleoprotein (a negative sense ssRNA + N protein) and the associated L and

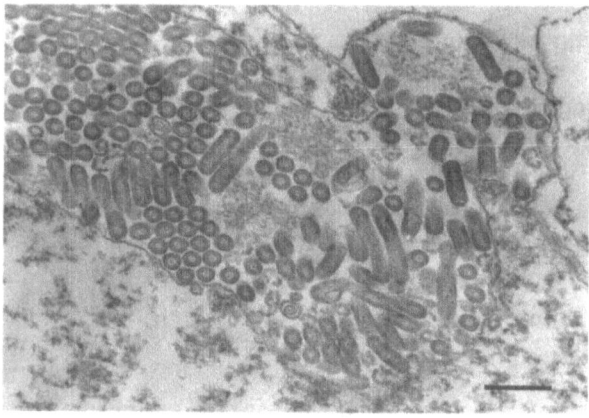

FIGURE 10.1. Thin-section of a leaf cell from a nasturtium (*Tropaeolum major*) plant showing symptoms of veination disease. The rhabdoviruslike particles are associated with a perinuclear space. Scale bar = 250 mμ.

NS proteins that are responsible for transcriptase activity. The nucleocapsid is contained in a protein matrix composed of one or two M proteins. Finally, during maturation, the entire particle becomes enclosed in a host-derived membrane into which the virus-derived projections of a G protein are inserted. As a staining artifact, unfixed PTA negatively-stained preparations most frequently yield virions that appear as bullet-shaped particles.

Five of 12 viruses, viz., BNYV (10), LNYV (19), SCV (28), SYNV (29), and SYVV (55) have been partially characterized physicochemically. The first three of these have been assigned by the International Committee on Taxonomy of Viruses (7) to subset A of the plant rhabdovirus group of the family Rhabdoviridae. The other two belong to subset B.

Members of subset A have a single matrix (M) protein. The in-vitro transcriptase activity is readily detectable. The virions mature by budding off cytoplasmic membranes, and frequently accumulate in cytoplasmic vesicles. Viruses of subset B have two M proteins (M_1 and M_2) and mature by budding off inner nuclear membranes. Mature particles are found predominantly in perinuclear cisternae. Based upon electron microscopy, RVCV (30, 31) would appear to be in subset A, while CFRVV (37, 38), CLV (41), and LEV (1) would be in subset B. Relevant information on NVV and PRV is lacking.

Aphids as Alternate Hosts for Rhabdoviruses

Whether the rhabdoviruses infecting plants are insect viruses that have become adapted to multiply in plants is a question that perhaps can be addressed by future molecular genetic studies. Up to the present the molecular virology that has been done on the aphid-borne rhabdoviruses is almost exclusively with virus from infected plants, and undoubtedly the viruses by convention will continue to remain in domain of "plant viruses."

Seed transmission does not occur, but transovarial passage in the aphid hosts does, and infections are frequently more symptomatically dramatic in plant hosts than in aphid vectors. These differences might be used to argue that pathogen/animal host adaptation is of a higher order, and therefore has evolutionary precedence over the pathogen/plant adaptation.

The case for replication of the aphid-borne rhabdoviruses in aphid vectors originally was made by Duffus (14). He found that SYVV had a long, temperature-sensitive, latent period in its vector, *Hyperomyzus lactucae* (L). Furthermore, in sequences of serial transfers of the insects, the distribution of the incubation period of the disease in the test plants had a skewed, U-shaped distribution. This was indicative that postacquisition variations in the inoculum dose were being provided by the aphids. Frazier (20) emphasized that a similar pattern of variation in the length of the plant incubation relationship existed in the transmission of SCV by *Chaetosiphon jacobi* (HLR.). An exceptionally long latent period of SCV, in another aphid species, *C. fragaefolii* (Cock.), had been reported and discussed previously (46).

Ultrastructure Studies of Infected Aphids

More evidence for rhabdovirus multiplication in aphids came from thin sectioning and electron microscopy. Uncoated particles of LNYV were found in the cytoplasm of a variety of tissue systems of infective *H. lactucae* (42) including muscle, alimentary, salivary, fat body, tracheal, and nervous. Since

TABLE 10.2. Infection of aphids with rhabdoviruses as evidenced by transmission electron microscopy.

Virus	Aphid host	Tissue Systems[a]											Reference
		A	S	T	M	My	F	N	R	H	B	E	
BNYV	*Brevicoryne brassicae*	+		+	+		+						23
LNYV	*Hyperomyzus lactucae*	+	+	+	+	+	+	+				+	42
RVCV	*Aphis idaei*	+	+		+			+					39
SYVV	*H. lactucae*	+	+		+	+	+	+	+				43, 59
	Macrosiphum euphorbiae	+		+	+?		+?						2
	Myzus persicae	+	+	+	+		+	+	+	+	+		
	Chaetosiphon jacobi		+	+	+							+	61
SCV	*C. jacobi*	+	+	+	+	+	+	+		+	+	+	49, 61
	H. lactucae	+	+	+	+			+		+	+	+	61
	M. ornatus	+	+	+	+			+		+	+	+	62
CFRVV	*Hyadaphis foeniculi*	+	+	+	+	+	+	+		+	+	+	

[a] The initials A, S, T, M, My, F, N, R, H, B, and E, indicate alimentary, salivary gland, tracheal, muscle, mycetome, fat body, nervous, reproductive, heart, blood cells, and epidermal, respectively.

then, the systemic infection of aphid vectors of several other plant-infecting rhabdoviruses has been reported (Table 10.2), including BNYV in *Brevicoryne brassicae* (L.) (23), RVCV in *Aphis idaei* (V.d.G) (39), SYVV in *H. lactucae* and *Macrosiphum euphorbiae* (Thom.) (2, 43, 59), and SCV in *C. jacobi* and *Myzus ornatus* Laing (49, 61, 62).

Examples of cellular infection of *M. persicae* (Sulz.) with SYVV and of *Hyadaphis foeniculi* (Pass.) with CFRVV are given in Figures 10.2 and 10.3. Although SYNV was reported as occurring in infected tissues of *Aphis coreopsis* Thom. (11) no photographs were published.

Only rudimentary data have been published concerning the transmission parameters found in aphid transmission of most of the viruses listed in Table

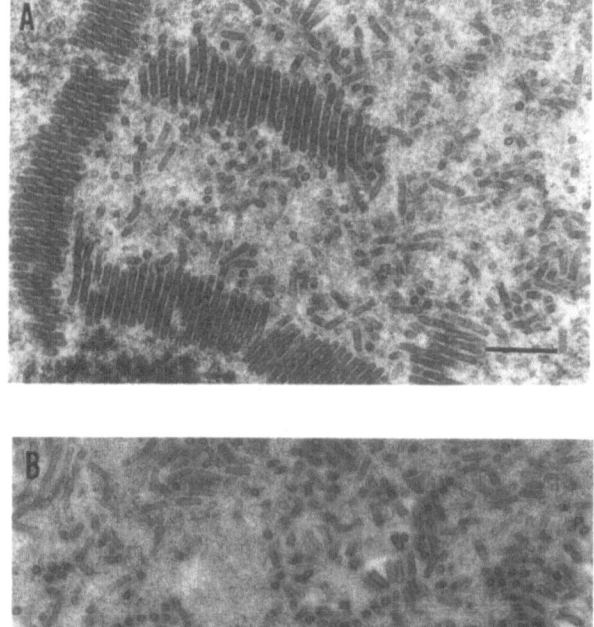

FIGURE 10.2. Thin-sections of tissues of *Myzus persicae* infected, by injection, with sowthistle yellow vein virus. (A) Nucleus of a salivary gland cell containing arrays of uncoated virus particles. Such arrays are also visible in sections made for light microscopy. (B) Nucleus of a cell from the ventriculus containing uncoated particles. Scale bars = 500 mμ.

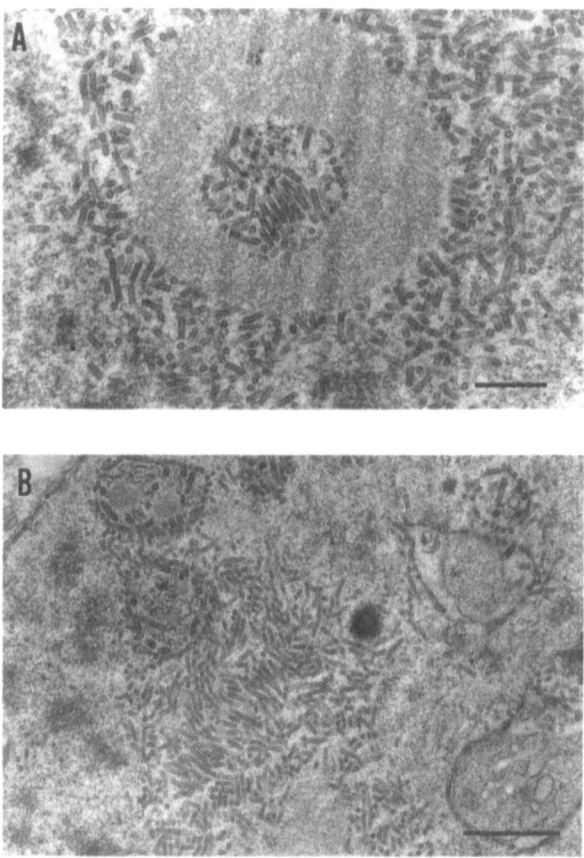

FIGURE 10.3. Infection of *Hyadaphis foeniculi* with coriander feathery red vein virus. The insects acquired virus by feeding on *Coriandrum sativum*. (A) Virions in the nucleus of a salivary gland cell (scale bar = 500 mμ) and (B) that of a cell from the ventriculus (scale bar = 1 μM).

10.1. Other than the fact that the virus has been transmitted by aphids, little has been reported on the vector relationships found with BNYV, SYNV, LEV, and PRV, and only fragmentary information is available on the aphid transmission of CLV and RVCV.

Transmission Parameters

Acquisition

The probability of obtaining an infective aphid increases with the length of time the insect feeds on a virus source plant. Most workers have used rela-

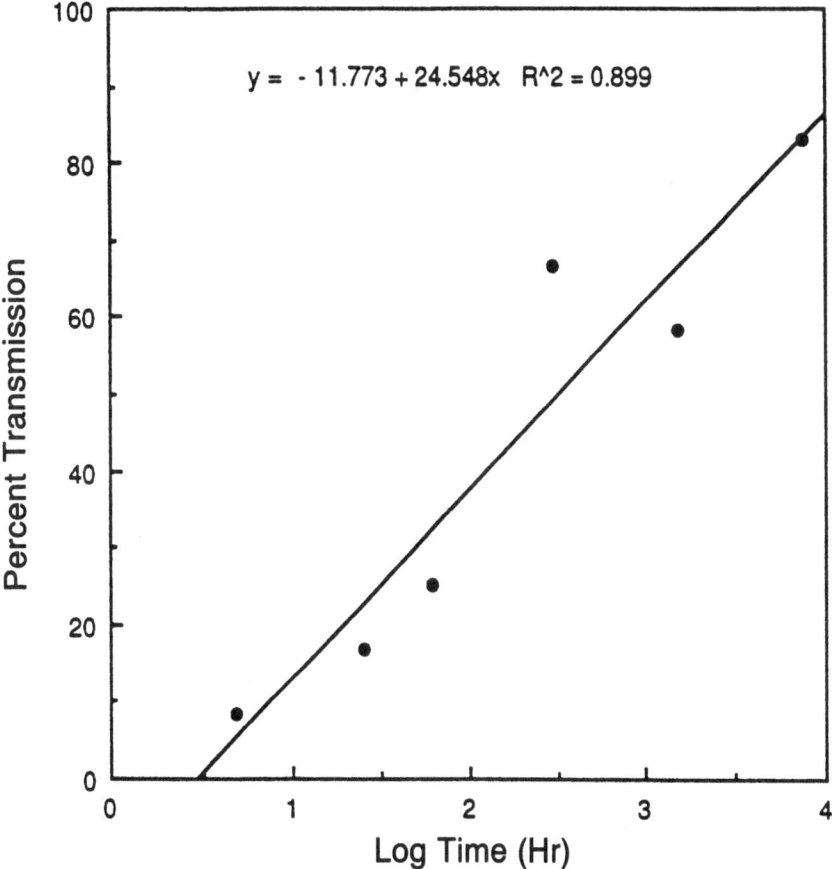

FIGURE 10.4. Efficiency of acquisition of sowthistle yellow vein from *Sonchus olera-ceous* fed upon by *Hyperomyzus lactucae*. Data from Duffus (14).

tively long acquisition access periods, 24 hrs or more, apparently in an attempt to insure adequate numbers of inoculative insects. SYVV can be acquired by *Hyperomyzus lactucae*, and CLV by *Semiaphis heraclei* (Taka.), within 2 hrs (14, 41). Although *Hyadaphis foeniculi* can acquire CFRVV by using an access period of 15 mins, the median acquisition access period appears to be approximately 4 hrs at either 20° or 25°C (37, 38), and that for *Hyperomyzus lactucae* in the acquisition of SYVV is approximately 12 hrs (Figure 10.4).

Data on the other viruses are minimal. LNYV can be acquired within 24 hrs (6). *A. idaei* (V.d.G) is said to need to feed for at least a day to acquire RVCV. The probability of obtaining an infective insect could be increased by using a 7-day acquisition access period, after which a maximum of about 46% of the tested insects transmitted (30). *C. fragaefolii* can acquire SCV within a 24-hr acquisition access period, but acquisition appears to be inefficient (46). Only a maximum of approximately 10% of *C. jacobi* aphids, given 14 days of access to SCV-infected plants, acquired virus (20). The only data on SLCV are those of Smith (53) where *C. fragaefolii* became infective after a 6-day, but not a 3-hr, acquisition access period.

Aphids are primarily phloem feeders, and thus virus acquisition presumably is from that tissue, but direct evidence is lacking. The inefficient acquisition of viruses such as RVCV and SCV may be because only low levels of virus circulate in the phloem.

Inoculation

There are few data on the inoculation threshold periods of the aphid-borne rhabdoviruses. Inoculative *Hyadaphis foeniculi* can transmit CFRVV in access periods of 15 mins and 1 hr at 25° and 20°C, respectively. Approximately

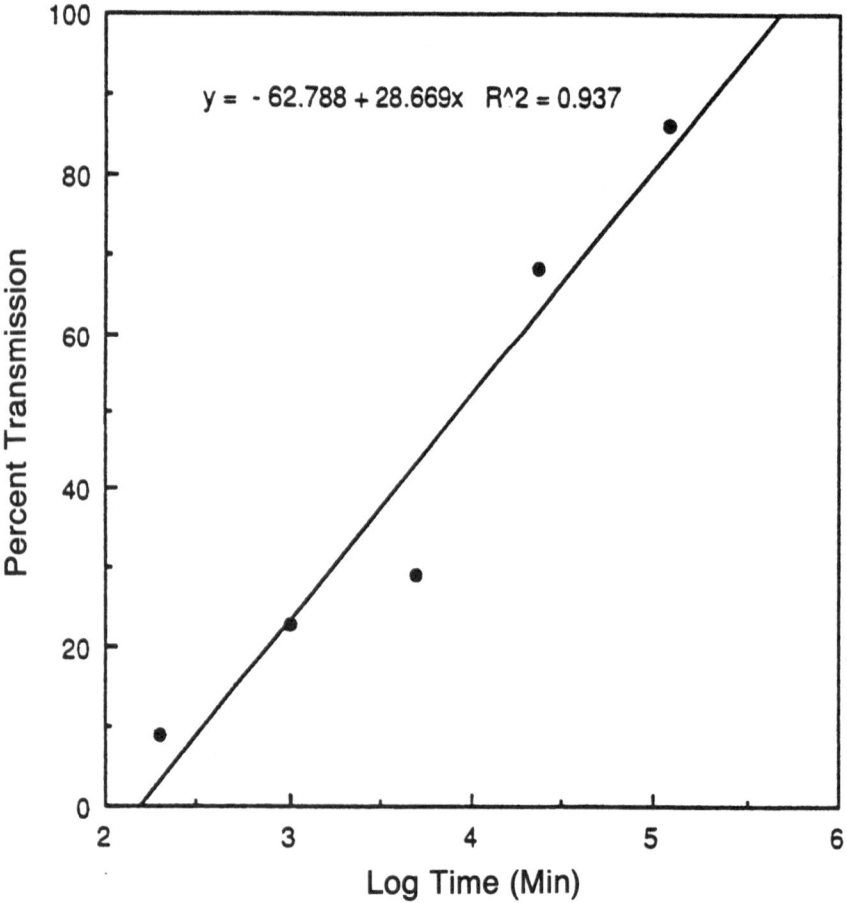

$$y = -62.788 + 28.669x \quad R^2 = 0.937$$

FIGURE 10.5. Rate of inoculation of strawberry crinkle virus by *Chaetosiphon fragaefolii* fed on *Fragaria vesca* (Alpine) seedlings. The aphids were infected by injection and the tests were done at 27°C and constant light.

50% of the inoculative test insects transmitted within 5.6 hrs (37, 38). The inoculation threshold period of LNYV by *Hyperomyzus lactucae* is reported to vary with the host plant, i.e., 1–5 mins for lettuce and 5–30 mins for *Sonchus*. In this work, the time commenced with placement of the rostrum in the feeding position (6). The brief inoculation threshold period found with lettuce indicates that penetration to vascular tissue is not required for the inoculation of LNYV. Recently we found that, at 27°C and constant light, *C. fragaefolii* could inoculate *Fragaria vesca* (Alpine) seedlings within a 10-min inoculation access period. The median inoculation access period was estimated to be about 50 mins. (Figure 10.5).

Latent Period

Aphid-borne plant rhabdoviruses characteristically have long latent periods in their vectors. Lettuce necrotic yellows virus has a temperature-sensitive latent period in *H. lactucae*, averaging 18, 9.2 and 5.4 days at 15°, 20°, and 28°C, respectively (6). Duffus (14) reported the minimum latent period of SYVV in *H. lactucae* to be independent of the duration of the acquisition access period, and the average length to be negatively correlated with temperature. His data showed the average 11-day latent period at 25°C, about doubled for each 10°C decrease in temperature. Thus the average latent period was about 21 and 40 days at 15° and 5°C, respectively. The latent period of SYVV in *H. lactucae* at 25°C was found to be dosage sensitive in injected aphids (65). The median latent period of 7 days increased by approximately 24 hrs for each 10-fold dilution of an inoculum dose that was estimated to contain about 10^5 virions.

The dosage effect on the latent period is dramatic with SCV, a case where acquisition by feeding aphids is quite inefficient. The literature reports the latent period of SCV, under variable temperature conditions, to range from 10–19 days in *C. fragaefolii* (46), and 14–59 days (20) in *C. jacobi*. However, using injected aphids, the median latent period at 25°C was estimated to be about 6.5 days in both species (64). The length of the latent period of SCV was found to be similar in the surrogate vector, *Macrosiphum euphorbiae* (66).

Estimates of the latent period can vary with the test conditions. Transmission of SCV had a higher probability of occurring during a sequence of interrupted feedings (daily transfers) than during a continuous inoculation feeding period of similar total length (24).

Latent periods reported for other aphid-borne rhabdoviruses in fed vectors include approximately 8 days for SLCV (53), 9–12 days for CLV in *S. heraclei* (41), and a median period of 8–10 days at 25°C for CFRVV in fed *Hyadaphis foeniculi*, and 7.2 days in injected adults (37, 38). There is some evidence that the duration of the latent period of CFRVV in *H. foeniculi* decreases as the length of the acquisition access period is increased, and when the virus is injected, the latent period may be shorter in larvae than in adults (37).

Retention of Inoculativity

Aphids, infected with rhabdoviruses, can retain the capacity to inoculate plants for a considerable period of time. SLCV is retained by *C. fragaefolii* for at least 9 days (53), and CLV by *S. heraclei* for at least 12 days (41). More generally, the statement is found in the literature, that "the virus is retained for life," with the caveat that loss of inoculativity may occur shortly before death (6, 14, 20). The implication is that, upon completion of the latent period, the age-specific transmission rate remains constant. The limited data of Duffus (14) on SYVV and *Hyperomyzus lactucae*, and those of Boakye and Randles (6) on LNYV by the same species, are convincing in support of this idea. Their data also show that both retention and vector longevity are temperature sensitive. Thus once the latent period is completed, the transmission curve primarily reflects that of survival.

However, other data indicate that the age-specific transmission rate declines in aging aphids. Such insects presumably are undergoing physiological senescence and decreased feeding competence. In work on SYVV at 25°C and using fed *H. lactucae*, the age-specific transmission rate was 2% on day 9, rose to a maxmimum of 93% on day 17, and then fell to 0% on day 27. At the time transmission ceased only 8% of the aphids were still living. The transmission expectancy curve, which is weighted for mortality, declined from a peak of 86% on day 15, to 51% and then to 3% on days 19 and 26, respectively (56).

Although Frazier (20) concluded that inoculative *C. jacobi* retained SCV essentially for life, his data show that as the insects aged, there was a continuous decline in transmission efficiency. In other work at 25°C, using aphids injected with SCV, both *C. jacobi* and *C. fragaefolii* had a distinct rise and fall in their patterns of transmission. Infected aphids lived for an average of about 24 ± 5 days after being injected. Following completion of a median latent period of approximately 5 days, *C. jacobi* and *C. fragaefolii* transmitted virus for an average of 5 and 8 days, respectively. The ratio of the retention period to longevity was approximately 50% for both species; thus both species were transmitting for only about half their post-injection survival period (64).

The short transmission period found in our experience with SCV may be an artifact resulting from the use of injection to obtain infective insects. We may have selected an isolate of virus with increased pathogenicity that effectively shortens the time that the salivary system will functionally replicate and release virus. We need an effective assay system that will quantitatively track virus within the vectors over time.

Transovarial Passage

Although seed transmission of plant rhabdoviruses has not been reported, the documented transovarial passage of some aphid-borne plant rhabdoviruses indicates that both vertical and horizontal transmission are processes in the endemic maintenance of these viruses. The reported rates of effective transovarial passage are low. An estimated 1% of the F_1 offspring of *H.*

lactucae apterae infected with either SYVV or LNYV are inoculative at birth, or shortly thereafter. The latent period apparently is completed during larval development within the maternal apterae. A much lower rate of passage to the F_2 generation has been reported (6, 58). In limited trials, the rate of transovarial passage of SYVV could not be increased by selection of subclones established from infected larvae (58).

A higher rate of transovarial passage (5–8%) was reported for *Hyadaphis foeniculi* infected with CFRVV (37, 38), but a much lower rate (0.03%) was found with SCV-infected *C. fragaefolii* inoculated by injection (64). Passage of virus via the true egg apparently has not been investigated with any species, and the rate of vertical transmission is probably too low to provide for virus maintenance in the absence of horizontal transmission among the plant hosts (16).

The impact of the infection of embryos on their survival has received little attention. Özel (43) found virions of SYVV in thin sections of embryos of infected *H. lactucae* and other work indicated that the probability of survival was reduced in *H. lactucae* larval cohorts containing individuals infected transovarially with SYVV (56). We found virions in PTA-stained preparations of aborted embryos of *C. jacobi* infected with SCV (64), but there is no information on whether there is a correlation between the incidence of abortion and maternal infection with the virus.

Virus Assay

Several qualitative methods can be used to detect plant rhabdoviruses in aphids, namely, transmission electron microscopy, injection, and singular instances of tissue culture, serology, and the use of cloned plasmid probes. All have their usefulness as well as their limitations. All methods have a probability of detection of less than one, and quantitative extensions of any of these methods are still crude.

Transmission

This is the fundamental method of establishing a relationship between a plant rhabdovirus and a vector. It is sensitive, informative, and essential for any epidemiological projection. In final analysis, all the other methods of detection may demonstrate the presence of virus, but that presence will have to be translated to a transmission expectancy, before it can be extrapolated to the field. Transmission studies are basic to the laboratory evaluation of such parameters as acquisition and inoculation threshold periods, latent periods, retention, transovarial passage, vector specificity, efficiency, and laboratory competency. However refined our experiments become, it is still a formidable undertaking to transfer such results into realistic probabilistic models of vector potential and performance in nature. There is much work to be done.

Transmission Electron Microscopy

Because of the unique shape, structure, and size of the rhabdoviruses, the electron microscope can be used not only for histological studies of ultra-structure, but also as an instrument for rapid detection of virions in simply prepared, negatively-stained crude extracts from both plant (27) and insect tissue (60). In our work, the inoculum used for injection was routinely checked for the presence of virions. In one series of experiments, we found that 90% of the injected and tested *H. lactucae* transmitted SYVV, and virions were found by EM in 70% of the crude extracts prepared from the same individuals. The corresponding figures for *C. jacobi* injected with SCV were 99% and 26%, respectively. In both cases transmission was superior, but neither method detected all the infected insects (61). Other data indicate that pre-latent-

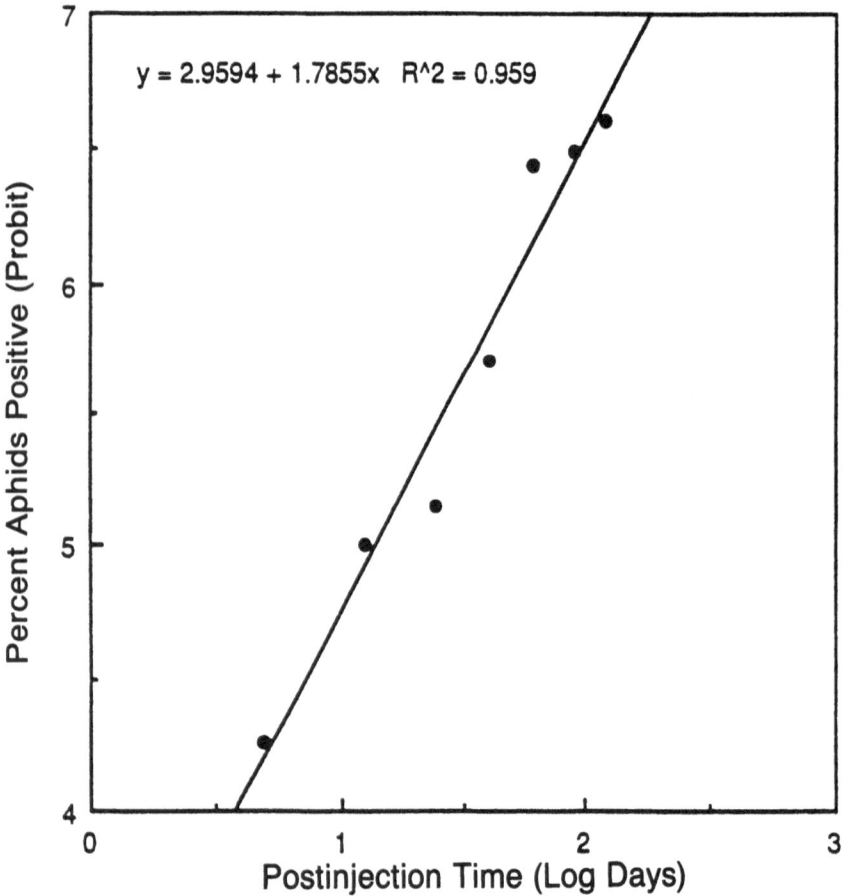

FIGURE 10.6. Efficiency of finding sowthistle yellow vein virions by TEM examination of PTA negatively-stained preparations from macerated heads of *Hyperomyzus lactucae* that had been injected with virus. Successive grids, up to a maximum of five, were examined until a virion was found.

period infections were more readily detected in SYVV by the electron microscopy assay (Figure 10.6), but as the insects aged, the procedure dropped in
efficiency. Thus there is a stage in the infection cycle where the probability of
detection by either method is maximized.

Injection

Injection is a useful mechanical method of inoculating aphids with rhabdoviruses, and as such it is a useful technique in vector transmission studies. Not
only is it efficient and precise as to the time of inoculation, but it can be used
effectively in serial passage and variable dosage experiments (60, 62, 65).
However, when used as an exclusive method of transfer, it can lead to the
selection of biotypes of the virus that are unique. Continuous serial passage of
SYVV in *H. lactucae* and in the inefficient vector *Macrosiphum euphorbiae*
gave rise to isolates that apparently had a very low probability of being
transmitted to plants (2, 60). In the case of the selection of the SYVV isolate in
H. lactucae there was also an increase pathogenicity to the vector as evidenced by a comparative reduction in longevity (60). The rapidity with which
biotypes can be selected by the continuous imposition of experimental proto-
cols, aimed at increasing the rigor of hypothesis testing, is impressive.

Tissue Culture

Continuous cell lines of aphid vectors have not been established. However,
Peters and Black (45) demonstrated that primary cultures of *H. lactucae* cells
could be infected with SYVV. Plaque formation did not occur, but infection
and viral increase were detected using fluorescent antibodies.

Serology

Although the aphid-borne rhabdoviruses are immunogenic, serology rarely
has been used to detect virus in aphids. Chu and Francki (12) used a double
antibody sandwich enzyme-linked immunosorbent assay (ELISA) to detect
LNYV in individual *H. lactucae*. The aphids were first tested for transmission
and then by ELISA. All transmitting aphids were ELISA-positive, and a
majority, but not all, of the individuals that were ELISA-positive, also transmitted. The test conditions were designed to eliminate any plant-derived
residual virus in the gut.

Cloned Plasmid Probes

Stenger et al. (55) recently demonstrated that a cDNA probe and dot hybridization technique could be used to detect viral specific RNA sequences of
SYVV in individual *H. lactucae*. All 25 aphids that had fed for 6 days on an
infected source plant and then were given three consecutive 48-hr inoculation

access periods on test plants tested positive when probed. Only 14 transmitted virus, but the latent period may not have been completed in all the individuals within the 12-day period of the experiment. For possible field survey work, it is important to know whether either the ELISA, or the cDNA probe, would detect nonreplicated virus that may accumulate in the gut during prolonged access periods on diseased plants.

Vector Range

The natural vector range of the individual aphid-borne plant rhabdoviruses, at least in the literature, is characterized by a high level of vector specificity. More than half of the viruses have only one reported vector. Presumptive evidence of a natural vector is acquisition and transmission by feeding aphids. Even in the other cases the number of vectors is quite limited. LNYV is transmitted by two species of *Hyperomyzus* (47), SCV by two species in the genus *Chaetosiphon* (20, 46), while CFRVV can be efficiently transmitted by *Hyadaphis foeniculi*, and inefficiently by *M. persicae* (37, 38). SYVV is efficiently transmitted by *Hyperomyzus lactucae* (14) and inefficiently by *M. euphorbiae* (2). However in each of these instances, based on current knowledge, only one species of aphid appears to be of major epidemiological significance. The data regarding NVV are too meager to conclude whether or not *A. fabae* (L.) and *Myzus persicae* are equivalent natural vectors.

The aphid host/vector range can be extended if the gut barrier is avoided. Thus, when inoculum is injected, containing either SYVV or SCV, either alone or in combination, both viruses, SYVV and SCV, will infect their reciprocal vectors, *H. lactucae* or *C. jacobi* (61). SYVV also has been shown to be capable of infecting *M. persicae* and SCV can infect both *M. ornatus* Laing and *Macrosiphum euphorbiae* (62, 66). Aphids infected by injection may or may not transmit the virus to plants. Behncken (2) found that even when infected by injection, *M. euphorbiae* remains an inefficient vector of SYVV, and although virus was found by TEM in various tissues of injected insects, it was not found in the salivary glands (2). However, this species is an efficient vector of SCV when injected. *Myzus ornatus*, injected with SCV, will replicate the virus, but rarely does it transmit, and although arrays of SYVV virions will accumulate in the salivary glands of injected *M. persicae* (Figure 10.2A), transmissions are rare.

Detailed studies are lacking, but the various results could be explained by transmission being moderated by barriers to virus replication and movement both in the gut and in the salivary glands.

Only *C. fragaefolii* is associated with the field spread of SCV in commercial strawberries. However, it may be that *C. jacobi* is a vector of consequence in the endemic maintenance of the virus among the native wild *Fragaria vesca* strawberries in the coastal forest areas where both the aphid and susceptible plant hosts are found.

Speculating further, a better host/pathogen adaptation may exist between SCV and *C. jacobi* than between SCV and *C. fragaefolii*. We did not detect any significant impact on the number of larvae produced by apterous *C. jacobi* infected with SCV. However, larviposition was reduced in infected *C. fragaefolii* (64).

Plant Host Range

Most of the aphid-borne rhabdoviruses appear to have a limited endemic host range among dicotyledonous plants. However, the host range can be considerably expanded through the use of mechanical transmission. Solanaceous hosts frequently prove to be susceptible, and apparently are preferred hosts from which the viruses are purified for molecular studies, e.g., LNYV from *Nicotiana glutinosa* (19) or SYNV from *N. clevelandii* (29). To date there have been no reports of extending the host range of SYVV beyond *Lactuca* species, but this may be because of the lack of a concerted effort. SCV was once considered to be restricted to *Fragaria* species, and not be juice inoculable (57). Then van de Meer (70) transmitted SCV both by aphids and juice inoculation to *N. occidentalis* and we infected both *Nicotiana* and *Physalis* species using injected *Macrosiphum euphorbiae* as a surrogate vector (63, 66). Once species of these plant genera were infected, SCV could be moved by juice inoculation. We also have obtained direct infection of *N. glutinosa* by juice inoculation from infected *Fragaria vesca* (unpublished).

Continuous passage of SCV by juice inoculation to *P. floridana* permitted the separation of at least two isolates that differed in the symptoms produced in plants and in their aphid transmissibility. One isolate (Clb) was similar, in its production of a relatively mild mottling and infrequent production of necrotic lesions, to the symptom type originally obtained by aphid transmission to *P. floridana*. The Clb isolate was maintained in further passages by continuously selecting source plants on the basis of symptom type. The other isolate (ANL) was selected, again on the basis of its symptomatology, after some 22 passages of SCV in *P. floridana* over a period of a year. Symptoms of infection by the ANL isolate were characterized by the production of primary local lesions on the inoculated leaves and a systemic necrosis, in addition to mottling, in the acute stages of the disease. Both isolates were maintained by source plant selection for another two years.

In 1989, aphid-transmission trials, using injected *C. fragaefolii*, were made comparing isolates Clb and ANL, as well as the isolate of SCV (Cf) that we had been using in our aphid transmission studies over the past several years. The Cf isolate was derived from the C10 isolate of Frazier (20) and had been kept, and periodically renewed, in injected *C. jacobi* and *C. fragaefolii* aphids frozen at −67°C.

The results of the trials are given in Table 10.3. The aphids injected with ANL died prematurely and no transmissions were obtained. Virions were

TABLE 10.3. Transmission of isolates of strawberry crinkle virus by injected *Chaetosiphon fragaefolii* aphids.

Trial[a]	Isolate[b]	N	Transmission (%)[c]	Postinjection longevity (days)[d]
I	CIb	50	133/483 (27.5)	23.1 ± 7.4
	ANL	57	0/166 (0.0)	9.6 ± 2.6
II	Cf	31	238/527 (45.2)	22.3 ± 5.9
	CIb	24	169/471 (35.9)	24.8 ± 4.7
	ANL	28	0/88 (0.0)	7.7 ± 2.3
III	Cf	59	309/517 (59.8)	22.4 ± 6.4
	CIb	56	197/468 (42.1)	21.4 ± 7.0
	ANL	66	0/186 (0.0)	10.3 ± 1.9

[a] In Trial I and III, the transfers to fresh Alpine *Fragaria vesca* test plants were made at 2-day intervals. In Trial II, following the first 4 transfers, the insects were moved daily.
[b] Isolate Cf was derived from Clone C10 of Frazier (20). Stock of the virus was kept in frozen (−67°C) injected aphids. Isolates CIb and ANL were derived from isolate Cf by selecting symptomatic variants during repeated passage of the virus by juice inoculation in *Physalis floridana*. ANL was a mottling isolate that produced both local and systemic necrotic lesions. CIb essentially was a nonnecrotic isolate that induced mottling symptoms.
[c] The test insects were fed in lots of 5 aphids per plant for three transfers, and then set up as individuals. The transmission data were from individual insects.
[d] Mean ± standard deviation.

found in a subsample of aphids that died within seven days after injection, and the short period of average longevity, would indicate that the ANL isolate was unusually pathogenic to the injected insects.

The CIb isolate was transmissible, though somewhat less efficiently than the Cf isolate. The aphids injected with either the CIb or the Cf isolate had a life span of more than twice that of the aphids inoculated with the ANL isolate.

Black (3, 4) found that vectorless strains of viruses could be selected if passage among plants, without the intervention of a vector cycle, was prolonged. Similarly, continuous passage in the insect host, without intervention of a plant cycle, can result in the selection of isolates that also have a reduced capacity to be transmissible by insects. However, in this latter case, it could be that the isolate no longer will replicate in plant tissue (2, 60).

Impact on Vectors

The statement, made in the case of the plant rhabdoviruses, that "there is no evidence that the vector insects suffer any disease" (Francki, et al., 18), is not correct. But the disease may be clinically inapparent. *H. lactucae*, infected with SYVV using the injection technique, exhibited reduced longevity under laboratory conditions (60), and there is some evidence that the length of life of larvae is shorter when they are infected by the transovarial route (58). Boakye and Randles (6) reported that *H. lactucae* aphids that transmitted LNYV were shorter lived than those that did not transmit. When injected *C. jacobi*, *C.*

fragaefolii, or *M. euphorbiae* were infected with SCV, longevity was reduced, and in the case of *C. fragaefolii*, fewer larvae per apterous female were borne when compared to virus-free inoculated control insects (64, 66).

However, in limited tests, using *Hyadaphis foeniculi* infected with CFRVV, there was no evidence that either longevity or reproduction were affected (37, 38). It may be that the deleterious effects are less apparent when the viruses are acquired by feeding, as compared to injection. Fed insects presumably ingest virus irregularly over time, the dosage can be variable, a gut barrier may exist, and the latent period is extended since a replication cycle in the gut presumably is needed to initiate infection. Injection avoids these restrictions.

CO_2 Sensitivity

Carbon dioxide has a variety of effects on insects (40), but one of particular interest is viral-induced CO_2 sensitivity. The classic example of the phenomenon is that found in *Drosophila melanogaster* infected with Sigma virus, an inherited rhabdovirus found in wild populations of the fruitfly (34). The mode of action of CO_2 is not explicitly known, but primary lethal effects are thought to be associated with the nervous system (8). Although the story is complex, in simplest terms, normal flies recover after being anaesthetized with CO_2 while Sigma-infected flies do not. Injection of *Drosophila* with certain serotypes of the vesicular stomatitis rhabdovirus (VSV) can induce a permanent, or temporary, sensitivity to CO_2 (8). The phenomenon is not restricted to infection with rhabdoviruses, for the Iota picornavirus has been shown to induce a CO_2 sensitivity in male *Drosophila* (32).

Sensitivity to CO_2 also has been found in virus-infected mosquitoes. Rosen and Shroyer (52) reported a naturally occurring virus-induced sensitivity in *Culex quinquefasciatus* but exposure to the gas had to be at 1°C. Laboratory studies (51) have indicated that intrathoracic injections of various VSV serotypes can induce a sensitivity in mosquitoes, but to detect the sensitivity it was important to expose the insects to the gas for 15 minutes at 13 to 15°C. Turell and Hardy (69) reported that a bunyavirus, California encephalitis, induced a temporary sensitivity to CO_2 in mosquitoes that had been infected using intrathoracic injection. In this case the exposure time to the gas was short (20 secs) and at room temperature.

Recently we have gathered experimental evidence that infection of aphids with rhabdoviruses can result in a sensitivity to CO_2. Serendipity played a major role in our becoming aware of this sensitivity. By way of background, when injected, *Chaetosiphon fragaefolii* and *C. jacobi* apterae cease to transmit SCV before mortality becomes the limiting factor. Our latest data indicated that at 25°C, transmission maximized about 7 days after injection and essentially ceased some 10 days later, even though approximately 80% of the vectors were still alive and reproducing (64). The reasons for loss of the ability to transmit are not known.

TABLE 10.4. Survival of, and transmission by, *Chaetosiphon fragaefolii* apterae injected once or twice (at 10-day intervals) with strawberry crinkle virus[a].

Treatment	Transmitting insects[b]	N	Posttreatment longevity (days, mean ± SD)	No. of deaths within 96 hrs after the time of the 2nd injection
Not injected	0/24	21	35 ± 10	1
Injected once:				
Initially	17/17	17	27 ± 7	0
10 days later	24/24	18	34 ± 5	1
Injected twice	18/18	17	13 ± 5	16[c]

[a] The test insects were serially transferred at 48-hr intervals to fresh Alpine *Fragaria vesca* test plants at 20°C and constant light.
[b] In ratios listed, the numerator is the number of insects transmitting; the denominator is the number injected. The difference between the value of the denominator and the N of the longevity category represents the number of aphids that were lost or missing before their time of death could be recorded.
[c] 16 out of the 17 aphids (94%) were dead within 72 hrs of the 2nd injection. One insect transmitted before the 2nd injection, and survived for 22 days after this injection. It inoculated the first two of its 11 postinjection test plants.

We attempted to determine if a second injection of virus would prevent the loss of inoculativity. Transmission by *C. fragaefolii* apterae injected twice with SCV, within a 10-day interval, was compared to that by aphids injected once. The two injections of SCV severely reduced the longevity of most of the insects (Table 10.4). It was not evident whether the high rate of mortality following the second injection was because the aphids had been *injected twice*, or simply because *infected* aphids were injected. However, it did not seem feasible to make further attempts to reestablish inoculativity by successive injections.

But the cause of the mortality was worth exploring. This we did in a series of experiments, the results of which are given in Table 10.5. In these experiments, with one exception, the aphids were transferred, until they died, to fresh Alpine *Fragaria vesca* test plant seedlings, at 2-day intervals under conditions of 20°C and constant light. In Experiment I, daily transfers were used.

The first experiment tested the consequences of two injections, with and without SCV. The shortened life (Table 10.5 Experiment I) was associated only with *two injections of SCV*. Injection alone was not associated with excessive early mortality. This conclusion was substantiated by a further experiment (Table 10.5, Experiment II) in which all insects were injected twice (within an 8-day interval), using reciprocal combinations of water and SCV.

Clearly the early mortality was associated with anaesthetization and injection when used in combination with an established infection of SCV, as evidenced by transmission. An additional experiment (Table 10.5, Experiment III) provided evidence that the early mortality of the infected insects was associated with the anaesthetization with CO_2, not the act of injection. Also, the early mortality was not associated with the fasting period that was routinely

TABLE 10.5. Survival of, and transmission of strawberry crinkle virus by, *Chaetosiphon fragaefolii* aphids injected once or twice at 7- or 8-day intervals with various combinations of treatments[a].

Treatment	Transmitting[b] insects	N	Posttreatment longevity (days, mean ± SD)	No. of deaths within 96 hrs after the time of the 2nd injection
EXPERIMENT I				
Injected once with:				
Water	0/26	25	41 ± 10	0
SCV	24/24	23	28 ± 7	0
Injected twice (7 days apart) with:				
Water then SCV	25/25	25	32 ± 6	0
SCV then SCV	20/26[c]	26	11 ± 3[d]	23
EXPERIMENT II				
Injected with (8 days apart):				
Water then water	0/43	41	42 ± 8	0
Water then SCV	39/42	36	36 ± 5	0
SCV then water	33/39	38	10 ± 13	36
SCV then SCV	39/42	39	10 ± 1[e]	37
EXPERIMENT III				
Injected with SCV, then 8 days later:				
Fasted 7 hrs	31/31	31	26 ± 6	1
Anaesthetized	32/34	32	14 ± 4	22
Injected with water	31/31	31	13 ± 2	23

[a] The insects were kept at 20°C and constant light and moved, at 24- (Experiment I) or 48-hr intervals, to fresh Alpine *Fragaria vesca* test plants.
[b] In ratios listed, the numerator is the number of insects transmitting; the denominator, the number injected. The difference between the value of the denominator and of N in the longevity column is due to aphids being lost or missing before their death could be recorded.
[c] The latent period may not have been complete in all the insects at the time of the 2nd injection (7 days after the 1st injection).
[d] One insect survived for 9, and one for 14 days, after the 2nd injection. Both transmitted before the 2nd injection, and they transmitted virus to the 5th and 8th of their postinjection test plants, respectively.
[e] One infective insect lived for 25 days after being injected the second time with SCV. However, it did not transmit until 6 days after the 2nd injection; presumably the 1st injection failed to inoculate the insect. Datum from this insect was omitted from the longevity calculations.

used to ready the insects for injection. In totality, we feel that there is strong experimental evidence that SCV induces a sensitivity to CO_2 in *C. fragaefolii*.

The results of others (34, 52) have indicated that the sensitivity is specific to CO_2, and the effect presumably is due to its action as an anaesthesia, and not anoxia. We compared the effect of N_2 versus CO_2 anaesthetization using *C. fragaefolii* apterae injected with SCV. Briefly, 13- to 14-day-old adults were injected, caged on Alpine seedlings, and transferred at 48-hr intervals under conditions of constant light and a min/max temperature averaging $19.2 \pm 0.5/21.9 \pm 1.0$ with a mean differential of $2.75 \pm 1.0°C$. Six days following injection, test insects were exposed, in groups of 10, for 2 mins to the test gas using a Buchner funnel through which a stream of pure, moistened treatment gas was flowing. The lots, and the order of treatment, were chosen

TABLE 10.6. Effect of carbon dioxide and nitrogen anaesthesia on the survival of *Chaetosiphon fragaefolii* aphids infected with strawberry crinkle virus.

Treatment[a]	Aphids transmitting[b]	N	Posttreatment longevity (days, mean $\pm$ SD)[c]
None	1/45[d]	45	42.0 $\pm$ 8.6a
Injected, then 6 days later:			
Untreated	50/50	48	29.6 $\pm$ 6.6b
Anaesthetized with:			
CO_2	29/50	44	12.8 $\pm$ 4.7c
N_2	48/50	46	28.5 $\pm$ 9.0b

[a] The insects were in lots of 5 per Alpine *Fragaria vesca* test plant for the first three transfers. Treatment was on day 6 (postinjection). The insects, after being treated, were individually tested using four successive transfers. The surviving insects were then recombined into their original lots, and each lot was transferred to fresh test plants until all aphids had died. All events, i.e., transmission or death were assumed to have taken place at the mid-point of the 48-hr test interval. Any difference between the number of aphids transmitting and N was due to the failure to find all aphids over the entire sequence of transfers.

[b] In the ratios listed, the numerator is the number of insects transmitting; the denominator, the number of insects tested.

[c] Means followed by the same letter were not separable using ANOVA and Duncan's multiple range test.

[d] In the untreated control (no injection, no anaesthesia) one insect died before the first three transfers were completed. It was inadvertently replaced with an insect from a pool of injected spare aphids. Thus transmission occurred in one of the control lots. The incubation period of the disease in the test plants precluded discovery of the error until after the insects had been recombined into a pool of five. It could not be readily determined which one of the 5 individuals was the infected one and therefore the mortality data on all insects in the lot were not used. The mean longevity with those 5 insects included was 41.0 $\pm$ 9.7 days.

at random. In total, samples of 50 insects per treatment were tested singly for transmission on day 6 (the day of treatment), 8, 10, and 12. The results of the experiments are in Table 10.6. An analysis of variance gave statistical support to the conclusion that injection and/or infection reduced longevity, and that infection with SCV induces sensitivity to CO_2, but not to N_2, when used as an anaesthesia. The virus-induced CO_2 sensitivity can be a transient condition and infected insects gradually lose their sensitivity (69). We checked this as follows. Groups of insects were injected and sequentially transferred to test plants at 48-hr intervals at 20°C and constant light. In the first trial, 20 days after injection, a random subsample of surviving aphids was given a 5 min exposure to CO_2. A second random subsample was anaesthetized 33 days after injection. In the second trial, the insects were anaesthetized 16 or 27 days after injection. The results are in Table 10.7. Aphids were sensitive for at least 20 days after injection; at 27 and 33 days post-injection, the short life expectancy made evidence of any sensitivity statistically insignificant.

A check was made upon the sensitivity to CO_2 of *Hyperomyzus lactucae* infected with SYVV. Adult aphids that had fed on SYVV-infected *Sonchus oleraceus* for up to 13 days were collected and divided into two groups. Twenty-four were anaesthetized using a 5-min exposure to pure CO_2, and 25 were not exposed. Following treatment each of the aphids was transferred to a fresh test plant and thereafter to fresh test plants at 2-day intervals until all

TABLE 10.7. Effect on survival of *Chaetosiphon fragaefolii* aphids exposed for 2 minutes to CO_2 at various intervals after being injected with strawberry crinkle virus[a].

Treatment	N	Posttreatment	
		Transmission rate[b]	Longevity[c]
Trial I			
20 days after injection:			
None	57	22/415 (5.3)	13.5 ± 6.6
CO_2 anaesthetization	98	6/216 (2.8)	4.2 ± 2.2
33 days after injection:			
None	32	4/118 (3.4)	6.3 ± 4.2
CO_2 anaesthetization	32	2/92 (2.1)	4.5 ± 2.8
Trial II			
16 days after injection:			
None	53	39/453 (8.6)	14.8 ± 6.0
CO_2 anaesthetization	96	4/236 (1.6)	2.9 ± 0.9
27 days after injection:			
None	40	2/199 (1.0)	8.2 ± 3.9
CO_2 anaesthetization	35	9/166 (5.4)	6.8 ± 2.6

[a] The test aphids were transferred to fresh Alpine *Fragaria vesca* test plants at 48-hr intervals at 20°C and constant light. All insects transmitted.
[b] In the ratios listed, the numerator is the number of plants infected; the denominator, the number tested. The per cent transmission is given in the parentheses.
[c] Days, mean ± SD.

had died. Seventeen of the anaesthetized aphids and 25 of the nonanaesthetized aphids transmitted virus. The anaesthetized aphids lived for an average of 4.4 ± 2.4 days, while the nonanaesthetized controls lived for an average of 7.9 ± 3.2 days ($t = 4.31$, with 47 df, $p = <0.01$). The results suggest that SYVV induces a sensitivity of CO_2 in *H. lactucae*.

A similar test was made with adult *Hyadaphis foeniculi* that had fed for approximately 2 weeks on coriander infected with CFRVV. Aphids were collected from the source plant, and divided at random into two groups of 25 individuals each. One group was then exposed to CO_2 for 6 mins, the other was not exposed. Each aphid was then caged on a healthy coriander seedling, and subsequently transferred, at 2-day intervals until it died, to a fresh test plant. One of the aphids in the anaesthetized group was missing, and one was accidently killed at the time of the first transfer. Eleven of the 23 anaesthetized aphids transmitted virus, compared to 25 out of the 25 nonanaesthetized control group. (Chi square = 9.25, df 1, $p < 0.01$). The anaesthetized aphids lived for an average of 7.0 ± 8.4 days, while the nonanaesthetized controls lived for an average of 18.3 ± 8.1 days ($t = 4.81$, with 45 df, $p = <0.01$). The results indicate that CFRVV also induces a sensitivity to CO_2 in infected *H. foeniculi*. In this trial, 13 out of the 23 aphids died within 42 hrs of being anaesthetized, but six lived for at least 16 days, transmitting virus intermit-

TABLE 10.8. Results of trials in which *Chaetosiphon fragaefolii* aphids were anaesthetized with CO_2 for 5 minutes at various times following injection with strawberry crinkle virus[a].

	Transmission[b]		Posttreatment survival (days, mean $\pm$ SD)	
Treatment	Trial I	Trial II	Trial I	Trial II
Not anaesthetized	39/39	26/26	21.3 ± 7.9	16.2 ± 6.3
Postinjection time of anaesthetization: (days)				
3	25/25	6/30	18.9 ± 7.9	13.7 ± 6.2
5	17/23	16/29	9.1 ± 9.1	4.3 ± 4.2
7	4/25[c]	7/27	1.6 ± 0.8	1.9 ± 1.1
9	22/24	26/30	1.7 ± 0.9	3.2 ± 2.4
11	25/25	24/26	1.5 ± 0.9	1.5 ± 1.6
13	24/25	18/18	1.2 ± 1.0	1.9 ± 1.1

[a] The injected aphids were moved to fresh Alpine *Fragaria vesca* test plants at 48-hr intervals at 20°C and constant light.
[b] In the ratios listed, the numerator is the number of insects transmitting; the denominator, the number injected.
[c] The low rate of transmission associated with this treatment time presumably was because the completion of the latent period and the development of sensitivity to CO_2, while correlated, are not completely coincident.

tently over that period. Thus there is variation in the sensitivity to CO_2 among the aphids that acquired virus by feeding.

Further tests on the CO_2 sensitivity induced by SCV in *C. fragaefolii* indicated that the effect was measurable about 5 days after injection (Table 10.8). Both the rate of transmission and the postinjection survival period were reduced. The effects maximized about 7 days after injection. Delaying anaesthetization with CO_2 until after completion of the latent period permitted the insects to transmit virus to test plants, but had little effect on the relatively brief posttreatment survival period.

The time of exposure to CO_2 was checked. A number of *C. fragaefolii* were injected, caged, in groups of five, on Alpine test plants and kept at 20°C and constant light. Subsequently the insects were moved to 4 sets of fresh test plants at 48-hr intervals. The insects were kept in lots of 5 for the first two transfers, then individually tested. At 9 days, the insects were treated. The treatments consisted of exposing the insects, at room temperature, to a stream of CO_2 for 10, 20, 40, 80, or 160 secs. A nonanaesthetized control also was used. Selection of the treatments was at random, and when all treatments were completed, using lots of 10 insects/treatment, the individual aphids were caged on test plants. The entire procedure was repeated three more times, using lots of 10, 10, and 5 aphids. Subsequently the individual aphids were moved, until all had died, to fresh test plants at 48-hr intervals. Mortality was checked daily. The results are given in Table 10.9. The sensitivity is a function

TABLE 10.9 Results of trials to determine the effect of varying the time of exposure to CO_2 on *Chaetosiphon fragaefolii* aphids injected with strawberry crinkle virus[a].

Duration of exposure to CO_2 (sec)	Infective insects[b]	Posttreatment	
		Transmission rate[c]	Longevity (days, mean ± SD)
0	31/31	3.4 ± 1.4	22.5 ± 9.0
10	34/35	3.4 ± 1.7	14.8 ± 8.3
20	34/35	3.7 ± 2.0	14.7 ± 8.4
40	35/35	1.3 ± 1.7	6.9 ± 5.2
80	32/35	0.03[d]	4.1 ± 1.1
160	33/34	0.03[d]	3.1 ± 1.0

[a] The injected aphids were moved to fresh Alpine *Fragaria vesca* test plants at 48-hr intervals at 20°C and constant light. The treatments were done 9 days after injection.
[b] In the ratio listed, the numerator is the number of insects transmitting; the denominator, the number injected.
[c] The mean ± SD number of plants inoculated per insect following treatment.
[d] Only one transmission occurred.

of dosage. A measurable reduction in transmission, a behavioral effect, occurred with exposures times of 40 or more seconds. But survival was reduced by both the 10 and 20-sec exposure times, and then increasingly so as the exposure time exceeded 20 seconds.

Epidemiology

STRAWBERRY CRINKLE

SCV is one of several viruses that alone, or in combination with other viruses, lead to the relatively rapid decline of the vigor and yield of commercial strawberries (21). It is assumed that, on a world-wide basis, *C. fragaefolii* is the field vector wherever it occurs. Commercial production of strawberries depends upon some type of certification program to insure growers an adequate supply of virus-free planting material derived from vegetatively-propagated, cloned, cultivars. If commercial production follows a scheme of annual planting of certified stock, then there is little evidence that SCV, with its inefficient acquisition, prolonged latent period in the vector, and limited host range, poses an economic threat. Presumably there are two sources of primary spread, viz., diseased planting stock and infected incoming alate aphids. Since the vector breeds on the host plant, secondary spread becomes important in the presence of high vector populations. As the time between the plantings of new stock is extended, the number of plants infected with aphid-borne viruses, including SCV, will accumulate as will the impact upon the sustainability of crop quality and yield.

SCV has been moved experimentally into solanaceous hosts, but we know of no reports of the occurrence of the disease in plants other than strawberry.

LETTUCE NECROTIC YELLOWS

This disease, first thought to be spotted wilt, has had a major impact in certain lettuce growing regions in Australia (54) and New Zealand (22). The identification of *Sonchus oleraceus* as the endemic host of LNYV, and of *Hyperomyzus lactucae* as the major vector (54), led to fruitful epidemiological investigations (47). The virus has a relatively long latent period in the aphid and a similar long incubation period in the plant. As in the case of SCV, both the virus and the vector have limited host ranges, but unlike SCV, lettuce is not a preferred (colonizing) host of the vector; thus secondary spread is considered negligible. However, alates are presumed to transmit the virus as they probe (?) lettuce in their search for additional summer hosts (47). The virus is not seed-borne and is vector-dependent for its maintenance in its annual hosts. Elimination of *Sonchus* from the proximity of newly planted lettuce crops is an effective method of control. The experimental hosts of LNYV in the Compositae, Solanaceae, Chenopodiaceae, and Amaranthaceae presumably play no part in the epidemiology in lettuce. Other plants in the Lactucae that are hosts of the virus and the vector may participate in the endemic cycle (48). These include *Sonchus hydrophilus*, *Embergeria megalocarpa*, and *Reichardia tingitana*. The other vector of LNYV, *H. carduellinus*, is not considered important in the epidemiology (47).

SOWTHISTLE YELLOW VEIN

A situation similar to that of LNYV exists in the case of SYVV, with *H. lactucae* the only vector of presumed consequence. Lettuce has been found naturally infected with the virus. The infection is characterized by vein clearing and yellow vein banding, with truncation of the leaf tips. The nonpreference of *H. lactucae* for immature (crop) lettuce appears to limit the extent of the disease to that resulting from primary spread from *Sonchus* into lettuce. Although field infection can be extensive, usually it is limited to peripheral areas of lettuce fields adjacent to *Sonchus* (15). Secondary spread in lettuce is probably limited to seed crops, where *H. lactucae* can be found infesting the seed stalks.

An interesting situation has occurred in Berkeley. Several years ago SYVV-infected *S. olerceus* plants were common, as were infestations of *H. lactucae*. In our relatively mild Mediterranean climate, *S. oleraceus* plants and their resident aphid populations will persist from one season to the next. In the last decade, another aphid has become noticeable as a species infesting *S. oleraceus*. It is a relatively large, dark brown species, very similar to *Uroleucon* (formerly *Dactynotus*) *sonchi* (L.). *U. sonchi* is more prolific than *H. lactucae*, and is frequently found on the stems, "below and between the flower heads" (25). *H. lactucae* prefers the base of the flower heads, but it will colonize the adjoining stem area. In the late winter and spring, mixed colonies of both species can be found, but *U. sonchi* soon dominates. Our attempts to transmit SYVV with *U. sonchi* failed. With the advent of the dominance of *U. sonchi* as the aphid species infesting the local populations of *S. oleraceus*, the frequency of plants showing symptoms of SYVV has declined to the point where diseased plants are relatively rare. This could be an example of the reduction in the incidence of a virus disease by the competitive displacement of a vector by a nonvector species.

Conclusions

Aphids are responsible for the horizontal transmission of most of the dicotyledonous plant rhabdoviruses for which vectors have been identified. More importantly, aphids can serve as invertebrate hosts of the viruses. Unlike virus infection in the plant hosts, infection in aphids results in a low rate of vertical transmission of the virus. The insect-rhabdovirus relationship may be a more fully evolved one between the host and the parasite, than that found in the plant-rhabdovirus relationship. However, from what we currently know, in the absence of virus maintenance by vegatative propagation, the endemic persistence of these rhabdovirus populations undoubtedly is dependent upon a seasonally repetitive plant-insect transmission cycle.

Clinically, the infection of the aphid host by a rhabdovirus is relatively inapparent. Some studies indicate that both longevity and the net reproductive rate of the host can be reduced, particularly if insects are infected by direct injection of virus. But if the virus is acquired by feeding, we know of no significant evidence that there is a reduction in the intrinsic rate of increase of the aphids. Thus from the standpoint of survival of the population, infection is relatively neutral.

Rhabdovirus infections can induce a carbon dioxide sensitivity in their aphid hosts. This sensitivity, as in the case of Sigma virus of *Drosophila*, may be a consequence of infection of the nervous system. Yet, in the absence of exposure to threshold concentrations of CO_2, there is no evident symptomatology of involvement of the nervous system.

We do not know to what degree the aphid-borne rhabdoviruses are related to the other invertebrate-vectored plant rhabdoviruses, or to the vertebrate-infecting rhabdoviruses, many of which will replicate in invertebrates. Answers to such questions perhaps could be found through integrated studies that cooperatively utilize elements of relevant disciplines that range from the molecular to the organismal. At one level, the invertebrates might be viewed as the key to evolution and diversity of the rhabdoviruses. If so, as the forbidding technological problems currently associated with small plant-feeding invertebrate vectors and their derived tissues are resolved, more attempts will be made to exploit that key.

References

1. Alliot, B, Giannotti, J, and Signoret, P.A., 1972, Mise en évidence de particules bacilliformes de associés à la maladie à énations de la luzerne (*Medicago sativa* L.), *C.R. Acad. Aci.* Serie D, Paris, **274**:1974–1976.

2. Behncken, G.M., 1973, Evidence of multiplication of sowthistle yellow vein virus in an inefficient aphid vector, *Macrosiphum euphorbiae*, *Virology* **53**:405–412.
3. Black, L.M., 1953, Loss of vector transmissibility by viruses normally insect-transmitted, *Phytopathology* **45**:208–216.
4. Black, L.M., 1958, Wound tumor, *Proc. Natl. Acad. Sci.* USA, **44**:364–367.
5. Blackman, R.L., Eastop, V.F., Frazer, B.D., and Raworth, D.A., 1987, The strawberry aphid complex, *Chaetosiphon* (*Pentatrichopus*) spp. (Hemiptera: Aphididae); taxonomic significance of variations in karyotype, chaetotaxy, and morphology, *Bull. Ent. Res.* **77**:201-212.
6. Boakye, D.B., and Randles. J.W., 1974. Epidemiology of lettuce necrotic yellows virus in South Australia. III Virus transmission parameters and vector feeding behaviour on host and non-host plants, *Aust. J. Agric. Res.* **25**:791–802.
7. Brown, F., Bishop, D.H.L., Crick, J., Francki, R.T.B., Holland, J.J., Hull, R., Johnson, K., Martelli, G., Murphy, F.A., Obijeski, J.C., Peters, D., Pringle, C.R., Reichmann, M.E., Schneider, L.G., Shope, R.E., Simpson, D.I.H., Summers, D.F., and Wagner, R.R., 1979, Rhabdoviridae, *Intervirology* **12**:1–7.
8. Brun, G., and Plus, N., 1980, The viruses of *Drosophila*, in Ashburner, M., and Wright, T.R.F. (eds): The Genetics and Biology of Drosophila, Volume 2d, Academic Press, New York, pp. 625–782.
9. Cadman, C.H., 1952, Studies in *Rubus* virus diseases II. Three types of vein chlorosis of raspberries, *Ann. Appl. Biol.* **39**, 61–68.
10. Campbell, R.N., and Lin, M.T., 1972, Broccoli necrotic yellows virus, *CMI/AAB Descriptions of Plant Viruses* **85**.
11. Christie, S.R., Christie, R.G., and Edwardson, J.R., 1974, Transmission of a bacilliform virus of sowthistle and Bidens pilosa, *Phytopathology* **64**:840–845.
12. Chu, P.W.G., and Francki, R.I.B., 1982, Detection of lettuce necrotic yellows virus by an enzyme-linked immunosorbent assay in plant hosts and the insect vector, *Ann. Appl. Biol.* **100**:149–156.
13. Demaree, J.B., and Marcus, C.D, 1951, Virus diseases of strawberries in the U.S., with specific reference to distribution, indexing, and insect vectors in the East, *Plant Dis. Rptr.* **35**:527–537.
14. Duffus, J.E., 1963. Possible multiplication in the aphid vector of sowthistle yellow vein virus, a virus with an extremely long insect latent period, *Virology* **21**:194–202.
15. Duffus, J.E., Zink, F.W., Bardin, R., 1970, Natural occurrence of sowthistle yellow vein virus on lettuce, *Phytopathology* **60**:1383–1384.
16. Fine, P.E.M., and Sylvester, E.S., 1978, Calculation of vertical transmission rates of infection, illustrated with data on an aphid-borne virus, *Amer. Natur.* **112**:759, 782–786.
17. Francki, R.I.B., Kitajima, E.W., and Peters, D., 1981, Rhabdoviruses, in Kurstak, E. 31 (ed): Handbook of Plant Virus Infections Elsevier/North Holland, Amsterdam, pp. 455–489.
18. Francki, R.I.B., Milne, R.G. and Hatta, T., 1985, Plant Rhabdoviridae: in An Atlas of Plant Viruses, Volume I, CRC Press, Boca Raton, Florida, pp. 73–100.
19. Francki, R.I.B., Randles, J.W., and Dietzgen, R.G., 1989, Lettuce necrotic yellows virus, *CMI/AAB Descriptions of Plant Viruses* **343**.
20. Frazier, N.W., 1968, Transmission of strawberry crinkle virus by the dark strawberry aphid Chaetosiphon jacobi, *Phytopathology* **58**:165–172.
21. Frazier, N.W., Sylvester, E.S., and Richardson, J. 1987, Strawberry crinkle, in

Converse, R.H. (ed): *Virus Diseases of Small Fruit*, U.S. Dept. Agric. Handbook 631, pp. 20–25.

22. Fry, P.R., Close, R.C., Procter, C.H., and Sunde, R., 1973, Lettuce necrotic yellows virus in New Zealand, *N. Z. Jour. Agric. Res.* **16**:143–146.

23. Garrett, R.G., and O'Loughlin, G.T., 1977, Broccoli necrotic yellows virus in cauliflower and in the aphid *Brevicoryne brassicae* L., *Virology* **76**:653–663.

24. Getz, W.M., Sylvester, E.S., and Richardson, J., 1982, Estimates of the latent period of strawberry crinkle virus in the aphid *Chaetosiphon jacobi* as a function of the experimental design, *Phytopathology* **72**:1441–1444.

25. Hille Ris Lambers, D., 1939, Contributions to a Monograph of the Aphididae of Europe II. The genera Dactynotus Rafinesque, 1818; Staticobium Mordvilko, 1914; Macrosiphum Passerini, 1860; Masonaphis Nov. Gen.; Pharalis Leach, 1826, *Temminckia* **4**:1–134.

26. Hills, G.J., and Campbell, R.N., 1968, Morphology of broccoli necrotic yellows virus, *J. Ultrastruct. Res.* **24**:134–144.

27. Hitchborn, J.H., and Hills, G.J., 1965, The use of negative staining in the electron microscopic examination of plant viruses in crude extracts, *Virology* **27**:528–540.

28. Hunter, B.G., Richardson, J., Dietzgen, R.G., Karu, A., Sylvester, E.S., Jackson, A.O., and Morris, T.J., 1990, Purification and characterization of strawberry crinkle virus, *Phytopathology* **80**:282–287.

29. Jackson, A.O., and Christie, S.R., 1979, Sonchus yellow net virus, *CMI/AAB Descriptions of Plant Viruses* **205**.

30. Jones, A.T., Murant, A.F., and Stace-Smith, R, 1977, Raspberry vein chlorosis virus, *CMI/AAB Descriptions of Plant Viruses* **174**.

31. Jones, A.T., Roberts, I.M., and Murant, A.F., 1974, Association of different kinds of bacilliform particle with vein chlorosis and mosaic diseases of raspberry (*Rubus idaeus*), *Ann. Appl. Biol.* **77**:283–288.

32. Jousset, F.-X., 1972, Le virus iota de "*Drosophila immigrans*" étudé chez "*Drosophila melanogaster*": symptôme de las sensibilité au CO_2, descriptions des anomalies provoquées chez l'hôte, *Ann. Inst. Pasteur Paris* **123**:275–288.

33. Leclant, F., Alliot, B., and Signoret, P.A., 1973, Transmission et épidémiologie de la maladie à énations de la luzerne (LEV). Primiers résultats, *Ann. Pathopathol.* **5**:441–445.

34. L'Heritier, Ph., and Teisser, G., 1937, Une anomalie physiologique héréditaire chez la drosophile, *C.R. Acad. Sci* (*Paris*) **205**:1099–1101.

35. Martelli, G.P., and Russo, M., 1977, Rhabdoviruses of plants, in Maramorosch K. (ed): The Atlas of Insect and Plant Viruses, Academic Press, New York pp. 181–213.

36. McGrew, J.R., 1987, Strawberry latent C, in Converse, R.H. (ed): *Virus Diseases of Small Fruits*, U.S. Dept. Agric. Handbook 631, pp. 29–31.

37. Misari, S.M., 1979, Vector relationships of the propagative plant rhabdovirus, coriander feathery red-vein to the aphid *Hyadaphis foeniculi* Passerini, Ph.D. Thesis, University of California, Berkeley, 181 p.

38. Misari, S.M., and Sylvester, E.S, 1983, Coriander feathery red-vein virus, a propagative plant rhabdovirus, and its transmission by the aphid *Hyadaphis foeniculi* Passerini, *Hilgardia* **51**:1–38.

39. Murant, A.F., and Roberts, I.M., 1980, Particles of raspberry vein chlorosis in the aphid vector *Aphis idaei*, *Acta Phytopathol. Acad. Sci. Hungaricae* **15**:103–106.

40. Nicolas, G., and Sillans, D., 1989, Immediate and latent effects of carbon dioxide on insects, *Ann. Rev. Entomol* **34**:97–116.

41. Ohki, S.T., Doi, Y., and Yora, K., 1978, Carrot latent virus: a new rhabdovirus of carrot, *Ann. Phytopath. Soc.* Japan **44**:202–204.

42. O'Loughlin, G.T., and Chambers, T.C., 1967, The systemic infection of an aphid by a plant virus, *Virology* **33**:262–271.

43. Özel, M., 1971, Vergleichende elektronenmikroskopische Untersuchungen an Rhabdoviren pflanzlicher und tierischer Herkunft 1. Erste elektronenmikroskopische Ergebnisse mit dem pflanzlichen Modell Sowthistle Yellow Vein Virus (SYVV) und seinem Vector Hyperomyzus lactucae (L.), *Zbl. Bakt. Hyg.*, I. Abt. Orig. A. **217**:160–175.

44. Peters, D., 1981, Plant Rhabdovirus Group, Descriptions of Plant Viruses *CMI/AAB Descriptions of Plant Viruses* **244**.

45. Peters, D., and Black, L.M., 1970, Infection of primary cultures of aphid cells with a plant virus, *Virology* **40**:847–853.

46. Prentice, I.W., and Woollcombe, T.M., 1951, Resolution of strawberry virus complexes IV. The latent period of virus 3 in the vector, *Ann. Appl. Biol.* **38**:389–394.

47. Randles, J.W., 1983, Transmission and epidemiology of lettuce necrotic yellows virus in Harris, K.F. (ed): Current Topics in Vector Research, Volume 1, Praeger, New York, pp. 169–188.

48. Randles, J.W., and Carver, M., 1971, Epidemiology of lettuce necrotic yellows virus in South Australia 11. Distribution of virus, host plants, and vectors, *Aust. J. Agric. Res.* **22**:231–237.

49. Richardson, J., Frazier, N.W., and Sylvester, E.S., 1972, Rhabdoviruslike particles associated with strawberry crinkle virus, *Phytopathology*, **62**:491–492.

50. Rorie, F.G., 1957, Resolution of certain strawberry viruses by the aphid vector *Capitophorus minor* Forbes, *Plant Dis. Reptr.* **41**:683–689.

51. Rosen, L., 1980, Carbon dioxide sensitivity in mosquitoes infected with sigma, vesicular stomatitis, and other rhabdoviruses, *Science* **207**:989–991.

52. Rosen, L., and Shroyer, D.A., 1981, Natural carbon dioxide sensitivity in mosquitoes caused by a hereditary virus, *Ann. Virol.* **132**:543–548.

53. Smith, H.E., 1952, Aphid transmission of strawberry viruses from commercial plants to Fragaria vesca L. (East Malling clone), *Phytopathology* **42**:20 (Abstr.).

54. Stubbs, L.L., and Grogan, R.G., 1963, Necrotic yellows: a newly recognized virus disease of lettuce, *Aust. J. Agric. Res.* **14**:439–459.

55. Stenger, D.C., Richardson, J., Sylvester, E.S., Jackson, A.O., and Morris, T.J., 1988, Analysis of sowthistle yellow vein virus-specific RNAs in infected hosts, *Phytopathology* **78**:1473–1477.

56. Sylvester, E.S., 1969, Evidence of transovarial passage of the sowthistle yellow vein virus in the aphid *Hyperomyzus lactucae*, *Virology* **38**:440–446.

57. Sylvester, E.S., Frazier, N.W., and Richardson, J., 1976, Strawberry crinkle virus, *CMI/AAB Descriptions of Plant Viruses* **163**.

58. Sylvester, E.S., and McClain, E., 1978, Rate of transovarial passage of sowthistle yellow vein virus in selected subclones of the aphid *Hyperomyzus lactucae*, *Jour. Econ. Entomol.*, **71**:17–20.

59. Sylvester, E.S., and Richardson, J., 1970, Infection of *Hyperomyzus lactucae* by sowthistle yellow vein virus, *Virology* **42**:1023–1042.

60. Sylvester, E.S., and Richardson, J., 1971, Decreased survival of *Hyperomyzus*

lactucae inoculated with serially passed sowthistle yellow vein virus, *Virology* **46**:310–317.

61. Sylvester, E.S., and Richardson, J., 1981, Inoculation of the aphids *Hyperomyzus lactucae* and *Chaetosiphon jacobi* with isolates of sowthistle yellow vein virus and strawberry crinkle virus, *Phytopathology* **71**:598–602.

62. Sylvester, E.S., and Richardson, J., 1986, Consecutive serial passage of strawberry crinkle virus in *Myzus ornatus* by injection and its occasional transmission to *Fragaria vesca, Phytopathology* **76**:1161–1164.

63. Sylvester, E.S., and Richardson, J., 1988, Successful juice inoculation of the aphid-vectored strawberry crinkle virus, *Calif. Agric.* **42**:6–7.

64. Sylvester, E.S., and Richardson, J., 1990, Comparison of vector-virus relationships of strawberry crinkle plant rhabdovirus in two aphids (*Chaetosiphon fragaefolii* and *C. jacobi*) infected by injection, *Hilgardia* **58**:1–23.

65. Sylvester, E.S., Richardson, J., and Behncken, G.M., 1970, Effect of dosage on the incubation period of sowthistle yellow vein virus in the aphid *Hyperomyzus lactucae, Virology* **40**:590–594.

66. Sylvester, E.S., Richardson, J., and Stenger, D.C., 1987, Use of injected *Macrosiphum euphorbiae* aphids as surrogate vectors for transfer of strawberry crinkle virus to *Nicotiana species, Plant Dis.* **71**:972–975.

67. Tomlinson, J.A., and Webb, M.J.W., 1974, Virus diseases of parsley. *Rep. Nat Veg. Res. Stn.*, Wellesbourne, 1973, pp. 100–101.

68. Tomlinson, J.A., Webb, M.J.W., and Faithfull, E.M. 1972, Studies on broccoli necrotic yellows virus, *Ann. Appl. Biol.* **71**, 127–134.

69. Turell, M.J., and Hardy, J.L., 1980, Carbon dioxide sensitivity of mosquitoes infected with California encephalitis virus, *Science* **209**:1029–1030.

70. van der Meer, F.A., 1989, Nicotiana occidentalis, a suitable test plant in research on viruses of small fruit crops, *Acta Horticulturae* **236**:27–36.

71. Vaughan, E.K., 1933, Transmission of the crinkle disease of strawberry, *Phytopathology* **23**:738–740.

72. Yoshikawa, N., Inouye, T., and Converse, R.H., 1986. Two types of rhabdovirus in strawberry, *Ann. Phytopath. Soc* Japan **52**:437–444.

73. Zeller, S.M., and Vaughan, E. K., 1932, Crinkle disease of strawberry, *Phytopathology* **22**:709–713.

Index